INTRODUCTION TO
QUANTITATIVE
CYTOCHEMISTRY-II

CONTRIBUTORS

NIELS B. ATKIN

S. BAILLY

LEON CARLSON

TORBJÖRN CASPERSSON

SALLY B. FAND

L. GALLIEN

ALFREDO MARIANO GARCIA

BARTON L. GLEDHILL

FREDERICK H. KASTEN

GÜNTER KIEFER

GÖSTA LOMAKKA

W. J. McCARTHY

BRIAN H. MAYALL

MORTIMER L. MENDELSOHN

KARL B. MORITZ

E. S. MOYER

RONALD J. PRZYBYLSKI

ELLEN M. RASCH

ROBERT W. RASCH

FRITZ RUCH

S. SILVERBÅGE

LUISE STANGE

WALTER E. STUMPF

M. VAN DER PLOEG

P. VAN DUIJN

SEYMOUR S. WEST

INTRODUCTION TO

QUANTITATIVE

CYTOCHEMISTRY-II

Edited by

GEORGE L. WIED

Department of Obstetrics and Gynecology
University of Chicago and the
Chicago Lying-In Hospital
Chicago, Illinois

AND

GUNTER F. BAHR

Biophysics Branch
Armed Forces Institute of Pathology
Washington, D.C.

1970

ACADEMIC PRESS   New York and London

ACADEMIC PRESS, INC.
111 Fifth Avenue, New York, New York 10003

United Kingdom Edition published by
ACADEMIC PRESS, INC. (LONDON) LTD.
Berkeley Square House, London W1X 6BA

LIBRARY OF CONGRESS CATALOG CARD NUMBER: 74-132769

PRINTED IN THE UNITED STATES OF AMERICA

CONTENTS

v

Cytophotometric Determination of the Nuclear DNA and RNA Content during Dedifferentiation of Certain Plant Cells

Luise Stange

Cytochemical Analysis of the Karyotype in the Amphibian Urodele *Pleurodeles waltlii* Michah Using the USMP-I

S. Bailly and L. Gallien

The Role of the Mounting Medium in UV Microphotometry

Günter Kiefer

A Precision Stage for Two-Dimensional Scanning Movement

Leon Carlson

A Rapid Scanning and Integrating Microinterferometer for Large-Scale Population Work

L. Carlson, T. Caspersson, G. Lomakka, and S. Silverbåge

Changes in Nuclear Stainability Associated with Spermateliosis, Spermatozoal Maturation, and Male Infertility

Barton L. Gledhill

Stoichiometry of Dye Binding versus Degree of Chromatin Coiling

Alfredo Mariano Garcia

Errors in Absorption Cytophotometry: Some Theoretical and Practical Considerations

Brian H. Mayall and Mortimer L. Mendelsohn

Recent Developments in Gallocyanine–Chrome Alum Staining

Günter Kiefer

Optical Rotatory Dispersion and the Microscope

Seymour S. West

Principles of Quantitative Autoradiography

Ronald J. Przybylski

Tissue Preparation for the Autoradiographic Localization of Hormones

Walter E. Stumpf

LIST OF CONTRIBUTORS

Numbers in parentheses indicate the pages on which the authors' contributions begin.

NIELS B. ATKIN, Department of Cancer Research, Mount Vernon Hospital, Northwood, Middlesex, England (1)

S. BAILLY, Laboratory of Embryology, Faculty of Sciences, University of Paris, Paris, France (87)

LEON CARLSON, Institute for Medical Cell Research and Genetics, Medical Nobel Institute, Karolinska Institutet, Stockholm, Sweden (113, 117)

TORBJÖRN CASPERSSON, Institute for Medical Cell Research and Genetics, Medical Nobel Institute, Karolinska Institutet, Stockholm, Sweden (27, 117)

SALLY B. FAND, Department of Pathology, Wayne State University School of Medicine and Veterans Administration Hospital, Allen Park, Michigan (209)

L. GALLIEN, Laboratory of Embryology, Faculty of Sciences, University of Paris, Paris, France (87)

ALFREDO MARIANO GARCIA, Department of Anatomy, State University of New York, Upstate Medical Center, Syracuse, New York (153)

BARTON L. GLEDHILL, Department of Clinical Studies, School of Veterinary Medicine, University of Pennsylvania, New Bolton Center, Kennett Square, Pennsylvania (125)

FREDERICK H. KASTEN,* Department of Ultrastructural Cytochemistry, Pasadena Foundation for Medical Research, Pasadena, California (263)

GÜNTER KIEFER, Department of Pathology, University of Freiburg, Freiburg, Germany (105, 199)

* Present address: Department of Anatomy, Louisiana State University Medical Center, New Orleans, Louisiana.

GÖSTA LOMAKKA, Institute for Medical Cell Research and Genetics, Medical Nobel Institute, Karolinska Institutet, Stockholm, Sweden (27, 117)

W. J. McCARTHY,* Department of Chemistry, West Virginia University, Morgantown, West Virginia (399)

BRIAN H. MAYALL, Department of Radiology, University of Pennsylvania, Philadelphia, Pennsylvania (171)

MORTIMER L. MENDELSOHN, Department of Radiology, University of Pennsylvania, Philadelphia, Pennsylvania (171)

KARL B. MORITZ, Zoology Institute, University of Munich, Munich, Germany (57)

E. S. MOYER,† Department of Chemistry, West Virginia University, Morgantown, West Virginia (399)

RONALD J. PRZYBYLSKI, Department of Anatomy, School of Medicine, Case Western Reserve University, Cleveland, Ohio (477)

ELLEN M. RASCH, Department of Biology, Marquette University, Milwaukee, Wisconsin (297, 335, 357)

ROBERT W. RASCH, Department of Physiology, Marquette School of Medicine, Inc., Milwaukee, Wisconsin (297)

FRITZ RUCH, Department of General Botany, Swiss Federal Institute of Technology, Zürich, Switzerland (431)

S. SILVERBÅGE, Institute for Medical Cell Research and Genetics, Medical Nobel Institute, Karolinska Institutet, Stockholm, Sweden (117)

LUISE STANGE, Institute of Botany, Technical University, Hannover, Germany (77)

WALTER E. STUMPF,‡ Department of Pharmacology, University of Chicago, Chicago, Illinois (507)

M. VAN DER PLOEG, Histochemical Laboratory of the Department of Pathology, University of Leiden, The Netherlands (223)

P. VAN DUIJN, Histochemical Laboratory of the Department of Pathology, University of Leiden, The Netherlands (223)

SEYMOUR S. WEST, University of Alabama in Birmingham Medical Center, Birmingham, Alabama (451)

* Deceased.
† Present address: Louisiana State University, New Orleans, Louisiana.
‡ Present address: Departments of Anatomy and Pharmacology, University of North Carolina, Chapel Hill, North Carolina.

PREFACE

This volume of "Introduction to Quantitative Cytochemistry" originated from the Second International Tutorial on Quantitative Cytochemistry held at the Center for Continuing Education of the University of Chicago.

Following requests from participants of the 1965 Tutorial, the faculty joined in the production of the first introductory volume. The wide acceptance of this publication among students of cytology and cell biology is reason for presenting in this second volume chapters on topics not previously included as well as on current progress in methodology and understanding.

Of particular interest to cytochemists are three discussions of changes in the interpretation of quantitative staining, notably because of the recently discovered effects of chromatin compaction. Also, fluorescence techniques are prominently represented, since they are finding rapidly increasing application because of both simplicity and sensitivity.

We hope that readers of these books will be convinced that quantitative cytochemistry continues to hold its own by providing unique data on the behavior of individual and populations of cells in growth, division, and differentiation, and that no other approach available can do the same. We further believe the reader will find the techniques as applicable and enjoyable as did each of the authors.

We wish to thank the contributing authors for expedient collaboration in the completion of this volume. We are especially grateful to Mrs. Dorothy E. Parrish and Mrs. Jacquelyn Strong for their assistance in editing and preparing the manuscripts.

PRINCIPLES AND APPLICATION OF THE DEELEY INTEGRATING MICRODENSITOMETER

Niels B. Atkin

DEPARTMENT OF CANCER RESEARCH, MOUNT VERNON HOSPITAL,
NORTHWOOD, MIDDLESEX, ENGLAND

Of the various approaches to the microspectrophotometry of natural or induced colored substances in cells, the scanning system introduced by Deeley in 1955 (*39, 40*) has perhaps proved of widest application in view of its relative simplicity, its speed of operation, and the accuracy that can be achieved provided that certain basic principles are adhered to. Since errors may be encountered where thick specimens, e.g., histologic sections, are used, it is desirable that measurements be made on smears in which the cells are already well flattened, or can be flattened immediately prior to measurement by the use of the crushing condenser. In order to acquaint the reader with the proper use of the Deeley integrating microdensitometer, it is necessary to discuss the above and other possible sources of error, first, by a consideration of the physical principles underlying the technique and, second, by emphasis of those features of the actual technique which will eliminate or at least minimize these errors.

I. Principles of Microspectrophotometry

The theoretical basis for the microspectrophotometric estimation of the amount of colored substances present in cells is the interaction between photons of specific wavelengths and the *chromophores* which are responsible for the phenomenon of color. Chromophores are atomic groupings such as the azo grouping, —N=N—, which in general have unsatisfied affinities for hydrogen, i.e., they are easily reducible. The specific interaction between photons and chromophores results in a loss of energy, or *absorption*, of the

1

light, which is transformed into heat or into fluorescent or phosphorescent light of wavelengths differing from its own; the amount of light absorbed is proportional to the number of chromophores in the medium through which the light passes.

However, when light, as well as other electromagnetic radiation outside the visible spectrum, passes through any transparent absorbing medium, only part of the light loss is due to absorption; part will also be lost by scattering or reflection at the surfaces, or scattering in the interior of the medium.

The quantitative estimation of absorption is based on two fundamental laws concerning the relationship between the intensities of the radiation incident on and transmitted by a homogeneous layer of absorbing substance.

(a) *Lambert's law:* As first stated by Bouguer in 1729 and enunciated by Lambert a few years later, "the proportion of radiation absorbed by a substance is independent of the intensity of the incident radiation" (this is a little different from the form in which the law is now usually stated; see below). It follows that each successive layer of thickness dl absorbs the same fraction dI/I of the radiation of incident intensity I_0, i.e.,

$$dI/I = -\mu\, dl$$

where μ is a constant. On integration,

$$I = I_0 \exp(-\mu l)$$

I and I_0 are the emergent and incident intensities, respectively, and l is the thickness.

Alternatively,

$$I = I_0 10^{-Kl}$$

hence

$$\log_{10} I_0/I = Kl$$

where $K = \log_{10} e\mu \ (= 0.4343\ \mu)$, μ is the absorption coefficient, and K the extinction coefficient first defined by Bunsen and Roscoe (*31*).

Thus it can be seen that according to Lambert's law the amount of light absorbed is related to the *thickness of the absorbing material through which it passes.* This relationship can be expressed in two ways. It will be seen first that the fraction of light absorbed, I/I_0, which is known as the *transmittance,* is equal to the negative exponential of μl. However, it is usually more convenient to consider the *absorbance* (also called the *extinction* or *optical density*), which is the common logarithm of the reciprocal of the transmittance i.e., $\log_{10} I_0/I$; this is *directly* proportional to the thickness.

(b) *Beer's law:* As originally stated by Beer in 1852, absorption depends only on the number of absorbing molecules through which the radiation passes. As a corollary of this, when an absorbing substance is dissolved in a nonabsorbing medium, the *absorbance is proportional to the concentration* (c) of the solution:

$$A = \log_{10} I_0/I = kcl$$

In the above formula, Beer's and Lambert's laws are combined; k is a constant known as the *absorptivity, absorbancy index,* or *specific extinction coefficient.*

The total mass M of absorbing substance is obtained by multiplying the concentration by the volume, i.e., $M = cal$, where a is the area (in microspectrophotometry, the area of the microscopic field); it follows that $M/A = a/k$ or $M = Aa/k$. Thus, M is proportional to the product of absorbance and area; neither the concentration nor the thickness appear in this formula.

II. Errors in Microspectrophotometry

Consideration will now be given to the special problems relating to measurements of objects at the microscopic level. There have been a number of extensive reviews concerned with errors in microspectrophotometry (*33, 37, 86, 99, 102*). As will be seen from the following brief account, which is relevant to the Deeley technique as used to estimate relative amounts of colored substances, e.g., stain, in single cells, several of the possible sources of error need not be a cause for undue concern.

A. DEPARTURE FROM BEER'S LAW

This implies a departure from the linear relationship between concentration and absorbance, and will need special consideration whenever a previously untried staining reaction is used; one possible reason for a departure from Beer's law is that chromophores may change their absorptivity (k) at the high concentrations likely to be encountered in cells, owing to molecular interaction. Unfortunately, the validity of Beer's law under these conditions cannot be directly tested since this would require an independent measurement of concentration, which is almost impossible in cells (*102*).

B. STRAY LIGHT

Light reflected from glass or metal surfaces and scattered by dust on the surfaces of lenses and further "nonspecific light loss" due to changes of refractive index or molecular scattering within the biological specimen may, under certain conditions, cause serious errors (*79*) but can be kept to within

1 % if the absorbances are less than 1.0 and if the area of the specimen illuminated by the condenser is only slightly larger than the measured area (*55*). Flattening of the specimen by means of the crushing condenser, as will be described later, serves to reduce scatter by the specimen which, it is important to note, increases with the absorbance of the material.

C. Nonparallel Light

The apparent increase in absorbance due to some rays from the condenser passing obliquely through the object is reduced to a minimum when the material is flattened (*40*).

D. Distributional Error

Light intensity can be conveniently and accurately measured by allowing the light to fall on the photocathode of a photomultiplier and recording the current produced. One can therefore estimate the transmittance (I/I_0) through an object such as a cell nucleus by measuring the transmitted light (I) and, after moving to an adjacent blank area, the light intensity without the object in the field which gives a measure of the incident light (I_0). However, the value for the transmittance so obtained is an average value which cannot be used as a measure of amount of absorbing substance if, as in most biological objects, the absorbing material is not evenly distributed. The reason for this is that the relation between transmittance and absorbance (the latter being directly related to the amount of absorbing material) is a logarithmic one.

The following numerical example will demonstrate that in certain circumstances nonuniform distribution of the absorbing material can introduce a *distributional error* into the calculation of absorbance from observed values of

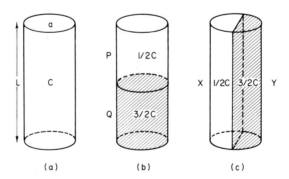

Fig. 1. Model cylindrical objects, to illustrate the numerical example indicating the conditions under which distributional error arises.

the transmittance. Two particular cases of nonuniform distribution will be compared with the case of uniform distribution.

Consider first the case where absorbing material is distributed at a uniform concentration c within a layer of depth l, and viewed in a field of area a (Fig. 1a). Suppose that the transmittance I/I_0 has been measured as $\frac{1}{4}$. Then the absorbance is

$$A = \log_{10} I_0/I = \log_{10} 4.0 = 0.6$$

Since $A = kcl$, the specific absorbance for this material is

$$k = A/cl = 0.6/cl$$

The total mass of absorber present is calculated from the relationship $M = Aa/k$ as

$$M = \frac{0.6a}{k} = \frac{0.6a}{0.6/cl} = cal$$

Now suppose the measuring volume divided into upper and lower halves P and Q (Fig. 1b) with the absorbing material redistributed at concentration $\frac{1}{2}c$ in P and $\frac{3}{2}c$ in Q. Noting that the pathlength in the region P is $l/2$ we calculate the absorbance of this region as

$$A_P = k\,(c/2)\,(l/2)$$

and substituting $k = 0.6/cl$,

$$A_P = \left(\frac{0.6}{cl} \cdot \frac{cl}{4}\right) = 0.15$$

The mass in this region can then be derived as

$$M_P = \frac{A_P a}{k} = \frac{0.15a}{0.6/cl} = \tfrac{1}{4}cal$$

and similarly for the region Q

$$A_Q = 0.45$$

and

$$M_Q = \tfrac{3}{4}cal$$

The transmittance can be derived by calculation as follows: for the region P, the transmittance is

$$T_P = \frac{1}{\log_{10}^{-1} 0.15} = \frac{1}{1.4} = 0.7$$

and for Q

$$T_Q = \frac{1}{\log_{10}^{-1} 0.45} = \frac{1}{2.8} = 0.35$$

The transmittance of the complete light path is the product

$$T_P T_Q = 0.7 \times 0.35 = \tfrac{1}{4} = I/I_0$$

This is the value that would be obtained by observation, and we note that it is the same as in the case of uniform distribution. The value of $M = cal$ deduced from this transmittance agrees with the sum of M_P and $M_Q = \tfrac{1}{4}cal + \tfrac{3}{4}cal$ arising from consideration of the separate absorbances of the two regions P and Q. Nonuniformity of distribution *in the direction of the light path*, therefore, introduces no error into the calculation of absorbance and mass from the observed transmittance.

Next, consider the measuring volume divided into halves X and Y by a plane parallel to the light path (Fig. 1c), with the region X filled at concentration $\tfrac{1}{2}c$ and Y at concentration $\tfrac{3}{2}c$. The absorbance of region X is

$$A_X = k\frac{c}{2}l = \frac{0.6}{cl}\frac{cl}{2} = 0.3$$

and similarly

$$A_Y = k\frac{3c}{2}l = \frac{0.6}{cl}\frac{3cl}{2} = 0.9$$

In calculating the masses of absorber present in X and Y we note that the area presented by these regions is $a/2$. Hence

$$M_X = \frac{A_X (a/2)}{k} = \frac{0.3a}{2 \times 0.6/cl} = \tfrac{1}{4}cal$$

and

$$M_Y = \frac{A_Y (a/2)}{k} = \frac{0.9a}{2 \times 0.6/cl} = \tfrac{3}{4}cal$$

For the transmittances of the two regions we have

$$T_X = \frac{1}{\log_{10}^{-1} 0.3} = \tfrac{1}{2}$$

$$T_Y = \frac{1}{\log_{10}^{-1} 0.9} = \tfrac{1}{8}$$

Since each region receives one-half of the total incident light, the emergent light from X is

$$\tfrac{1}{2}I_0 \times \tfrac{1}{2} = \tfrac{1}{4}I_0$$

and from Y,

$$\tfrac{1}{2}I_0 \times \tfrac{1}{8} = \tfrac{1}{16}I_0$$

The observed transmittance of a field containing the two regions is thus $\tfrac{5}{16}$.

If one attempts to use this observation to deduce absorbance and mass, the values obtained will be

$$A' = \log_{10} \frac{16}{5} = 0.50$$

$$M' = \frac{0.50a}{k} = \frac{0.50a}{0.6/cl} = 0.83cal$$

Comparing this with the correct value $M_X + M_Y = cal$ derived from consideration of the separate regions, it is seen that in this case the nonuniformity has introduced a distributional error of 17%.

In summary, therefore, while nonuniformity in the direction of the light path is without effect, distributional errors arise from nonuniformity across the field of view. To eliminate such errors it is necessary to subdivide the field into elements sufficiently small to present only negligible transverse variation of concentration. The transmittances of these elements must be measured separately and converted to absorbances, which are summed to derive the total mass of absorber present. This principle is the basis of the scanning method.

E. OUT-OF-FOCUS ERRORS

Errors may arise if part of the absorbing material lies outside the depth of focus of the microscope objective; to avoid these, it will therefore be necessary to ensure that the material is sufficiently flat. In the "two-wavelength" method (see Section III), as opposed to scanning methods, errors due to material being out of focus are less serious.

III. Microspectrophotometric Systems Which Minimize Distributional Error

From the above discussion it will be seen that the total absorbance of an object cannot be estimated from its average transmittance if the material is irregularly distributed; it is possible, however, to measure the transmittance of small portions of the object, and if the dimensions of these areas are small enough the distributional error within each area will be negligible. (In practice the limit will be reached when the dimensions approach the limit of resolution of light.) One may therefore estimate the total absorbance of an object by measuring the transmittances of a number of representative "plugs" through the object and converting each of these to absorbance.

There are other approaches to the problem of reducing distributional error; that most frequently used is the two-wavelength method of Ornstein (*83*) and Patau (*84*). Other methods include the one-wavelength, two-area method of Garcia (*47*) and a photographic method developed by Adams (*1*). Micro-fluorometry offers a further approach (*92*). [The plug and two-wavelength methods are described elsewhere in this volume (*48*).]

However, the most satisfactory solution to the problem of distributional error would appear to be provided by the scanning methods, where (as in the Deeley system) the transmittance for each small area of the object as it is scanned is converted to absorbance and an integrated value for the total absorbance over the whole field automatically obtained.

Besides the Deeley system, scanning systems have been used in the pioneering studies of Caspersson (*33*), whose design, in which the object is scanned relative to the light beam by moving the microscope stage, forms the basis of the Zeiss Universal Microspectrophotometer, which operates in both UV and visible light. Other scanning systems include that of Jansen (*58, 59*). A scanning and integrating microdensitometer, the Optoscan, is in the course of development by Vickers Instruments Ltd., York, England; in this instrument the object itself is scanned by a small moving spot, and an additional feature is that the total area of regions of the specimen which have absorbances exceeding any selected value up to 1.3 can be determined.

IV. The Deeley Integrating Microdensitometer

This system has been extensively used for microspectrophotometric studies on biological material. Many of the studies have been made with the commercially available instrument (Fig. 2) manufactured by Barr and Stroud Ltd., Glasgow W.3, Scotland (U.S. representatives: National Instrument Laboratories, 12300 Parklawn Drive, Rockville, Maryland 20852).

In the Deeley system, a small moving aperture scans an image of the object. Total absorbance in arbitrary units is obtained by taking the difference of two meter readings: from a reading obtained with the object in the field is subtracted a "blank" (clear field) reading obtained after the object has been moved outside the field. Each meter reading can be read immediately after the scan, which takes about 3 sec. The readings represent the integral of the

FIG. 2. The Barr and Stroud integrating microdensitometer, type **GN2**.

logarithm of the intensity of each fraction of the field as seen by the aperture and recorded by a photomultiplier, the logarithmic conversion and integration being accomplished by appropriate electronic circuits.

A. THE OPTICAL SYSTEM

The optical system is shown schematically in Fig. 3. The light source, which was a 125-watt mercury vapor lamp in the instrument originally described by Deeley (*39*), now conveniently takes the form of a 75-watt tungsten filament, or an iodine vapor lamp, which provides an adequate and easily stabilized supply of light. A field lens images the lamp filament in the plane of the substage diaphragm according to the Köhler principle. Field stops are provided by a rotatable disk containing suitable apertures. The stop should be matched to the objective being used: In order to reduce glare, the stop, in use during measurement, should limit the area of the microscopic field illuminated to little more than the area that is being scanned. It is necessary to be able to control the amount of light reaching the specimen, so as to ensure that both readings are within the range of the meter; altering the light will affect the

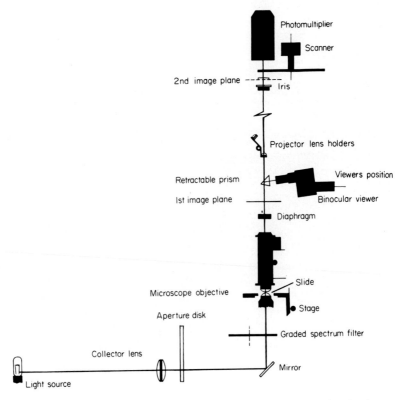

FIG. 3. Schematic arrangement of the Barr and Stroud integrating microdensitometer.

meter reading, but will not affect the *difference* between the readings for object and blank field, provided, of course, that the light is not altered *between* the two readings. This may be done by means of a graded neutral filter or, as on the current Barr and Stroud instrument, by altering the current to the lamp. In order to restrict the wavelength to a suitable narrow range a graded spectrum (interference wedge) filter can be used; Barr and Stroud provide a filter which gives a bandwidth of 150 Å at 50% of peak transmission over a range of 4000 to 7000 Å. Alternatively, a prism mono-chromator or suitable interference filters can be used. It may be noted that for most purposes it is not necessary to have strictly monochromatic light. It may also be noted here that the wavelength of the light used need not corre-spond to the peak of the absorption curve of the particular colored substance being measured, and it may in fact be an advantage to measure at a wave-length which is off the peak, where absorption is less intense (but proportional to that at the peak). Thus, for the Feulgen reaction, where the peak of the absorption curve, which incidentally shows slight variation under different conditions, is at about 575 mμ (*98*), a wavelength of 546 mμ might be used; this was in fact the wavelength used for this purpose in early studies with the Deeley apparatus (*40*) when the light source was a mercury vapor lamp, since adequate light could be obtained at this wavelength (the green line of the mercury spectrum).

Although a conventional condenser can be used, it is frequently desirable to have a crushing condenser; the use of the latter is described below. In the Barr and Stroud model, the same condenser can be used either conventionally or for crushing by changing the top lens.

For viewing the specimen, the light is reflected by a prism into a binocular viewer. During measurement this prism is displaced out of the beam so that the light rays which have formed an image of the specimen at the first image plane, just below the prism, now pass through a projector lens and form a further image at the second image plane. An iris diaphragm or a series of fixed stops are situated in the region of the first image plane; with a suitably sized stop, an object can thereby be isolated from other objects in the field for measurement. The scanning mechanism is situated at the level of the second image plane. The rays which pass through the scanning aperture are collected by a field lens and fall on the photocathode of a photomultiplier; they do not form an image at this level.

B. The Scanning Mechanism

A mechanical system is used in which the light passes through a scanning aperture formed by the intersection of two slits 0.2 mm in width situated approximately at right angles to one another. One slit is one of a series cut into the circumference of a disk 10 cm in diameter. As this disk rotates the aperture formed by the intersection of the two slits travels the length of the second slit (15 mm); as rotation continues a further slit on the disk comes into apposition and again forms an aperture travelling the length of the second slit. Since during one scan the disk rotates three times and since there are 30 radially orientated slits at the periphery of the disk, a scanning spot effectively traverses a series of 90 straight lines. During the scan the synchronous electric motor which rotates the disk also actuates a cam-operated mechanism which moves both the disk as it is rotating and the plate with the second slit together in a direction at right angles to the scan lines; each successive scan line is, therefore, in a slightly different position so that eventually the whole of the required area is scanned.

In the Barr and Stroud instrument the area of the image scanned (when using the 10 $\times$ projector lens) corresponds to an area of the actual microscopic field of 10 μ in diameter, and the aperture itself corresponds to an area of 0.25 μ in diameter. It will be noted that successive scan lines partially overlap one another.

C. ELECTRONIC CIRCUITS

Although a detailed description of the circuits is outside the scope of this chapter, there are certain aspects which deserve consideration here. As we have seen, absorbance (A) is equal to the common logarithm of the reciprocal of the transmission (i.e., to $\log_{10} I_0/I$), and the total absorbance due to an inhomogeneous object can be obtained by summing the separate absorbances of each small fraction of the field seen by the scanning aperture. Since log I_0/I equals log I_0 minus log I, the total absorbance for inhomogeneous objects can be obtained by taking the difference of two values, the integrated values for the logarithms of the light intensities at each instance during the scan, with and without the object in the field, respectively. A voltage-integrating circuit gives a signal proportional to $\int(-\log i)dt$ (i being the current generated at the photocathode, which is proportional to the intensity of the light falling on it). The amount of absorbing material in the field is then proportional to the meter reading representing the output voltage obtained with the object in position less the reading obtained without the object in position.

When no light falls on the photomultiplier cathode the absorbance is theoretically infinite, and the voltage rises to its maximum value. However, it is obvious that integration should proceed only when light is falling on the photocathode; this is ensured by means of a diode switch which disconnects the input of the integrator when the voltage reaches a certain value. The operator by turning the switch brings in one of a series of resistors, thereby selecting the limiting value so that integration ceases if the absorbance rises above this level (e.g., 0.8).

When measuring an object, it will be placed in the center of the field and a suitable stop at the first image plane selected to enclose the object. Until the scanning aperture reaches the edge of the image of this stop no light will reach the photomultiplier and no integration will occur. As the aperture crosses the edge of the stop it will begin to become illuminated and when this illumination reaches a certain level, depending on the setting of the diode switch, integration will commence. Each time the aperture crosses the edge of the stop during the scan a short period of partial illumination will therefore result in a contribution to the final signal. The same contribution from the stop will appear in the blank reading obtained with the object out of the field, and so when this value is subtracted from that obtained with the object in the field, the contribution from the stop will be cancelled out; it is nevertheless desirable that this contribution should be relatively low, and this can be ensured by selecting a low " maximum absorbance " setting of the diode switch.

If the absorbance of any parts of the object, or of any particles of extraneous matter in the field, exceeds the maximum set by the diode switch, these parts will not be recorded. An object containing regions which are too dense will therefore give a value for total absorbance which is too low.

D. The Preparation of Material for Measurement and the Use of the Crushing Condenser

1. *Control Cells*

Since values are in arbitrary units, it is usually desirable that some control objects in the form of readily identifiable cells, to which the measurements may be related, should be included in the preparation; these may already be present in the material [for instance, leukocytes or fibroblasts present in tumor tissue (*16*)], or they may be added to the preparation. The use of controls enables allowance to be made, when comparing cells from two different preparations, for differences in stain intensity due to technical factors which may be superimposed on those representing real differences in the substance being estimated. Thus, in order to compare cells of type A with cells of type B, which are on separate slides or coverslips, the same control cells X_A and X_B are added to each; it is then merely necessary to obtain by measurement, mean values for A, X_A, B, and X_B and then to estimate the ratios $A:X_A$ and $B:X_B$. Since the control cells X_A and X_B are in fact identical cell types, any difference between their mean values, apart from those arising from instrumental variation, which can be reduced by taking sufficiently large samples, are presumed to be due to technical factors relating to the staining reaction, and the estimated true relative values for the substance being measured in cells A and B can thus be obtained by dividing the ratio $A:X_A$ by the ratio $B:X_B$, which (assuming now that $X_A = X_B$) gives $A:B$, i.e., an estimate of the relative amounts of the substance in cells A and B. It is important to realize that in general the intensity of staining reactions depends on a number of factors, some of which may be difficult or impossible to control. Thus, as regards the Feulgen reaction, the intensity of color depends on the type of fixation, time of hydrolysis, and the pH of and time of exposure to Schiff's reagent; the ensuing variability, according to Richards (*89*) "... virtually precludes the comparison of amounts of Feulgen colour in situations other than between cells in the same area of the slide."

2. *Flattening of the Material*

If the biological material is too thick, errors will be introduced which may or may not be acceptable depending on the standard of accuracy required; the extent of the errors will depend partly on the range of absorbances of the material, and will be less at low absorbances. Flattening brings about the following desirable results: The whole of the object can be simultaneously brought into focus; absorbance can be reduced to within the optimal range (thus reducing errors from glare); and compression of the object and surrounding material into a flat layer reduces errors from light scattering. Flattening may be achieved in the course of making the slide preparation or at the time of measurement by the use of the crushing condenser.

3. *Slide Preparations of Flattened Material*

It is difficult or impossible to make suitably flattened preparations after fixation in conventional fixatives such as ethyl alcohol or acetic alcohol; squashing by application of pressure to the coverslip rarely results in satisfactory flattening and, moreover, these fixatives tend to produce clumping of cellular material, e.g., chromatin, resulting in areas of high absorbance.

However, one method which has been found to be satisfactory for the measurement of the Feulgen stain content of cells in solid tissues or suspensions is as follows (for solid tissues a suspension is first made by mincing the tissue in physiologic saline with scissors). The cells in suspension are centrifuged, exposed to a hypotonic solution, and then fixed in 3:7 glacial acetic acid:methanol. Air-dried slide preparations are subsequently made. This procedure is similar to that used for making chromosome preparations from solid tumor tissue (*14*); both the metaphase chromosome groups and the interphase nuclei are well flattened and even though many of the cells still remain in clumps, individual nuclei in the clumps are frequently sufficiently well separated to be suitable for measurement. Also, the staining of the nuclei tends to be rather homogeneous, since the hypotonic pretreatment blurs the chromatin pattern so that there are no dense areas; despite the hypotonic pretreatment the total amount of stain is comparable to that of material not so pretreated, and, more important, the ratio of stain content for different cell types (e.g., epithelial cells and lymphocytes) is not altered (*13*).

4. *Use of the Crushing Condenser; Methods of Fixation*

The Boddy crushing condenser, introduced by Davies, Wilkins, and Boddy (*38*), enables the operator to flatten cells under direct observation. Cells fixed by most conventional fixatives are usually too tough for satisfactory crushing; the procedure of choice is fixation by methanol freeze substitution. A description of this procedure, which involves quenching coverslip preparations (necessary when using the crushing condenser) in a propane–isopentane mixture cooled in a liquid nitrogen bath followed by quick transfer to methanol at −76°C and gradual warming up to room temperature, is given by Richards (*89*). Smears can be made from pieces of solid tissue by tapping the fresh material with a flat-ended metal rod; the resultant isolated cells or nuclei and small clumps of cells adhere well to the glass following this fixation procedure, which results in a minimum of artifactual clumping of the chromatin. (It may be an advantage that clumps of cells are present in the preparation since the interrelationship of the cells and the amount of cytoplasm can be seen, and these may be an aid to identification of the cell-type.) Although the cells in a clump may overlap one another, it will often be found that they can be sufficiently spread out by application of the crushing condenser so that many of the nuclei will be separated from their neighbors and can be isolated by the appropriate diaphragm for individual measurement.

One other method of preparation has been found to be suitable for subsequent use of the crushing condenser: Smears of cells on coverslips obtained by scraping the buccal surface of the cheek (18) and the cervix uteri (6) were allowed to dry in the air and, without being immersed in liquid fixatives, subsequently stained by the Feulgen method. These smears were found to be satisfactory in that the cells were soft and could be crushed easily. (If many erythrocytes are present they can be lysed by immersing the smear in 10% acetic acid for a few minutes prior to staining.) A possible disadvantage of air-drying procedures is that there is some cellular distortion which, while not interfering with the measurements (the stain content being unaffected), may render identification of cell types less easy than in preparations fixed by freeze substitution.

The above two methods of preparation may be combined; thus, air-dried coverslip preparations of nucleated animal erythrocytes were made, and fresh control cells (human lymphocytes or epithelial cells) were subsequently added, the preparation then being immediately put through the freeze-substitution procedure (20).

5. The Technique of Crushing

The crushing condenser is an oil immersion condenser with a special top lens having a convex upper surface. When racked up this makes contact with the cells over an area of approximately 350 μ in diameter. In order that a high-power objective may be used, the cells must be mounted on a No. 1 thickness coverslip (thinner coverslips tend to break easily during the crushing process). The cells, which when the coverslip is in position will be situated on its *under* surface, are mounted in glycerol and covered by a slip of cellophane about 25 μ thick. (Coverslip preparations stained by the Feulgen procedure can be conveniently stored without risk of deterioration over a period of months or even years when mounted in immersion oil on a slide and kept in a refrigerator at 4°C or a deep freeze unit. The cells, however, cannot usually be crushed satisfactorily when mounted in oil.) With the coverslip mounted on a metal plate which is placed on the microscope stage (the latter should have micrometer controls to facilitate exact positioning of the preparation), the cells are crushed through the cellophane. Sufficient counterforce may be provided by the weight of the plate, or the latter may be fixed to the microscope stage by clips. Flattening can be aided by a sideways movement of the preparation while the pressure is being applied. When sufficiently flattened, it will be possible to bring the whole of the object into focus simultaneously; there should not be any densely stained regions.

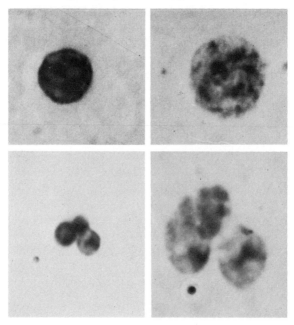

FIG. 4. Nuclei before and after crushing (reproduced at the same magnification). Feulgen stain. Top: Lymphocyte. Bottom: Polymorph. The small dense extraneous particle seen just below the nucleus after crushing has a high absorbance which probably is above the maximum that will be recorded, as determined by the setting of the diode switch.

For measurement, the object is placed in the center of the field and isolated from any other objects in the field by the smallest-sized diaphragm (at the first image plane) which will still leave a rim of clear field all around the object. To increase the accuracy of measurement, the mean of three consecutive readings is taken and from this is subtracted the mean of three measurements with the diaphragm still in position but with the object moved sideways so that it is just out of the field. This " blank " field should not, of course, include any colored material, but if any uncolored material such as cytoplasm is seen by the observer to have been unavoidably included in the field with the object, a similar amount of this material should if possible be included in the " blank " field (rather than that the field be moved some distance from the object to a completely clear area).

6. *The Diode Switch Setting*

The diode switch which determines the upper limit of absorbance that will be recorded should be set to suit the material. If this is faintly colored, a low setting is desirable, in order to reduce the contribution from the edge of the diaphragm, which adds a "dead weight" to both the reading with the object in the field and to that of the "blank" field. Whether the setting is too low for a given object may be determined by taking measurements at decreasing settings: A drop in the measurements will indicate that the setting is too low and that material is not being recorded. If material is too dense to be recorded even with the switch at its maximum setting this test cannot be used, but with experience the operator can judge that an object with dense areas which gives an unexpectedly low reading is probably being underrecorded. However, a possible means of determining that a field contains areas that are too dense to be recorded at any given setting of the diode switch is as follows (suggested by F. S. Stewart):

A pulse counter (which may conveniently be of the decatron type) is actuated by the potential which operates the diode switch, and records a count each time the switch changes from the closed to the open state. In scanning a blank field, this occurs each time the scanning aperture passes out of the field, so that the number of counts recorded is equal to the number of scanning lines sweeping the field. With the object in the field, the number of counts recorded will remain the same provided the absorbance everywhere within the field is below the setting of the maximum absorbance control. If this condition is not satisfied, either because of the choice of an unduly low setting for the maximum absorbance control, or the presence of an artifact in the specimen, the consequent operation of the diode switch at points inside the field will be revealed by an increase in the count recorded.

E. RESULTS OBTAINED WITH THE DEELEY INTEGRATING MICRODENSITOMETER

1. *Accuracy*

Small variations in readings occur as a result of photomultiplier fluctuations and slight irregularities in the operation of the scanning mechanism, and give rise to a random error which can be reduced if the mean of three readings of the object and of the blank field are taken. Deeley and his colleagues (42) found a coefficient of variation of 2.8% for 24 independent determinations of the Feulgen stain in a single ram sperm head; measurements on 45 different sperm heads from the same field gave a coefficient of variation

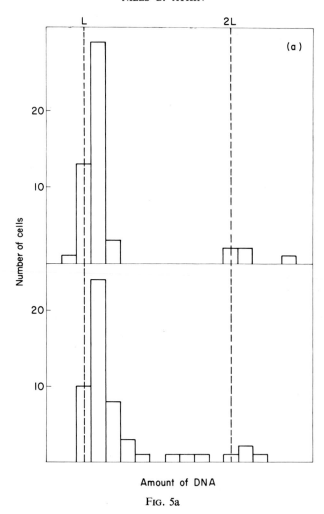

FIG. 5a

of 3.9 %. James (56), replacing the condenser by a fluorite objective, measured rat liver nuclei which had been flattened by manual pressure at the moment of fixation, and found very little variation, the coefficient of variation being less than 1 %.

In general, measurements on a population of cells which show very little real variation in stain content may be expected to show a coefficient of variation of 3 % or less, although when the cells are spread over a wide area of the slide or coverslip, more variation may be encountered; Hale (51) found a difference in Feulgen stain content between cells at opposite ends of the same slide which was attributed to a temperature gradient in the bath ($\sim 60°$C) during acid hydrolysis (this was overcome by agitating the bath). Obviously, where such variation in intensity of staining due to technical factors may occur over

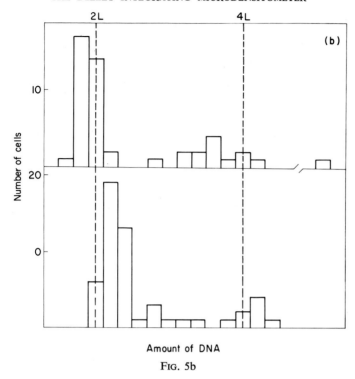

Amount of DNA

Fig. 5b

Fig. 5. (a and b) Histograms of DNA values of interphase epithelial cells obtained by measurement of their Feulgen stain content. (a) Top: Adenomyomatous polyp of the corpus uteri. Bottom: Normal endometrium (proliferative stage). (b) Two carcinomas *in situ* of the cervix uteri. L, Mean value of the control cells (leukocytes and fibroblasts); 2L and 4L, twice and four times this value, respectively.

a single preparation, cells which are being used as controls should be situated in the same regions of the preparation as the cells that are being investigated.

2. *A Comparison with Other Methods*

A particular advantage of the Deeley system is that large numbers of cells can be measured in a reasonable time; a rate of one cell per minute or better can often be achieved. If, when using the crushing condenser, it is not considered necessary to examine each cell before crushing, the condenser can simply be left in the raised position with a constant pressure being applied. When the field is moved and new cells are brought in, these will then be automatically crushed.

Figure 5 shows histograms of measurements obtained on normal endometrial cells, cells from a polyp of the corpus uteri with presumed normal karyotypes, and abnormal epithelial cells from two carcinomas *in situ* of the cervix uteri. The first two show modes for the epithelial cells which are about

TABLE I

STUDIES USING THE DEELEY INTEGRATING MICRODENSITOMETER

Objects of the studies	References
1. *Estimation of the DNA content*	
(a) Of normal cells, e.g., as a parameter related to the karyotype of a species, or to demonstrate the occurrence of different degrees of ploidy or of mitotic activity	
(i) In animals	15, 20, 27, 65, 75, 77, 78, 80–82, 97
(ii) In plants	22, 23, 28, 30, 36, 41, 44, 62, 66, 67, 87
(b) Of nonmalignant cells, in relation to the appearance of the sex chromatin body in different parts of the cell cycle	61, 73, 74, 76
(c) Of tumor cells, as an indication of their karyotypes or mitotic activity	
(i) In man	4, 5, 7, 9–12, 16, 17, 19, 21, 68, 90, 91, 94, 101
(ii) In the mouse	69, 70, 85, 88, 93
(d) Of human tumor cells, relating the presence of sex chromatin bodies to the karyotypes of the tumors	3, 8
(e) Of exfoliated human normal and abnormal (including premalignant and malignant) cells, in cervical and other smears	6, 13, 18, 71, 103, 104
(f) Of lymph node, leukemic and lymphoma cells, in relation to their karyotypes or mitotic activity	25,[a] 35,[a] 46,[a] 52–54,[a] 63, 64
(g) Of human or animal cells under other pathologic or experimental conditions	57, 95, 100
2. *Estimation of other substances*	
(a) Protein, in animal cells	2, 49, 70[b]
(b) Protein, in plant cells	26
(c) Arginine, in plant cells	66
(d) Enzymes, in cells or tissues	34, 43, 45
3. *Studies which are largely concerned with various aspects of technique, including staining reactions*	24, 29, 32, 49, 51, 56, 60, 93, 96

[a] Radioactive isotope incorporation was studied in conjunction with DNA content.
[b] In tumor cells.

10% greater than those of leukocytes and fibroblasts in the preparation, in agreement with previous results (*16, 19, 51*): the carcinomas *in situ* show modes which are a little below and a little above *twice* the control values, in agreement with chromosome studies which indicated a hypotetraploid and a hypertetraploid mode, respectively. Most of the cells in both the normal and abnormal populations lie close to the mode, although the latter show a little more variation, which probably reflects a moderate degree of variation in chromosome number.

A strict comparison with another microspectrophotometric system using the same material necessitates, of course, that the material be suitable for both. Mendelsohn and Richards (72) compared measurements on air-dried mouse ascites tumor cells obtained with a Deeley integrating microdensitometer and using the two-wavelength method, and found no significant difference in the results obtained with the two methods.

However, when comparing results obtained using two different methods on similar but differently prepared material, e.g., on histologic sections using the plug or two-wavelength method and on smears using a scanning method, a wide spread in the values obtained with only one of the techniques would suggest that there was an appreciable degree of random variation associated with this technique, but caution should be exercised in assigning the variation to any particular aspect of the technique (for instance, the variation could perhaps represent a real variation in stain content rather than errors in measurement). Also, if a different ratio between two cell types (e.g., epithelial cells and leukocytes) was obtained with the two techniques, this difference could arise from the presence of a systematic error associated with one of the measuring techniques; it could also be due to a difference in stain intensity of technical origin.

3. *Some Applications of the Deeley Integrating Microdensitometer*

A useful bibliography was published in 1964 relating to microspectrophotometry in general and the estimation of DNA content in particular; it is contained in a paper by Garneau (50). Some of the studies that have been made with the Deeley integrating microdensitometer are listed in Table I.

ACKNOWLEDGMENTS

The author wishes to thank Messrs. Barr and Stroud Ltd. for permission to use the photograph and drawing of their Integrating Microdensitometer and Mr. F. S. Stewart, A.M.I.E.E., for his advice and for formulating the numerical example illustrating distributional error.

REFERENCES

1. Adams, L. R., A photographic cytophotometric method which avoids distributional error. *Acta Cytol.* **12**, 3–8 (1968).
2. Alvarez, M. R., and Cowden, R. H., Cytophotometric study of acid dye binding by proteins of isolated frog liver nuclei in relation to nuclear size. *Histochemie* **7**, 22–27 (1966).

3. Atkin, N. B., Sex chromatin and chromosomal variation in human tumours. *Acta Unio Intern. Contra Cancrum* **16**, 41–46 (1960).

4. Atkin, N. B., Some general considerations on the chromosomes of human tumours. *In* "Human Chromosomal Abnormalities" (W. M. Davidson and D. Robertson Smith, eds.), pp. 135–142. Staples Press, London, 1961.

5. Atkin, N. B., The relationship between the deoxyribonucleic acid content and the ploidy of human tumours. *Cytogenetics (Basel)* **1**, 113–122 (1962).

6. Atkin, N. B., The deoxyribonucleic acid content of malignant cells in cervical smears. *Acta Cytol.* **8**, 68–72 (1964).

7. Atkin, N. B., The chromosomal changes in malignancy; an assessment of their possible prognostic significance. *Brit. J. Radiol.* **37**, 213–218 (1964).

8. Atkin, N. B., Die chromosomale Basis von Sexchromatinabweichungen in menschlichen Tumoren. *Wien. Klin. Wochschr.* **76**, 859–862 (1964).

9. Atkin, N. B., Sex chromosome studies in trophoblast. *In* "The Early Conceptus, Normal and Abnormal" (W. Wallace Park, ed.), pp. 130–134. Univ. of St. Andrews, 1965.

10. Atkin, N. B., The influence of nuclear size and chromosome complement on prognosis of carcinoma of the cervix. *Proc. Roy. Soc. Med.* **59**, 979–982 (1966).

11. Atkin, N. B., DNA content of breast lesions. *Lancet* **1**, 1372 (1966).

12. Atkin, N. B., Perimodal variation of DNA values of normal and malignant cells. *Acta Cytol.* **13**, 270–273 (1969)

13. Atkin, N. B., The use of microspectrophotometry. *Obstet. Gynecol. Surv.* **24**, 794–804 (1969).

14. Atkin, N. B., and Baker, M. C., Chromosome abnormalities as primary events in human malignant disease: Evidence from marker chromosomes. *J. Natl. Cancer Inst.* **36**, 539–557 (1966).

15. Atkin, N. B., and Ohno, S., DNA values of four primitive chordates. Chromosoma **23**, 10–13 (1967).

16. Atkin, N. B., and Richards, B. M., Deoxyribonucleic acid in human tumours as measured by microspectrophotometry of Feulgen stain: A comparison of tumours arising at different sites. *Brit. J. Cancer* **10**, 769–786 (1956).

17. Atkin, N. B., and Richards, B. M., Clinical significance of ploidy in carcinoma of cervix: Its relation to prognosis. *Brit. Med. J.* **II**, 1445–1446 (1962).

18. Atkin, N. B., Boddington, M. M., and Spriggs, A. I., Deoxyribonucleic acid measurements on buccal cell nuclei in megaloblastic anaemias. *Nature* **195**, 394–395 (1962).

19. Atkin, N. B., Mattinson, G., and Baker, M. C., A comparison of the DNA content and chromosome number of fifty human tumours. *Brit. J. Cancer* **20**, 87–101 (1966).

20. Atkin, N. B., Mattinson, G., Beçak, W., and Ohno, S., The comparative DNA content of 19 species of placental mammals, reptiles and birds. *Chromosoma* **17**, 1–10 (1965).

21. Atkin, N. B., Richards, B. M., and Ross, A. J., The deoxyribonucleic acid content of carcinoma of the uterus: An assessment of its possible significance in relation to histopathology and clinical course, based on data from 165 cases. *Brit. J. Cancer* **13**, 773–787 (1959).

22. Avanzi, S., and D'Amato, F., New evidence on the organization of the root apex in leptosporangiate ferns. *Caryologia* **20**, 257–264 (1967).

23. Avanzi, S., Brunori, A., D'Amato, F., Ronchi, V. N., and Mugnozza, G. T. S., Occurrence of 2C (G_1) and 4C (G_2) nuclei in the radicle meristems of dry seeds in *Triticum durum* Desf. Its implications in studies on chromosome breakage and on developmental processes. *Caryologia* **16**, 553–558 (1963).

24. Bahr, G. F., and Wied, G. L., Cytochemical determinations of DNA and basic protein in bull spermatozoa. Ultraviolet spectrophotometry, cytophotometry and micro-fluorometry. *Acta Cytol.* **10**, 393–412 (1966).
25. Balfour, B. M., Cooper, E. H., and Meek, E. S., DNA metabolism of the immunoglo-bulin-containing cells in the lymph nodes of rats. *Res. J. Reticuloendothelial Soc.* **2**, 379–395 (1965).
26. Barnard, E. A., Quantitative cytochemical observations on a specific nucleoprotein reaction in cell nuclei. *Nature* **186**, 447–449 (1960).
27. Benirschke, K., and Malouf, N., Chromosome studies of *Equidae. In* " Equus " (H. Dathe, ed.), pp. 253–284. Tierpark, Berlin, 1967.
28. Bennici, A., Buiatti, M., and D'Amato, F., Nuclear conditions in haploid *Pelargonium in vivo* and *in vitro. Chromosoma* **24**, 194–201 (1968).
29. Böhm, N., and Sandritter, W., Feulgen hydrolysis of normal cells and mouse ascites tumor cells. *J. Cell Biol.* **28**, 1–7 (1966).
30. Brunori, A., and D'Amato, F., The DNA content of nuclei in the embryo of dry seeds of *Pinus pinea* and *Lactuca sativa. Caryologia* **20**, 153–161 (1967).
31. Bunsen, R. W., and Roscoe, H. E., Photochemische Untersuchungen. *Ann. Physik* [2] **117**, 529 (1862).
32. Burks, J. L., and Bakken, A. H., Measurements of deoxyribonucleic acid on mam-malian germ cells using the integrating microdensitometer. *Nature* **203**, 325–326 (1964).
33. Caspersson, T., "Cell Growth and Cell Function." Norton, New York, 1950.
34. Coimbra, A., and Tavares, A. S., Deoxyribonuclease II, acid ribonuclease and acid phosphatase activities in normal and chromatolytic neurons. Histochemical and photometric study. *Histochemie* **3**, 509–520 (1964).
35. Cooper, E. H., Peckham, M. J., Millard, R. E., Hamlin, I. M. E., and Gerard-Marchant, R., Cell proliferation in human malignant lymphomas. Analysis of labelling index and DNA content in cell populations obtained by biopsy. *European J. Cancer* **4**, 287–296 (1968).
36. D'Amato, F., and Avanzi, S., The shoot apical cell of *Equisetum arvense*, a quiescent cell. *Caryologia* **21**, 83–89 (1968).
37. Davies, H. G., and Walker, P. M. B., Microspectrophotometry of living and fixed cells. *Progr. Biophys. Biophys. Chem.* **3**, 195–236 (1953).
38. Davies, H. G., Wilkins, M. H. F., and Boddy, R. G. H. B., Cell crushing: A technique for greatly reducing errors in microspectrometry. *Exptl. Cell Res.* **6**, 550–553 (1954).
39. Deeley, E. M., An integrating microdensitometer for biological cells. *J. Sci. Instru.* **32**, 263–267 (1955).
40. Deeley, E. M., Scanning apparatus for the quantitative estimation of deoxyribonucleic acid content. *Biochem. Pharmacol.* **4**, 104–112 (1960).
41. Deeley, E. M., Davies, H. G., and Chayen, J., The DNA content of cells in the root of *Vicia faba. Exptl. Cell Res.* **12**, 582–591 (1957).
42. Deeley, E. M., Richards, B. M., Walker, P. M. B., and Davies, H. G., Measurements of Feulgen stain during the cell-cycle with a new photo-electric scanning device. *Exptl. Cell Res.* **6**, 569–572 (1954).
43. Felgenhauer, K., and Glenner, G. G., Quantitation of tissue-bound renal aminopepti-dase by a microdensitometric technique. *J. Histochem. Cytochem.* **14**, 53–63 (1966).
44. Fernández-Gómez, M. E., Rate of DNA synthesis in binucleate cells. *Histochemie* **12**, 302–306 (1968).
45. Filipe, M. I., and Dawson, I. M. P., Qualitative and quantitative enzyme histochemis-try of the human endometrium and cervix in normal and pathological conditions. *J. Pathol. Bacteriol.* **95**, 243–258 (1968).

46. Foadi, M. D., Cooper, E. H., and Hardisty, R. M., DNA synthesis and DNA content of leucocytes in acute leukaemia. *Nature* **216**, 134–136 (1967).
47. Garcia, A. M., A one-wavelength, two-area method in microspectrophotometry for pure amplitude objects. *J. Histochem. Cytochem.* **13**, 161–167 (1965).
48. Garcia, A. M., Stoichiometry of dye binding versus degree of chromatin coiling. *In* "Introduction to Quantitative Cytochemistry-II" (G. L. Wied and G. F. Bahr, eds.). Academic Press, New York, 1970.
49. Garcia, A. M., and Iorio, R., Studies on DNA in leucocytes and related cells of mammals. V. The fast green-histone and the Feulgen-DNA content of rat leucocytes. *Acta Cytol.* **12**, 46–51 (1968).
50. Garneau, R., Analyses quantitatives cytospectrophotométriques de L'ADN "in situ" dans la thyroïde humaine. *Laval Med.* **35**, 71–211 (1964).
51. Hale, A. J., The leucocyte as a possible exception to the theory of deoxyribonucleic acid constancy. *J. Pathol. Bacteriol.* **85**, 311–326 (1963).
52. Hale, A. J., and Wilson, S. J., The deoxyribonucleic acid content of leucocytes in normal and in leukaemic human blood. *J. Pathol. Bacteriol.* **77**, 605–614 (1959).
53. Hale, A. J., and Wilson, S. J., The deoxyribonucleic acid content of the leucocytes in human blood, bone-marrow and lymph-glands. *J. Pathol. Bacteriol.* **82**, 483–501 (1961).
54. Hale, A. J., Cooper, E. H., and Milton, J. D., Studies of the incorporation of pyrimidines into DNA in single leukaemic and other proliferating leucocytes. *Brit. J. Haematol.* **11**, 144–161 (1965).
55. Howling, D. H., and Fitzgerald, P. J., The nature, significance and evaluation of the Schwarzschild-Villiger (SV) effect in photometric procedures. *J. Biophys. Biochem. Cytol.* **6**, 313–337 (1959).
56. James, J., Constancy of nuclear DNA and accuracy of cytophotometric measurement. *Cytogenetics (Basel)* **4**, 19–27 (1965).
57. James, J., Feulgen-DNA changes in rat liver cell nuclei during the early phase of ischaemic necrosis. *Histochemie* **13**, 312–322 (1968).
58. Jansen, M. T., A simple scanning cytophotometer. *Histochemie* **2**, 342–347 (1961).
59. Jansen, M. T., and Leeflang, C. W., Feulgen-DNA content of individual bull spermatozoa. *Exptl. Cell Res.* **44**, 614–616 (1966).
60. Kiefer, G., Kiefer, R., and Sandritter, W., Cytophotometric determination of the binding of basic dyes by DNA following acetylation. *Histochemie* **14**, 65–71 (1968).
61. Klinger, H. P., Schwarzacher, H. G., and Weiss, J., DNA content and size of sex chromatin positive female nuclei during the cell cycle. *Cytogenetics (Basel)* **6**, 1–19 (1967).
62. La Cour, L. F., Deeley, E. M., and Chayen, J., Variations in the amount of Feulgen stain in nuclei of plants grown at different temperatures. *Nature* **177**, 272–273 (1956).
63. Lampert, F., DNS (Feulgen)- und Chromosomenuntersuchungen bei Mongolismus mit akuter Leukämie. *Klin. Wochschr.* **45**, 512–516 (1967).
64. Lampert, F., Kerntrockengewicht, DNS-Gehalt und Chromosomen bei akuten Leukämien im Kindesalter. *Arch. Abt. B. Zellpath.* **1**, 31–48 (1968).
65. Lapham, L. W., Tetraploid DNA content of Purkinje neurons of human cerebellar cortex. *Science* **159**, 310–312 (1968).
66. McLeish, J., Comparative microphotometric studies of DNA and arginine in plant nuclei. *Chromosoma* **10**, 686–710 (1959).
67. McLeish, J., and Sunderland, N., Measurements of deoxyribose nucleic acid (DNA) in higher plants by Feulgen photometry and chemical methods. *Exptl. Cell Res.* **24**, 527–540 (1961).

68. Meek, E. S., The cellular distribution of deoxyribonucleic acid in primary and secondary growths of human breast cancer. *J. Pathol. Bacteriol.* **82**, 167–176 (1961).
69. Meek, E. S., Deoxyribonucleic acid content of mouse ascites tumour cells in interphase and mitosis. *Brit. J. Cancer.* **15**, 162–167 (1961).
70. Meek, E. S., Quantitative cytochemical analysis of protein and deoxyribonucleic acid in ascites tumour cells. *Brit. J. Cancer* **16**, 157–162 (1962).
71. Meek, E. S., Application of the Feulgen reaction to smears in exfoliative cytology. *J. Clin. Pathol.* **15**, 387–388 (1962).
72. Mendelsohn, M. L., and Richards, B. M., A comparison of scanning and two-wavelength microspectrophotometry. *J. Biophys. Biochem. Cytol.* **5**, 707–709 (1958).
73. Mittwoch, U., Barr bodies in relation to DNA values and nuclear size in cultured human cells. *Cytogenetics (Basel)* **6**, 38–50 (1967).
74. Mittwoch, U., Atkin, N. B., and Ellis, J. R., Barr bodies in triploid cells. *Cytogenetics (Basel)* **2**, 323–330 (1963).
75. Mittwoch, U., Kalmus, H., and Webster, W. S., Deoxyribonucleic acid values in dividing and non-dividing cells of male and female larvae of the honey bee. *Nature* **210**, 264–266 (1966).
76. Mittwoch, U., Lele, K. P., and Webster, W. S., Relationship of Barr bodies, nuclear size and deoxyribonucleic acid value in cultured human cells. *Nature* **205**, 477–479 (1965).
77. Mittwoch, U., Lele, K. P., and Webster, W. S., Deoxyribonucleic acid synthesis in cultured human cells and its bearing on the concepts of endoreduplication and polyploidy. *Nature* **208**, 242–244 (1965).
78. Muramoto, J., Ohno, S., and Atkin, N. B., On the diploid state of the Fish order *Ostariophysi*. *Chromosoma* **24**, 59–66 (1968).
79. Naora, H., Microspectrophotometry of cell nuclei stained with the Feulgen reaction. IV. Formation of tetraploid nuclei in rat liver cells during post-natal growth. *J. Biophys. Biochem. Cytol.* **3**, 949–976 (1957).
80. Ohno, S., and Atkin, N. B., Comparative DNA values and chromosome complements of eight species of fishes. *Chromosoma* **18**, 455–466 (1966).
81. Ohno, S., Muramoto, J., Christian, L., and Atkin, N. B., Diploid-tetraploid relationship among old-world members of the fish family *Cyprinidae*. *Chromosoma* **23**, 1–9 (1967).
82. Ohno, S., Wolf, U., and Atkin, N. B., Evolution from fish to mammals by gene duplication. *Hereditas* **59**, 169–187 (1968).
83. Ornstein, L., The distributional error in microspectrophotometry. *Lab. Invest.* **1**, 250–265 (1952).
84. Patau, K., Absorption microphotometry of irregular-shaped objects. *Chromosoma* **5**, 341–362 (1952).
85. Pearson, A. E. G., and Atkin, N. B., Changes in the deoxyribonucleic acid content of mouse Sarcoma 37 cells following serial irradiation. *Nature* **186**, 647–648 (1960).
86. Pollister, A. W., and Ornstein, L., The photometric chemical analysis of cells. *In* "Analytical Cytology" (R. C. Mellors, ed.), pp. 431–518. McGraw-Hill, New York, 1959.
87. Rees, H., Deoxyribonucleic acid and the ancestry of wheat. *Nature* **198**, 108–109 (1963).
88. Richards, B. M., Deoxyribose nucleic acid values in tumour cells with reference to the stem-cell theory of tumour growth. *Nature* **175**, 259 (1955).
89. Richards, B. M., Cytochemistry of the nucleic acids. *In* "Protoplasmatologia" (L. V. Heilbrunn and F. Weber, eds.), pp. 2–38. Springer, Vienna, 1966.

90. Richards, B. M., and Atkin, N. B., DNA content of human tumours: Change in uterine tumours during radiotherapy and their response to treatment. *Brit. J. Cancer* **13**, 788–800 (1959).
91. Richards, B. M., and Atkin, N. B., The differences between normal and cancerous tissues with respect to the ratio of DNA content to chromosome number. *Acta Unio. Intern. Contra Cancrum* **16**, 124–128 (1960).
92. Ruch, F., Principles and some applications of cytofluorometry. *In* "Introduction to Quantitative Cytochemistry-II" (G. L. Wied and G. F. Bahr, eds.), Academic Press, New York, 1970.
93. Sandritter, W., and Böhm, N., Atypische Hydrolysekurve bei der Feulgenreaktion von Mauseascitestumorzellen. *Naturwissenschaften* **51**, 273 (1964).
94. Sandritter, W., Carl, M., and Ritter, W., Cytophotometric measurements of the DNA content of human malignant tumors by means of the Feulgen reaction. *Acta Cytol.* **10**, 26–30 (1966).
95. Sandritter, W., Hilwig, I., Engelbart, K., Kiefer, G., and Kiefer, R., Versuche zur Carcinogenese in vitro. Chromosomenanalysen und cytophotometrische Messungen des DNS-Gehaltes. *Z. Krebsforsch.* **67**, 57–68 (1965).
96. Sandritter, W., Jobst, K., Rakow, L., and Bosselmann, K., Zur Kinetik der Feulgenreaktion bei verlängerter Hydrolysezeit. Cytophotometrische Messungen im Sichtbaren und Ultravioletten Licht. *Histochemie* **4**, 420–437 (1965).
97. Sparshott, S. M., and Meek, E. S., The quantitative distribution pattern of deoxyribonucleic acid in cell cultures. *Exptl. Cell Res.* **44**, 521–526 (1966).
98. Swift, H., Cytochemical techniques for nucleic acids. *In* "The Nucleic Acids" (E. Chargaff and J. N. Davidson, eds.), pp. 51–92. Academic Press, New York, 1955.
99. Swift, H., and Rasch, E., Microphotometry with visible light. *In* "Physical Techniques in Biological Research" (G. Oster and A. W. Pollister, eds.), pp. 353–400. Academic Press, New York, 1956.
100. Tavares, A. S., Maia, J. C., Pereira, J. M. R., and da Costa, J. G., Estudos de Citologia do Cancro. IV. Variações do A. D. N. nuclear nas cervicites e no carcinoma *in situ* do colo uterino. *Acta Gynaecol. Obstet. Hispano Lusitana* **12**, 245–267 (1963).
101. Tavares, A. S., Pereira, J. M. R., and Branco, J. B., Estudios de Citologia do Cancro. II. A determinação do A.D.N. no citodiagnóstico do cancro vesical. *J. Med.* **50** 361–368 (1963).
102. Walker, P. M. B., and Richards, B. M., Quantitative microscopical techniques for single cells. *In* "The Cell" (J. Brachet and A. E. Mirsky, eds.), pp. 91–138. Academic Press, New York, 1959.
103. Wied, G. L., Messina, A. M., Meier, P., and Rosenthal, E., Deoxyribonucleic acid assessments on Feulgen-stained endometrial cells and comparison with fluorometric values on acridine-orange stained material. *Lab. Invest.* **14**, 1494–1499 (1965).
104. Wied, G. L., Messina, A. M., and Rosenthal, E., Comparative quantitative DNA-measurements on Feulgen-stained cervical epithelial cells. *Acta Cytol.* **10**, 31–37 (1966).

RECENT PROGRESS IN QUANTITATIVE CYTOCHEMISTRY: INSTRUMENTATION AND RESULTS*

Torbjörn Caspersson and Gösta Lomakka

INSTITUTE FOR MEDICAL CELL RESEARCH AND GENETICS, MEDICAL NOBEL INSTITUTE,
KAROLINSKA INSTITUTET, STOCKHOLM, SWEDEN

The purpose of this survey is to present views on suitable ways to promote instrument development for quantitative cytochemical work in general and to trace the general tendencies in the work at present. This will be done against the background mainly of the work being undertaken in this field at our laboratory in Stockholm.

During the last year three tendencies have been evident in the biophysical work on instrumentation for quantitative cytochemical work with optical techniques, defined in a wide sense; that is, absorbometric techniques in general for the visible, ultraviolet, and also the X-ray regions and interferometric techniques, working in the visible spectral range.

One trend concerns the efforts to carry the high-resolution techniques to the physical limit with regard to object size. One reason for doing this is the desire to get different kinds of information about the nucleic acid components and the different proteins of the individual nuclear structures—namely, chromosomes and nucleoli in normal and pathologic conditions.

Another trend is to improve the procedures for measuring the ultimate results of the main synthetic processes—the bulk of the cytoplasmic processes —a field with a series of special technical problems caused mainly by the irregular shapes and optical inhomogeneity met in practically every type of

* The development of the different types of instrumentation described has been supported mainly by the Wallenberg Foundation and the Swedish Medical Research Council. The biomedical working lines which have induced the various construction projects have been supported from several Swedish sources and from the U.S. National Institutes of Health (C 3082 and C 4716) and the Damon Runyon Memorial Fund.

cellular material. Here considerable technical developments have been made in recent years.

The third tendency is to extend the automation of measurement techniques and data processing in order to facilitate instrumental work and make it possible to operate with a large series of biological objects.

In spectrophotometric as well as in interferometric and X-ray absorbometric work it is convenient to distinguish between two types of measurements (*3–13*). One is the spot measurement and the other the integrating method of measuring. Spot measurement is the determination of optical constants of a very small spot in the specimen—as a rule of a magnitude corresponding to the limit of resolution of the optical system used. An absorption spectrum, for instance, of such a minute spot in the cell gives information about the composition of this small part of the cell. In the study of dynamic biomedical problems, which entirely dominates the field of modern cytochemistry, spot measurements are, as a rule, of only limited value; the aim of most investigations is to determine total amounts of substances within defined regions of tissues or cells. This requires integrating measurements, which predominate in biological work at present. The major complication in such modes of work is the optical inhomogeneity of the biological materials —a factor that entirely dominates the technical arrangements of the measuring instrumentation. Integrating measurements of inhomogeneous objects can be made in different ways. The most direct but in principle most laborious way is the measuring of a very great number of individual spots, which are so small that they can be regarded individually as homogeneous, and then the integration of the data over the desired part of the tissue. So great are the advantages of this technique over any others proposed to date that the present review will be confined to such procedures. They are the only general methods by which practically any type of inhomogeneous object can be mastered, up to such extreme cases of inhomogeneity as, for instance, metaphase plates. It should be noted that the need to measure a very great number of individual spots and to process the different measurements individually before integrating them is no complication nowadays because the procedures for measuring as well as those for data conversion can easily be partially or entirely automated. When developing cytochemical lines of work one point, regrettably often overlooked, is that techniques have to be available which guarantee that the optical conditions in the specimens are such that the data in the measuring instrumentation can be correctly analyzed. This requires techniques for determining not only the optical conditions in the specimens to be measured but also the optical constants of the substances to be determined under the conditions met in the specimens (*6–8, 15, 16, 18, 19, 22*). Devices for such determinations can be incorporated with any high-class microspectrophotometer.

I. The Need for Different Types of Instruments

Planning for general equipment for quantitative cytochemical work involves weighing complexity in instrumentation against costs and operational convenience and the desired penetrative power of the techniques in the biomedical field to be approached. The following are general comments on this question:

(1) It is necessary to accept that instrumentation for quantitative cytochemical work is bound—because of the difficulties inherent in work with such very small objects—to be relatively complex and expensive as compared with analogous techniques working in the macroscale. As a result, when planning instrumentation, reasonable efforts should always be made to reach the goal with the simplest possible instrumentation (4, 5, 18).

(2) The foremost strength and penetration of the field of quantitative cytochemistry stems from the opportunity which available techniques afford of obtaining information about several cell parameters rather than being limited to individual ones. This obviously calls for more complex instrumentation, but also raises the question of the extent to which instruments can be made to serve different purposes, thus saving expense and possibly time as well (4, 5, 18).

(3) The fields of application in biology and medicine are very wide; there is thus a multitude of different types of biologic materials and metabolic problems to be studied. Consequently the most suitable instrumentation is the one in which work can be performed on many different types of materials and in which different types of measurements can be done. In practice, however, the choice is often between different instruments, each specialized and

	Microspectro-photometry in UV and visible	Interferometry	Microfluorimetry
Multipurpose recording instruments	Universal MSP	Recording interferometer	Recording ultra-microfluorimeter, fluorescence, and excitation spectra
Rapid instruments for population studies	Rapid MSP (rapid Feulgen MSP)	Rapid interferometer	
Rapid instruments with preselection of measuring field	Rapid MSP with preselection	Rapid interferometer with preselection	Rapid utramicro-fluorimeter with preselection

FIG. 1. Survey of the main classes of instruments.

possibly very efficient for specific materials or specific tasks, and techniques that are more generally applicable to different types of biologic materials, but less rapid and convenient.

Such questions have been in the minds of all planners and builders of cytochemical instrumentation. In our work these considerations have led to

FIG. 2. The universal microspectrophotometer (UMSP).

three groups (Fig. 1) of instrumentation, which will be dealt with below. First, there is the more general, universal instrument, applicable to a variety of different objects and types of problems, but slow in operation—the "multipurpose" instrument. Second, we have the group of very fast instruments for metabolic studies on specialized materials—the rapid scanning and integrating instruments. Third, there is a group of instruments, not quite as fast as the second group, but permitting a much wider scope in the work on cell metabolic problems—the methods for automatized scanning and integrating work with preselection of the measuring field.

II. Multipurpose Instruments

The "multipurpose" instruments must necessarily be rather slow in their work, besides being relatively complex and expensive. The trend is undoubtedly for more and more work to be done with the other two groups of instruments. With access to the larger instruments of the multipurpose type—which also permit calibration and control experiments—rapid instruments can be built relatively simply in a laboratory and thus supplement the multipurpose instruments for large-scale work.

The first instrument of this general type to be developed was the universal microspectrophotometer (UMSP), which has reached the stage indicated in Fig. 2. This permits work with a resolution close to the physical limit in middle and long-wave ultraviolet and in the visible spectral region. The instrument is built for spot measurements as well as for integrating measurements. In the case of integrating work, the instrument records the absorption of the specimen along scanning lines swept by very small measuring spots in a regular pattern. At the same time, recordings are made of the integrated value of the extinction over the area measured. A commercial model of this instrument developed by our staff in very constructive collaboration with the Zeiss Company, Oberkochen, West Germany, is presented in Fig. 3. An important contribution by Zeiss to the development of this instrument was the construction of achromatic lenses for the ultraviolet regions, the Ultrafluars. These were developed specifically for this instrument but have since found many other uses.

Because of its aim—high-resolution studies with spot as well as integrating measurements on a variety of types of biological objects, including very heterogeneous specimens—the instrument is built as a recording machine and is thus rather slow. It can also be applied to certain other modes of work, such as analyses of plates from photographic X-ray absorbometry, and, with minor supplementary additions, also to model substance measurements of different types.

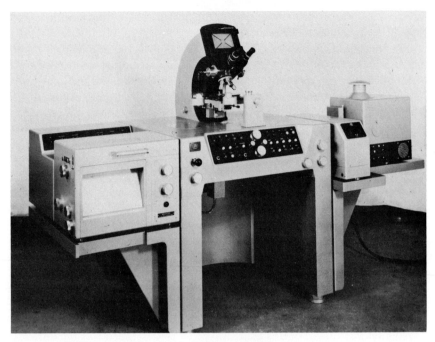

FIG. 3. The UMSP-I built by C. Zeiss.

The ultramicrointerferometer (*18*, *23*), which scans and records in a way similar to the UMSP, is presented in Fig. 4. It records the optical increment of small measuring spots along scanning lines arranged in a regular pattern and, on the same paper, the running integrals of the measurements. The latter parameter is proportional to the total amount of dry mass in the specimen passed by the measuring ray during its scanning. The mode of measurement is suitable for biological objects of different types and different sizes. This, together with its use in combination with the UMSP for model substance measurements, explains why it is classified among the multipurpose recording instruments.

Figure 5 shows a fluorimeter built for high-resolution measurements in biological material (*12*, *17*). It can be used for recording fluorescence as well as excitation spectra and covers the range from 240 nm to the long-wave visible region. A built-in reference system makes absolute measurements possible. It is combined with an ultramicrospectrophotometer, as these two types of measurements are often needed in the same object. Furthermore, the rapid quenching of fluorescence in many objects makes it undesirable to work with the same specimen in a series of different and separate instruments, since this inevitably increases the ultraviolet exposure.

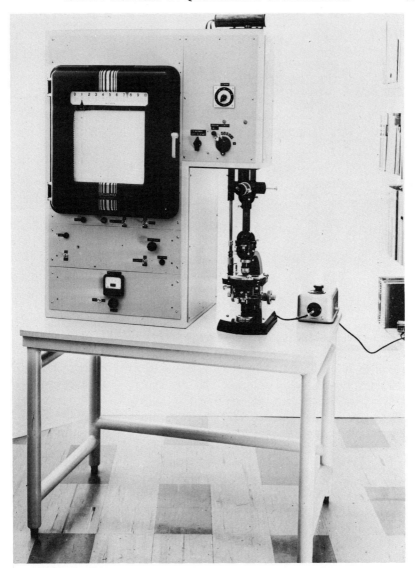

FIG. 4. A scanning, recording, integrating ultramicrointerferometer.

III. Instruments for Cell Population Studies

The rapid advances in nucleic acid and protein biochemistry in the late 1950's and early 1960's presented a whole new group of problems for quantitative cytochemistry and also incited the trends described in the introduction.

FIG. 5. High-resolution ultramicrofluorimeter for recording of fluorescence and excitation spectra, combined with an UMSP.

Some of these cell metabolic problems, mainly concerning protein and nucleic acid syntheses, could be approached by changes in existing instrumentation; others, however, could not, and thus called for new developments (6, 7, 13, 14, 16, 19). The first field to come into focus was the kinetics of cellular synthetic processes, such as protein synthesis, and of disturbances in these processes. In general such approaches call for measurements of whole cells—the cell being the metabolic unit in growth and differentiation. Kinetic studies call for measurements of very large numbers of individual cells in different stages of growth or function. Determination of total contents of substances in whole cells in such large series is hopelessly slow in instruments of the type described above. This led to the development of the rapid instruments for population studies indicated in Fig. 1. An additional reason for these efforts came out of the earlier observations in our laboratory of a very pronounced variability between the individual cells in tumor cell populations with regard to different cell parameters, such as total amounts of DNA, RNA, and protein in whole cells, in nuclei, and in nucleoli.

The idea behind the construction of rapid instruments was as follows: The primary measuring problem concerns the determination of total amounts of substances in whole cells. This rules out preparation techniques involving sectioning of tissues. It is necessary to have the whole cell on one slide. Using suitable biological material and applying simple cell isolation techniques, it

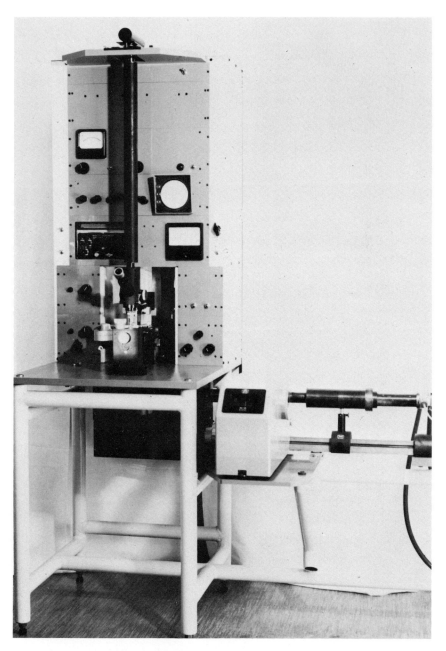

FIG. 6. A rapid microspectrophotometer for ultraviolet and/or visible light.

FIG. 7. A rapid microspectrophotometer especially for Feulgen work.

is almost always easy to obtain specimens in which the cells lie free from each other. This is also a very favorable situation for automated measurements: The cell and its nearly free surroundings simply have to be swept by a measuring beam, the necessary conversions and integrations being made electronically. This entirely eliminates the time-consuming recording and the still more time-consuming analysis of curves. The time required for the work is thus 100–1000 times less than with recording machines.

When building rapid machines the working speed has to be chosen carefully. A higher speed than that mentioned above can easily be achieved but will complicate the instrumentation. In all our instruments we have selected as a suitable speed a measuring time for an average mammalian cell of the order of 5 sec. The reason is that in all cytologic work the observer needs to select the cells to be measured individually; even in large-scale routine work, experience has shown that this calls for about ten times longer and thus no practical purpose would be served by making the measurements still faster.

A rapid ultramicrospectrophotometer for the wavelength range 240–700 nm is presented in Fig. 6. The size and shape of the rectangular scanning field can be regulated by eye and the measuring spot can be as small as the optics permit in the different wavelength ranges used.

The very great demand for measurements of total nuclear DNA in Feulgen preparations led to the construction of the instrument shown in Fig. 7. It can be used even in such extremely inhomogeneous objects as whole metaphase plates. For this work in the visible regions the construction can be considerably simpler than that of the above-mentioned instrument.

Rapid interferometric measurements are especially important as they give the total amount of dry mass in the cells, which can serve as a reference system for several other types of cytochemical work. The construction in Fig. 8, originally developed by G. Lomakka, permits very precise measurements on fixed as well as living cell material. A special practical advantage is that measurement is independent of irregularities in the source of light. A commercial model of this instrument has recently become available.

The X-ray absorption–mass determination method of Engström and Lindström has been modified and developed by L. Carlson so as to be suitable for mass determinations in large series of cells. The interferometric procedure is usually more convenient but certain objects present difficulties for interferometric work and in these the X-ray absorption technique is of advantage.

Figure 9 gives an example of the type of information the rapid instruments can obtain from large cell populations—histograms of the amounts of various compounds in individual cells.

These techniques have been used by many research groups on a wide variety of problems within normal and pathologic cell physiology. Extensive

FIG. 8. A rapid microinterferometer.

studies have been made of the kinetics of nucleic acid and protein synthesis in the normal interphase—a period of the cell cycle that was earlier practically inaccessible for such work.

The work on cytochemical tumor variability has also been quite extensive, engaging groups in our laboratory and at the Children's Cancer Research

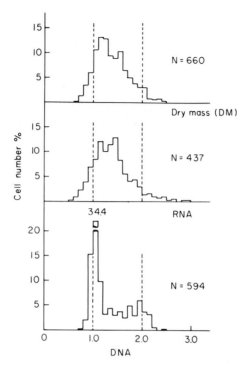

FIG. 9. Examples of frequency diagrams from a cell population (mouse fibroblasts *in vitro*). Determinations in individual cells of relative dry mass, RNA and DNA. *N* is number of cells measured. The broken lines indicate the postmitotic (1.0 relative units) and the premitotic (2.0 relative units) values. From Zetterberg and Killander (*25*).

Foundation in Boston. Figure 10 shows the differences in variability in populations of normal cells and of different strains of tumor cells for one parameter. Figure 11 illustrates that in a great number of cell strains studies the cells without pathologic growth characteristics in general show comparatively little variability (they lie close to the top of the curve in the figure), while those capable of invasive growth show very large variability (far down the curve). The relation between malignancy and variability is further illustrated by Fig. 12, in which the uppermost curve gives the mass distribution histogram of a malignant sarcoma strain, which is much broader than the distribution histogram of the normal strain (bottom curve). The middle histogram is from a spontaneously arisen strain of the sarcoma culture that has lost its capacity of invasive growth, as tested in hamster cheek pouch. The variability is correspondingly lower.

In the collaboration between the two institutes mentioned above, these techniques have been extensively used for work on tumor chemotherapy. The

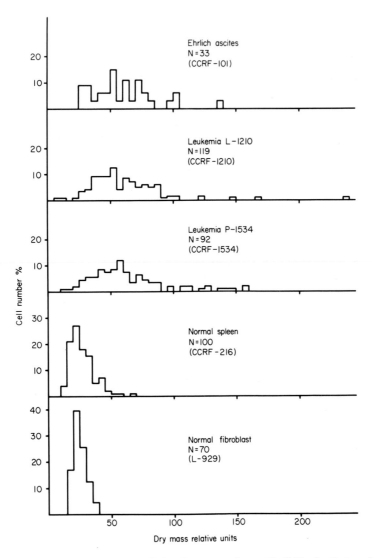

FIG. 10. Frequency diagram of relative dry mass values of individual cells in cell lines derived from neoplastic and normal mouse tissues. From Caspersson *et al.* (*11*).

rapid cytochemical techniques permit very fast determinations of the re- action on different external agents of DNA, RNA, and overall protein- synthetic processes in the cell. This makes the methods suitable for screening purposes as well as for rapid studies of the effects on complex cell populations of combinations of different antitumor agents with different points of attack in the cell.

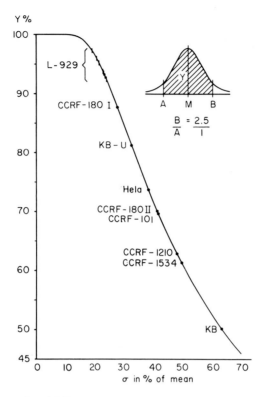

Fig. 11. Degree of variability (standard deviation in percent of mean) in dry mass content of individual cells in different cell lines: Comparison of cell lines derived from normal (L-929) and neoplastic (all other cell lines) sources. The CCRF-180 I and KB-U cell lines have lost their malignant properties of growth in the Syrian hamster cheek pouch. The relationship between the standard deviation in any set of observations and the percentage of the total number of cells which falls between the limit dry mass values A and B when $A:B = 1:2.5$ is demonstrated. From Caspersson et al. (11).

Figure 13 draws attention to a field of some promise, the identification of precancerous cells by cytochemical approaches. The figure shows histograms of cell populations from scrapings of human cervix in a normal state, in the cancer *in situ* situation, and in a cancerous state. The differences in the RNA and protein distribution patterns between normal and cancerous cell populations are particularly evident and it is of interest to note that the cancer *in situ* picture shows deviations from the normal pattern even before distinct signs of invasiveness can be detected. This situation has been confirmed in about 20 different cancer *in situ* cases and work is in progress on the study of other precancerous lesions.

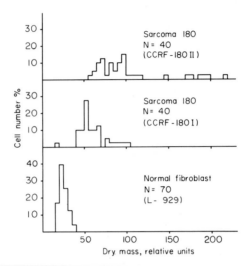

FIG. 12. Frequency distributions of relative dry mass values of individual cells in cell lines derived from normal (L-929) and neoplastic (Sarcoma 180) mouse tissues. The CCRF-180 I strain has lost its property of malignant growth in the hamster cheek pouch, whereas CCRF-180 II has retained its malignant properties. From Caspersson *et al.* (*11*).

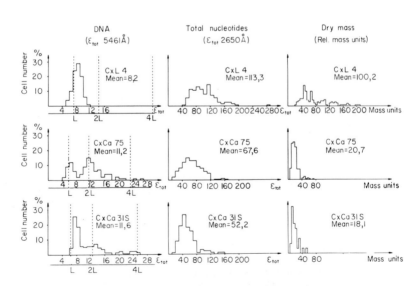

FIG. 13. Frequency diagrams from scrapings from human portio uteri. CxL 4 normal mucosa. CxCa 75 cancer. CxCa 31 S cancer *in situ*. From Caspersson (*2*).

IV. Instruments with Preselection of Measuring Field

The applicability of the rapid instruments to different types of problems concerning normal and pathologic cell growth, differentiation, and function was found to be so wide that in the last few years it became desirable to obtain

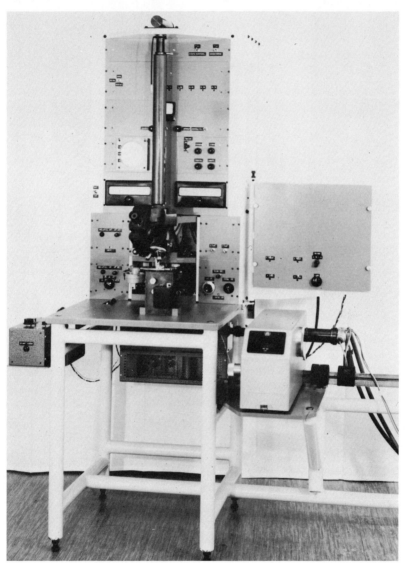

FIG. 14. Rapid microspectrophotometer for UV and/or visible light, with preselection of measuring field.

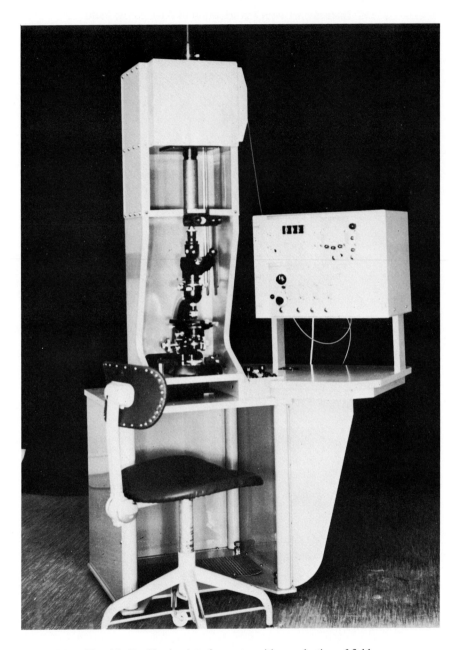

FIG. 15. Rapid microinterferometer with preselection of field.

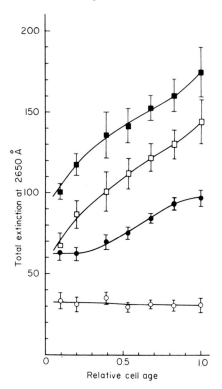

FIG. 16. Amounts of (from above) total cellular RNA, cytoplasmic RNA, total nuclear acids, and nuclear RNA, plotted against interphase time. From Zetterberg (*24*).

similar types of information from selected parts of the cells, such as the cell nucleus, for the study of cytoplasmic interactions, but also from individual nuclear elements, chiefly chromosomes and nucleoli. This has led to development of the rapid techniques with preselection of field (Fig. 1). In these instruments, which are necessarily considerably more complex than those described above, the observer can select the size and shape of the field and the instrument makes integrating measurements over this selected field only.

The instrument shown in Fig. 14 is an ultramicrospectrophotometer for ultraviolet and visible light of the type that has proved very suitable for the study of nucleus–cytoplasmic interactions and also of small endocellular elements. Figure 15 shows an ultramicrointerferometer that is similar in principle to the instruments in Fig. 8, but with a special arrangement for preselection of field. Together these instruments permit separate determinations of nuclear and cytoplasmic proteins and nucleic acids (Fig. 16).

Figure 17 shows a high-resolution ultramicrospectrofluorimeter with a special arrangement for selecting the measuring field from whole cells down

Fig. 17. High-resolution microfluorimeter with preselection of measuring field.

to the limit of resolution, for instance, for studying details in chromosomes. High-resolution fluorimetry on biological objects is faced with two major difficulties. One is to avoid stray light in the optics during measurement and the other concerns the need to keep the radiation load on the object at a very low level. The instrument has been built up with these points in mind. It permits determinations on areas under 1 mμ^2 and work with quite radio-sensitive substances. The bulk of work with this third group of instruments (Fig. 1) has been developed to such an extent during the last few years that they may become the predominant instruments in the near future.

In the study of the complex cell populations often met in biomedical work, especially in studies of carcinogenesis and cell differentiation or of such tissues as spleen or bone marrow, a special problem arises when individual cell types have to be measured according to two or more different methods, for instance, spectrophotometry and interferometry. If these measurements have to be made on different instruments, the identification of individual cells in the

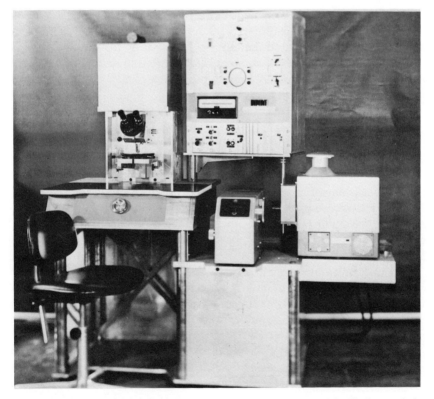

Fig. 18. Instrument for rapid measurements of two parameters (extinction and dry mass) in large and complex cell populations.

second instrument is normally so time-consuming that it may jeopardize work on sensitive objects. An instrument has therefore been developed (Fig. 18) for simultaneous measurements of extinction in UV and/or visible and optical increment, thus eliminating the need for time-consuming identification.

V. Measurements on Very Small Objects

The instruments described above represent two of the trends mentioned in the introduction, techniques for large-scale integrating measurements and extension of automation. The main reason for the third trend towards working close to the resolution limits of the techniques, is the desire to study the detailed organization of the cell nucleus, a field of rapidly growing importance for the study of cell and tissue differentiation and the mechanisms regulating gene function.

Extensive studies have been made of the possibilities of pressing the resolution of ultramicrospectrophotometry (in the first place in UV), interferometry, and fluorimetry very close to the physical limits set by the wavelength used. By developing a whole series of minor modifications in several of the instruments described above it has proved possible to come very close to this goal (1, 10, 20, 21).

Because this field is relevant for several branches of biological research and in view of the relative complexity of the methodological side, descriptions of the latter will be omitted and instead a series of examples will be presented to serve as demonstrations of what is at present within reach. The examples will be taken from work on metaphase chromosomes in our laboratory.

The dimensions of most mammalian chromosomes are in the range from one to a few wavelengths of visible light and thus the larger ones should be amenable for studies with reasonable accuracy of, for instance, their DNA content, while others are beyond reach or only possible to measure with low accuracy (1).

This is demonstrated by Fig. 19, which gives measurements of individual chromosomes in Chinese hamster metaphase plates, carried out in the ultraviolet and in the visible in Feulgen preparations with a modified high-resolution UMSP. Each diagram represents 10 male and 10 female metaphase plates. The individual spots give measurements of individual chromosome pairs and thus the spread of the points gives the dispersion of the measurements and—assuming that the amount of DNA is constant and the same for each individual chromosome in different metaphase plates—the accuracy attainable with the methods on the very borderline of the physically permissible region. The diagrams demonstrate that in the larger chromosomes, changes

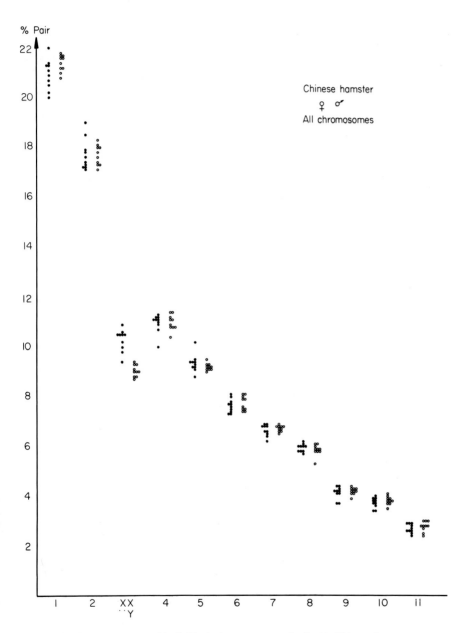

FIG. 19. Measurements of individual chromosome pairs in 20 Chinese hamster meta-phase plates, by aid of a modified UMSP for very high resolution.

of the order of 5% should be observable, for instance, during a virus transformation process.

Human chromosomes are obviously of special interest. Because of the greater number per metaphase plate and the corresponding smallness of most of the chromosomes, the situation is not quite as favorable as in the Chinese hamster. Figure 20 illustrates the situation with the different groups

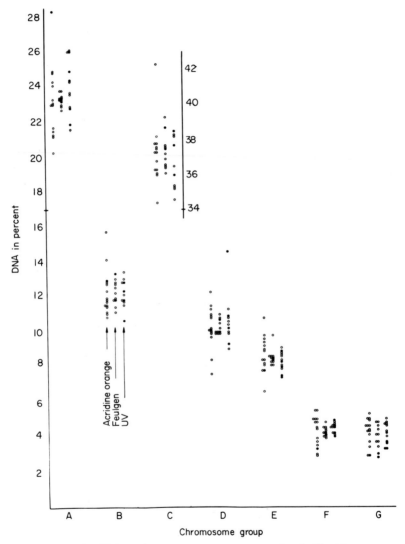

FIG. 20. Human chromosome groups, measured as in Fig. 19.

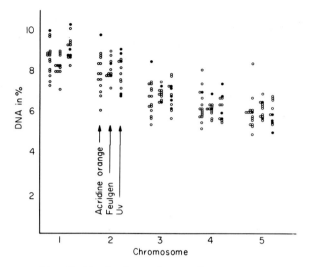

FIG. 21. The five largest human chromosome pairs.

of human chromosomes and Fig. 21 shows that the five largest chromosomes are easily within reach, with a reproducibility in the measurements similar to that attained for the larger Chinese hamster chromosomes. However, working at the very limit of resolution it is possible to get further information, namely, on the distribution of DNA along the chromosomes. Figure 22 illustrates the DNA distribution along the chromosomes of rye, from the work by Heneen and others at our institute. The plant has a set of seven chromosome pairs similar in size to the larger human chromosomes. Morphologically it is very difficult to differentiate between the chromosomes, which are of the same length and differ only in a few morphologic details. In a scanning high-resolution UMSP, consecutive measuring sweeps were made across the chromosome at intervals of 0.25 μ along its total length. The total extinction values (which are proportional to the amount of DNA swept by the ray) for each sweep across the chromosome were plotted against chromosome length to give a diagrammatic presentation of the DNA distribution patterns along the chromosome. Figure 22 shows that these DNA patterns are reproducible for the individual chromosomes in different metaphase plates but differ distinctly between the chromosomes of different numbers. Such patterns can also be obtained from several of the Chinese hamster chromosomes and the largest of the human chromosomes.

The possibility of obtaining such detailed DNA distribution patterns in small metaphase chromosomes led to the use of the DNA determinations as a base line for more penetrating studies of the chemical differentiation of the

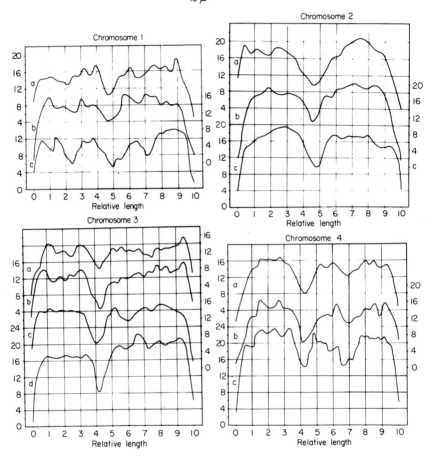

FIG. 22.

metaphase chromosomes (*10, 20, 21*). The very great sensitivity of the ultra-microfluorimetric procedures makes an approach with fluorescing DNA-binding compounds appear especially promising. At first, efforts were made with alkylating fluorescing agents belonging to the quinacrine group, as alkylating agents are known to react preferentially with the guanine moeity of the DNA and consequently might provide information about the statistical distribution of bases along the chromosome. Model experiments showed that amounts of 10^{-16} gm DNA can be measured fluorimetrically when bound to quinacrine derivatives. This in an amount corresponding to only about 100

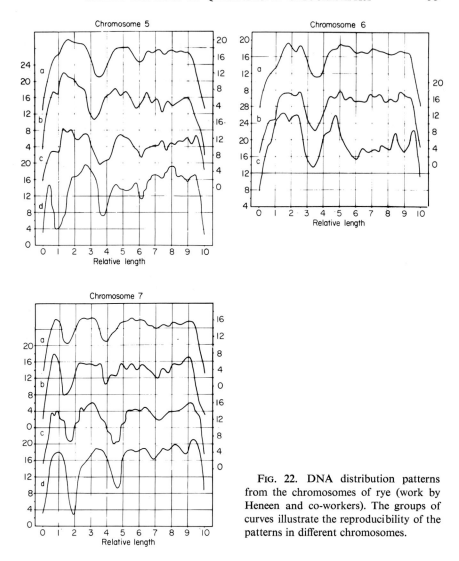

FIG. 22. DNA distribution patterns from the chromosomes of rye (work by Heneen and co-workers). The groups of curves illustrate the reproducibility of the patterns in different chromosomes.

genes, estimating the size of the gene to be 1000 nucleotide pairs. Selective binding of these compounds has been demonstrated, as has the preferential binding of other types of DNA reagents to certain chromosome regions. Of special interest are the results on *Vicia*, for which it had previously been shown that alkylating agents give chromosome breakages at well-defined regions. These are the regions which react preferentially with the compounds mentioned, indicating also a preference for heterochromatic regions, as defined by cold treatment. The regions preferentially broken in the M chromosome and which are strongly fluorescent show the cold starvation phenomenon of Darlington and La Cour. Comparison of the DNA distribution pattern along the chromosome with the fluorescence picture (Figs. 23 and 24) shows that the binding of the fluorescence compound is attributable not to differences in the DNA amounts of different parts, but to a qualitative difference between the chromosome regions in question. The high-resolution ultramicrospectrofluorimeter (Fig. 17) permits measurements of the fluorescence in such narrow chromosome bands.

FIG. 23. Quinacrine mustard-treated M chromosome from *Vicia faba*. 2000×. From Caspersson *et al.* (*20*).

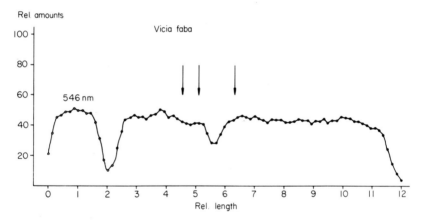

FIG. 24. DNA distribution pattern (Feulgen preparation) of chromosome as in Fig. 23. The arrows mark the places for the strongly fluorescent bands in the quinacrine mustard preparations. From Caspersson *et al.* (*20*).

This investigation has been extended to include several other plants as well as mammalian material. The larger chromosomes from Chinese hamster and from man can also be used to follow the relationship between fluorescence intensity and DNA amounts in spite of the small size of the objects, which well illustrates the penetration of the present high-resolution techniques.

This is not the place to present these studies in greater detail. The examples are intended to illustrate that the high-resolution techniques for ultramicrospectrophotometry in ultraviolet and visible light, for interferometry and for fluorimetry can be used very close to the physical limit set by the wavelength of the light, besides being applied in routine work on a wide variety of biological material.

Summarizing, it can be said that in recent years great improvements have been made in the quantitative cytochemical tools for the study of cell metabolic processes by the use of large cell series and especially for the work on nucleus–cytoplasmic interactions. Considerable progress has also been made in the study of the details of the nuclear structures by improving techniques for work at very high resolutions.

REFERENCES

1. Carlson, L., Caspersson, T., Foley, G. E., Kudynowski, J., Lomakka, G., Simonsson, E., and Sören, L., The application of quantitative cytochemical techniques to the study of individual mammalian chromosomes. *Exptl. Cell Res.* **31**, 589–594 (1963).
2. Caspersson, O., Quantitative cytochemical studies on normal, malignant and atypical cell populations from the human uterine cervix. *Acta Cytol.* **8**, 45–60 (1964).
3. Caspersson, T., "Cell Growth and Cell Function." Norton, New York, 1950.
4. Caspersson, T., Quantitative cytochemical methods for the study of cell metabolism. *Experientia* **11**, 45–60 (1955).
5. Caspersson, T., A project for the development of quantitative methods for the endocellular range. *Exptl. Cell Res.* Suppl. 4, 3–8 (1957).
6. Caspersson, T., Quantitative cytochemistry for the study of normal and abnormal growth. *Federation Proc.* **20** (1961).
7. Caspersson, T., Cytochemical aspects of the problem of tumour growth. The morphological precursors of cancer. *Proc. Intern. Conf. Univ. Perugia*, 1961.
8. Caspersson, T., Physikalisch-optische Methoden in der Histochemie. *Acta Histochem.* Suppl. 6, 21 (1965).
9. Caspersson, T., Farber, S., Foley, G. E., and Killander, D., Cytochemical observations on the nucleolus-ribosome system. Effects of actinomycin D and nitrogen mustard. *Exptl. Cell Res.* **32**, 529–552 (1963).
10. Caspersson, T., Farber, S., Foley, G. E., Kudynowski, J., Modest, E. J., Simonsson, E., Wagh, U., and Zech, L., Chemical differentiation along metaphase chromosomes. *Exptl. Cell Res.* **49**, 219–222 (1968).
11. Caspersson, T., Foley, G. E., Killander, D., and Lomakka, G., Cytochemical differences between mammalian cell lines of normal and neoplastic origins. Correlation with heterotransplantability in Syrian hamsters. *Exptl. Cell Res.* **32**, 553–565 (1963).
12. Caspersson, T., Hillarp, N. A., and Ritzén, M., Fluorescence microspectrophotometry of cellular catecholamines and 5-hydroxitryptamine. *Exptl. Cell Res.* **42**, 415–428 (1966).
13. Caspersson, T., and Lomakka, G. M., Scanning microscopy techniques for high resolution quantitative cytochemistry. *Ann. N.Y. Acad. Sci.* **97**, 449–463 (1962).
14. Caspersson, T., and Lomakka, G., Microscale spectroscopy. *In* "Instrumentation in Biochemistry" (T. W. Goodwin, ed.), p. 25. Academic Press, New York, 1966.
15. Caspersson, T., Lomakka, G., and Carlson, L., Eine Instrumentausrüstung für quantitative Cytochemie. *Acta Histochem.* **9**, 139–162 (1960).
16. Caspersson, T., Lomakka, G., and Caspersson, O., Quantitative cytochemical methods for the study of tumor cell populations. *Biochem. Pharmacol.* **4**, 113–127 (1960).

17. Caspersson, T., Lomakka, G., and Rigler, R., Jr., Registrierender Fluoreszenzmikrospektrograph zur bestimmung der Primär und Sekundärfluoreszenz verschiedener Zellsubstanzen. *Acta Histochem.* Suppl. 6, 23 (1965).
18. Caspersson, T., Lomakka, G., and Svensson, G., A coordinated set of instruments for optical quantitative high resolution cytochemistry. *Exptl. Cell Res.* Suppl. 4, 9–24 (1957).
19. Caspersson, T., Vogt-Köhne, L., and Caspersson, O., Relations between nucleus and cytoplasm in normal and malignant growth. *In* " Cell Physiology of Neoplasia". Univ. of Texas Press, Austin, Texas, 1960.
20. Caspersson, T., Zech, L., Wagh, U., Modest, E. J., and Simonsson, E., *Exptl. Cell Res.* (1969) (in press).
21. Caspersson, T., Zech, L., Wagh, U., Modest, E. J., and Simonsson, E., *Exptl. Cell Res.* (1969) (in press).
22. Lomakka, G., Computers for microphotometric data analyses. *Exptl. Cell Res.* Suppl. 4, 54–57 (1957).
23. Svensson, G., Scanning interference microphotometry. *Exptl. Cell Res.* Suppl. 4, 165–171 (1957).
24. Zetterberg, A., Nuclear and cytoplasmic nucleic acid content and cytoplasmia protein synthesis during interphase in mouse fibroblasts *in vitro*. *Exptl. Cell Res.* 43, 517–525 (1966).
25. Zetterberg, A., and Killander, D. Quantitative cytochemical studies on interphase growth. II. Derivation of synthesis curves from the distribution of DNA, RNA and mass values of individual mouse fibroblasts *in vitro*. *Exptl. Cell. Res.* 39, 22–32 (1965).

QUANTITATIVE ASPECTS OF CHROMOSOMAL COMPOSITION IN ASCARIS MEGALOCEPHALA

Karl B. Moritz

ZOOLOGY INSTITUTE, UNIVERSITY OF MUNICH, MUNICH, GERMANY

I. Introduction

A great number of experiments in developmental biology demonstrate that differentiation of embryonic cells is controlled by specific cytoplasmic factors of the zygote; biochemical data, however, are very scarce (*10*). Historically, there are three experiments which show that in the differentiation process especially the nuclei undergo alterations: First, Boveri (*4*) discovered "chromatin diminution" in the *Ascaris* egg. During mitosis the long chromosomes in the presoma cells break into a number of small segments. Only the intercalary segments are incorporated in the daughter nuclei, whereas the ends are resorbed into the cytoplasm. This process does not take place in the germline blastomeres. The fate of the nuclei depends upon the cytoplasmic milieu into which they enter (*7*). Second, Seidel (*18*) found in the dragon fly's egg a cytoplasmic region, the "Bildungszentrum," which influences the nuclei so that they initiate the formation of the "Keimanlage." Third, the transplantation experiments of King and Briggs (*12*) indicated stable alterations of nuclei during egg cleavage.

It is extremely difficult to find out which cytoplasmic components act on the nuclei. In the *Ascaris* egg, as an example, the chromosomes were studied in germline and presoma cells before chromatin diminution (*14*). We hoped to find alterations in the chromosomes which could perhaps in turn give us a clue to their cause. In our first experiments we did find a very interesting variation in chromosomal DNA content. These observations inspired the investigations reported in Section IV.

57

II. Nuclear Behavior during Cleavage

The *Ascaris* egg is unique because of its determinative cleavage, i.e., the blastomeres resulting from the early divisions of the zygote are all different from one another in their developmental fate (5). Among them are the cells of the germline, which are easily identifiable because of their heterochromatic nuclei.

Both egg pronuclei are Feulgen negative. In the prophase of the first cleavage mitosis the long chromosomes become visible just beneath the nuclear membrane. In this state the organization of the plurivalent chromosomes can be seen (Fig. 1). The ends are strongly heterochromatic. They

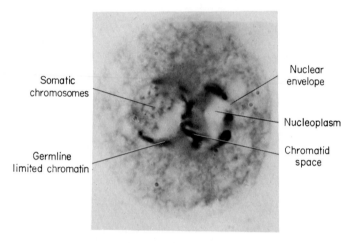

FIG. 1. Pronuclei of the zygote in prophase. The chromosome ends are curling just beneath the nuclear envelope. In the left nucleus the knoblike swellings of the intercalary chromatin, the somatic chromosomes, are easily identifiable. Stained with Heidenhain hematoxylin.

consist of the so-called germline-limited chromatin, which persists only in the germline. The intercalary region is euchromatic. Its thin threads have knoblike swellings. These swellings are probably the somatic chromosomes which appear as independent segments after chromatin diminution in the somatic cells has taken place. In metaphase the intercalary regions show no localized swellings because of their highly condensed state. There are two races in *Ascaris*, univalens and bivalens. In univalens there are two chromosomes, in bivalens four. The metaphase chromosomes differ from one another in size and shape (Fig. 2). They can therefore hardly be grouped in pairs of two at all. Similar observations were made on the heterochromatic germline chromosomes in *Cyclops* (1). In addition to this structural heterozygosity,

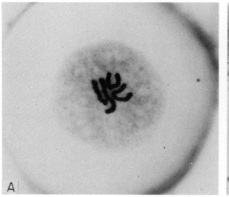 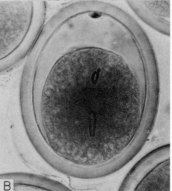

FIG. 2. Metaphase of the first cleavage mitosis in the *Ascaris* egg. In both eggs the chromosomes are in focus. (A) All four chromosomes differ from one another in size and shape; race bivalens. Section stained with hematoxylin. (B) The homologs differ in length; race univalens. Whole mount stained with acetocarmine.

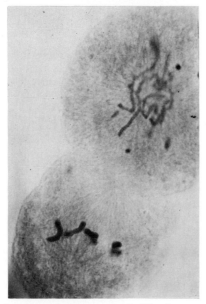

FIG. 3. Metaphase of the second cleavage mitosis. The chromosomes in the AB cell (top) are much less condensed than those of the P_1 cell. Note the different spindle orientations. Unstained cryostat section at wavelength 280 nm. Agfa Agepe film. UMSP-I equipment.

the chromosomes are characterized by their holokinetic organization. In anaphase all segments are capable of actively moving to the poles (Fig. 4). The spindle is longitudinally oriented in the polarity axis of the egg. After cytokinesis two similar blastomeres are formed but each will have a different

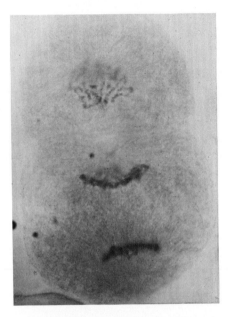

FIG. 4. Anaphase of the second cleavage mitosis. The configuration of the P_1 cell chromosomes in the anaphasic plates indicate their holokinetic property. *Top*: Note the nucleus of one daughter blastomere of AB in telophase. The cytoplasmic absorption at the vegetative pole of P_1 is very high, indicating ribosomal RNA concentration in this spindle pole. Technical data as in Fig. 3.

developmental fate. The cell at the animal pole is a presoma cell, called AB, the cell at the vegetative pole is a germline cell, called P_1.

During the second cleavage cycle the difference between the two cell lines becomes apparent: The centromeres of P_1 arrange themselves again longitudinally in the egg axis, in AB they arrange at right angles to it (Fig. 3). It is not known what kind of cytoplasmic factors direct centrosome movement, which in turn determines the furrow planes and therefore the distribution of the heterogeneous cytoplasm. Only the existence of such factors could be demonstrated (9).

The chromosomes are strikingly different in the two sister cells in the following respects: In all mitotic phases the AB cell chromosomes are less condensed than those of P_1 (Fig. 3), although in most of the cases the AB blastomere undergoes mitosis earlier than the sister cell (20). From this state of incomplete spiralization the anaphase movement starts. The distal chromosomal parts have lost their capability for active anaphase movement, as the cytologic appearance of the chromosome daughter plates indicate. The ends are passively dragged by the active intercalary parts (17). Sometimes ends of the sister chromatids stick together; then they are broken off and not incorporated in the resulting daughter nuclei. The complete diminution is, however, shifted

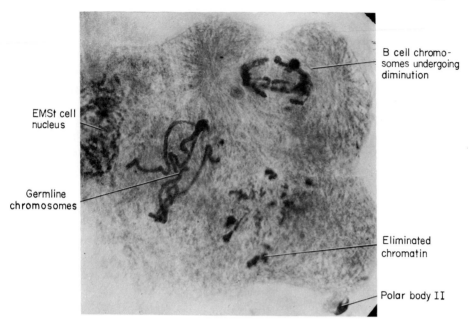

B cell chromo-
somes undergoing
diminution

EMSt cell
nucleus

Germline
chromosomes

Eliminated
chromatin

Polar body II

FIG. 5. Third cleavage mitosis. Squashed unstained cryostat section at wavelength 280 nm. Agfa Agepe FF film. UMSP-I equipment.

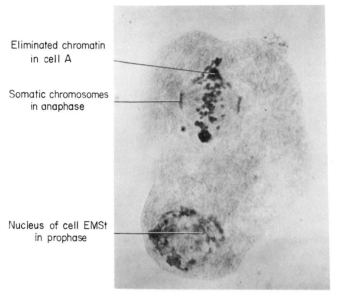

Eliminated chromatin
in cell A

Somatic chromosomes
in anaphase

Nucleus of cell EMSt
in prophase

FIG. 6. Third cleavage mitosis. Anaphase in blastomere A (top) with chromatin diminution. Race bivalens, unstained cryostat section at wavelength 280 nm. Agfa Agepe FF film. UMSP-I equipment.

to the third mitosis. In the sister cell P_1 the chromosomes maintain their holokinetic property (Fig. 4). The result of cytokinesis is the T state, with its cells A, B, and the daughter cells of P_1, the so-called blastomeres EMSt and P_2.

During the third mitotic cycle chromatin diminution takes place in the cells A, B, and EMSt. The extremely decondensed chromosomes break up into a number of segments (Fig. 5). One part of these, the somatic chromosomes, are capable of moving to the poles in anaphase (Fig. 6), whereas the ends rest in the equatorial plane and are distributed to the daughter cells at random.

III. Comparative Microphotometric Measurements on Germline and Soma Chromosomes

So far, it has been demonstrated that chromosomes of sister cells show a difference in their behavior, even before chromatin diminution. Spectrophotomicroscopic measurements have shown that this qualitative difference is accompanied by quantitative differences in chromosomal composition. As spot measurements have shown (Table I), the extinctions at the wavelengths

TABLE I

SPOT EXTINCTIONS OF DIFFERENT CHROMOSOMAL REGIONS DETERMINED IN UNSTAINED METAPHASIC CHROMOSOMES OF THE SECOND CLEAVAGE MITOSIS, $N = 12$

Wavelength (nm)	Mean extinction of chromatin in cell P_1		Mean extinction of chromatin in cell AB	
	Germline-limited chromatin	Intercalary chromatin	Germline-limited chromatin	Intercalary chromatin
260	0.454	0.247	0.385	0.213
280	0.310	0.150	0.240	0.114
$Q_{260/280}$	1.46	1.65	1.60	1.87

280 and 260 nm of the P_1 germline chromosomes are altogether higher than those of AB cell chromosomes. This is easily understood because of the higher degree of chromosomal condensation in P_1. In both blastomeres the quotients of the extinction values at the two wavelengths are higher in intercalary chromatin than in germline chromatin. These quotients are also different if one compares corresponding chromosomal regions of the two cells.

These differences are not based on differences in DNA quantities. Before chromatin diminution sister cells have identical DNA contents, as shown by scanning measurements of Feulgen-stained chromosomes of the second cleavage mitosis (Table II). The differences in the UV absorption curves, therefore,

TABLE IIA

EXTINCTION INTEGRALS OF FEULGEN-STAINED MITOTIC CHROMOSOMES AND THEIR LENGTHS IN THE TWO-CELL CLEAVAGE STAGE[a]

Integral of extinction		Chromosomal length in arbitrary units	
P_1	AB	P_1	AB
73.4	73.0	$22.0 + 8.0 = 30.0$	$18.0 + 8.2 = 26.2$
70.9	69.3	—	—
68.6	71.5	$10.0 + 11.6 = 21.6$	$24.6 + 18.5 = 43.1$
67.6	72.0	$12.3 + 18.0 = 30.3$	$13.5 + 14.0 = 27.5$
67.5	64.75	$16.8 + 13.2 = 30.0$	$16.6 + 11.8 = 28.4$
65.5	68.0	$8.0 + 13.4 = 21.4$	$9.0 + 14.0 = 23.0$
59.0	60.0	$17.0 + 15.0 = 32.0$	$22.0 + 15.5 = 37.5$
55.25	52.0	$9.0 + 10.0 = 19.0$	$18.6 + 10.2 = 28.8$

[a] Between sister cells P_1 and AB there are no significant differences in total extinction values. Among the eggs the differences are significantly higher than those between sister cells. Statistics: Variance between sister cells $s_0^2 = 3.53$, variance among eggs $s_1^2 = 40.33$. F test: $F = 23.85$, $F_{0.01} = 4.14$. There is no relation between the extinction integrals of chromatin and chromosomal length. All values are derived from eggs on one slide. Race univalens.

TABLE IIB

EXTINCTION INTEGRALS OF FEULGEN-STAINED METAPHASE HOMOLOGS OF SISTER CELLS P_1 AND AB IN AN UNIVALENS EGG[a]

Cell	Chromosome I	Chromosome II	Sum	Quotient
P_1	28	21	49.0	1.33
AB	34.8	15	49.8	2.32

[a] The sums are identical for both blastomeres, the quotients are different.

may be attributed to differences in the chromosomal protein fraction. In conformity with these data the AB cell chromosomes bind less Fastgreen FCF at pH 8.1 after DNA extraction than the P_1 cell chromosomes do. These observations indicate that the chromosomes that will undergo diminution contain in the intercalary regions as well as in the germline-limited material less basic protein than the corresponding segments of the germline chromosomes.

Taking the influence of localized different cytoplasmic components on the nuclei so far as a fact (7), the question still remains whether the cytoplasm acts on primarily identical chromosomes during nuclear division, i.e., after breakdown of the nuclear envelope, or whether it already acts on interphasic chromosomes during their replication. There is some evidence for the first hypothesis (15, 16): germline and soma spindles differ from one another in their ribosomal RNA content (Fig. 4). But the aforementioned data on chromosomal composition indicate that changes take place before the AB cell chromosomes are directly exposed to the soma cytoplasm.

If one tries to compare the results of our measurements with the earlier described cytologic observations, one can state a correlation between the loss of basic protein and the decrease of spiralization of the AB cell chromosomes as well as their stickiness in anaphase. And since there is also a correlation between the degree of metaphasic condensation and anaphasic activity—strongly condensed chromosomes behave holokinetically, whereas the ends of weakly condensed ones have no activity—there may also be an interdependence between the amount of protein and the spindle activity of chromosomes.

IV. The Variation in DNA Content of Germline Chromosomes

The Feulgen photometric measurements given in Table II had shown that DNA replication is complete in chromosomes that will undergo diminution. These same measurements yielded two additional results.

(1) There are differences in DNA content among the eggs produced by one female, differences which cannot be explained by variations in technique. Immediately connected with this result is another one. In all those cases where it had been possible to measure the extinction integrals of both chromosomes within one cell separately, the quotient of the integrals was different from egg to egg.

(2) In one egg of this same material each of the two chromosomes in both blastomeres was measured separately (Table IIB). The structural heterozygosity of the chromosomes was apparent. The sums of the extinction integrals were identical in both sister cells. The quotients of the extinction integrals, however, were different. This means that a recombination of chromomatin must have taken place during interphase. If such a recombination is possible in the germline, the differences in DNA content between the eggs from one female become intelligible. They can arise by fusion of gametes, each carrying chromosomes different in their DNA content. These differences are due to

different amounts of germline-limited chromatin. This is indicated by the variation in the quantities of the material eliminated in different eggs.

These results lead us to three questions:

(1) What is the chemical nature of the germline-limited chromatin?
(2) What is the function of this chromatin in the germline?
(3) What are the mechanisms that cause the changes in DNA content of the plurivalent chromosomes?

To find an answer to this last question the following measurements were done. The DNA content of Feulgen-stained spermatids was measured microphotometrically (Fig. 7). The material was again *Ascaris megalocephala* race univalens. The data from the measured sperm, which represent one chromosome each in this race, show a continuous variation (Fig. 8). Since all sperm

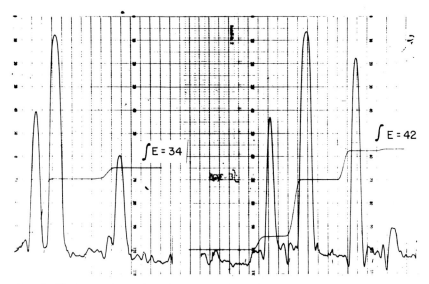

FIG. 7. The integrating registration curves of two Feulgen-stained spermatids from one male. The differences in the integrated values are not due to measuring error resulting from a heterogeneous background. Race univalens.

chromosomes of one male are derived from either the maternal or the paternal chromosome of the zygote, one could have expected to substantiate the basis of earlier mentioned data, two groups of sperm with different DNA content. The deviation from this expectation, that is, the continuous variation of the extinction integrals, indicates that there is also a recombination of DNA in the germline among the germline chromosomes. Evidence that variation in the extinction integrals is not due to variations in preparation of the material and/or the technique of measurement comes from three sources.

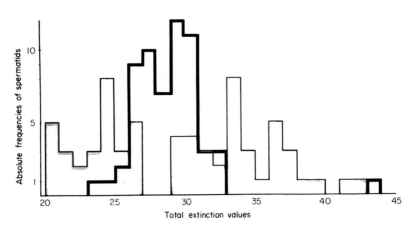

Fɪɢ. 8. Frequency distributions of the total extinctions of Feulgen-stained *Ascaris* spermatids from two males. There is a wide continuous variation in one case ($N = 60$); in the other, there are two blocks ($N = 66$). Race univalens.

(1) Application of the two-wavelength method shows that unspecific components of extinction can be excluded. For that purpose the absorption spectrum in a less-condensed nucleus was determined. Then, at the wavelengths of half-maximal and maximal extinction the spermatids were measured in the scanning procedure. The total extinction values showed a relation of one to two.

(2) We could estimate technical errors by repeated measurements. Before repetition of each measurement the object was reoriented in the microscope. The scanning of the spermatids has been made overlappingly.

(3) To test the reliability of the method, microphotometric measurements on other material were made for comparison. Material was chosen in which the DNA content of the structures measured can safely be considered constant. We used the X chromosome of pachytene and diplotene stages in the spermatogenesis of Locusta migratoria.

This locust chromosome is easily identifiable as a single heterochromatic, rod-shaped element (Fig. 9). Its DNA content is in the same order of magnitude as that of *Ascaris* spermatids. The objects were a few microns in diameter. Scanning with the 0.5-μ and the 1.0-μ diaphragm gave results that were not significantly different from one another. To test whether the shape of the chromosome plays a role in the extinction measured, each locust X chromosome was measured both crosswise and lengthwise. There were no differences in the extinction integrals (Fig. 10). For the locust X chromosomes variance analysis (Table III) shows a variation due to experimental error and in addition a variation between X chromosomes. The variation coefficient for

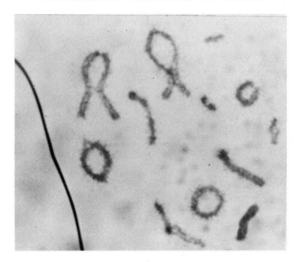

FIG. 9. Diplotene of the locust spermatogenesis. There are 11 lampbrushlike autosomes showing chiasmas to some extent and one heterochromatic element, the X chromosome (bottom right). Feulgen-stained squash preparation.

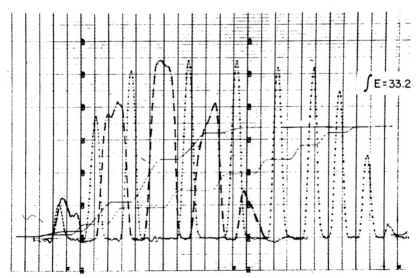

$$\int E = 33.2$$

FIG. 10. The integrating curves of one Feulgen-stained locust X chromosome. The scanning registrations were made crosswise and lengthwise. The integrals are identical.

repetitions is $v_r = 2.19\%$, the one for the X chromosomes $v_n = 1.49\%$, with a mean of $M = 16.44 \int E$.

Applying the F test, the quotient of the mean squares of the integrated extinctions between the nuclei and of integrated extinctions for repeated

TABLE III

VARIANCE ANALYSIS OF CYTOPHOTOMETRIC DATA ON 24 FEULGEN-STAINED MEIOTIC LOCUST
X CHROMOSOMES[a]

Variation	Degrees of freedom	Sum of squares	Mean square	F Test
Total	95	17.86	—	—
Between X chromosomes	23	8.47	0.368	$F = 2.83$
Between readings	72	9.39	0.130	$F_{0.01} = 2.07$
Variances	$s_r^2 = 0.13$	$s_n^2 = 0.06$		
Coefficients of variation	$v_r = 2.19\%$	$v_n = 1.49\%$		

[a] The variance among X chromosomes is different from zero ($s_n^2 = 0.06$). The F test shows a significant F value. The mean total extinction is $M = 16.44$ $\int E$.

measurements is $F = 2.83$; i.e., it indicates a significant difference. This means that the differences between the X chromosomes of different nuclei cannot be explained exclusively by experimental error, that is, by variations in the measuring process.

However, the quotient is not too different from $F_{0.01} = 2.07$. Therefore, I do not believe that the inhomogeneity of the data has any biological significance, that is, I do not believe that it is based on DNA differences between X chromosomes. The only conclusion from these data is that the measuring procedure can be carried out with high accuracy and that it reveals only the degree of our insufficiency in preparing the material. In support of this interpretation I may add that no investigator has succeeded as yet in furnishing proof with the same technique and similar material for an absolute constancy of DNA in nuclei (11).

Considering the data from these control experiments when judging the great variation of extinctions found in *Ascaris* sperm as expressed by the highly significant F values (Table IV) one has to conclude that technical errors cannot be the sole source of the deviation. As shown by variance analysis, the experimental error in sperm of all investigated males is generally larger than that in the DNA determination of locust X chromosomes. This is certainly due to the heterogeneous background in the germline cells of *Ascaris*. Nevertheless, the differences between spermatids can under no circumstances be explained on the basis of experimental errors.

The means of extinction integrals for sperm of different males also deviate strongly. This comparison, however, is subject to another, probably relatively small, error: The slides carrying sperm from different males have been made at different times.

TABLE IV

STATISTICAL ANALYSIS OF VARIATION OF CYTOPHOTOMETRIC DATA ON FEULGEN-STAINED SPERMATID NUCLEI FROM UNIVALENS MALES[a]

| Ascaris | N | Mean∫E | Variance | | F Value | $F_{0.01}$ | Coefficient of variation | |
			Between readings	Between nuclei			Residual (%)	Between nuclei (%)
IV	31	21.45	0.32	1.93	13.2	2.03	2.62	6.48
V	31	18.74	0.29	2.05	15.1	2.03	2.89	7.63
II	16	32.90	0.47	8.28	36.2	2.62	2.08	8.75
III	30	31.44	0.51	6.13	25.0	2.04	2.28	7.88

[a] Applying the *F* test for each male there are significant differences in DNA content among the spermatids. In all cases the reading error is greater than in the measurement of the locust X chromosomes (Table II). Nevertheless, the differences between the spermatids are not due to chance.

Furthermore, there are more than two classes of sperm within one male. In one case, the critical difference for the extinction integrals of sperm of a male was determined to be $D_k = 1.18 \int E$. This is the difference in extinction between two spermatids, which can still be interpreted as a chance deviation. The empirical variation was, however, $I = 9.8 \int E$. Into this interval one could fit without any trouble more than four classes of spermatids. The existence of more than two classes calls for some kind of new combination of DNA. Since there are more than even four classes, one is forced to conclude that not only whole ends of chromosomes, but sections of the distally located germline chromatin are exchanged between the homologs, in some kind of an unequal crossover event.

We wondered whether the stage of exchange during germline development could be determined more precisely. As mentioned earlier, mitotic exchange could be involved. A meiotic exchange mechanism, however, could also account for the DNA variation of spermatids, either exclusively or in addition to a mitotic exchange. In order to distinguish between these different possibilities, other stages of spermatogenesis were investigated, that is, primary spermatocytes, metaphase and anaphase of both meiotic divisions within one male of race univalens.

For metaphase-I there were no differences in DNA content between different tetrads. (It should be emphasized here that each tetrad represents one spermatocyte, since each spermatocyte has only one tetrad; Table V.) This uniformity in DNA content excludes mitotic exchange at the four-strand

TABLE V

STATISTICS OF FEULGEN-PHOTOMETRIC DATA ON TETRADS IN *Ascaris* SPERMATOCYTES[a]

Variation	Degrees of freedom	Sum of squares	Mean square
Between nuclei	18	577.27	32.07
Between readings	19	209.27	11.01
Total	37	786.54	
Variances	$s_r^2 = 11.01$	$s_n^2 = 10.53$	
Coefficients of variation	$v_r = 3.12\%$	$v_n = 3.05\%$	

[a] The F values, $F = 2.9$ and $F_{0.01} = 3.1$, mean that the variation may be due to technical errors. The mean extinction integral is $M = 106.5$ ∫E. Race univalens.

stage. We have to consider this possibility on the basis of structural heterozygotes participating in the event. And at random segregation of chromatids half of all four-strand-stage exchanges would lead to daughter nuclei differing in DNA content. Mitotic exchange which we have to consider here, therefore, can only take place at the two-strand stage.

It should be mentioned that Cytophotometric proof of DNA constancy is not possible in all cases. Spermatocytes of *Ascaris* are characterized during their growth phase by large nuclei (*13*). These regularly contain a Feulgen-positive chromocenter, representing germline-limited chromatin. The rest of the nuclear volume is almost entirely Feulgen negative. Photometry of these nuclei leads to strong variations of extinction integrals.

Taking the aforementioned uniformity of DNA content at this stage as a fact, we would have to conclude that sections of the germline-limited chromatin, all different in extent, have been so strongly despiralized that they slip below the threshold of detection with this technique. This interpretation finds support in the observation that in the subsequent pronucleus stage, the originally entirely Feulgen-negative nuclei regain their stainability simultaneously with chromosomal condensation. It is reasonable to assume that at these stages of despiralisation the germline-limited chromatin is genetically active.

In first meiotic divisions the members of dyad pairs were measured separately (Table VI). As in the extinction integrals of whole tetrads, the sums of dyad pairs were identical in different cells. The differences between the members of pairs, however, were highly significant. If one excludes meiotic recombination for a moment, the result would point to prereductional segregation of structurally heterozygous bivalents. It should be pointed out

TABLE VI

VARIANCE ANALYSIS OF FEULGEN-PHOTOMETRIC DATA ON FIRST MEIOTIC DIVISIONS IN *Ascaris* SPERMATOGENESIS[a]

Source of variation	Degrees of freedom	Sum of squares	Mean square	F Value
Dyads	1	178.09	178.09	15.43**
Cells	9	75.31	8.37	1.37
Repetition	20	53.31	2.67	4.32**
Discrepance	9	103.85	11.54	
Total	39	410.56		

[a] The right column shows that the preparation error is significantly (**significant at the 1% level) higher than the reading error, and that the differences between the dyads cannot be explained by experimental errors. The differences between cells are due to chance. The mean extinction integral is $M = 52.61$ ∫E. Race univalens.

here that also the difference between the residual mean square and the mean square of the repetition is significant, which means that the measuring error is considerably smaller than the preparation error. This is easily explained by the fact that the background in the stages is heterogeneous.

If one compares the experimental errors for the slides of metaphase I (Table V) and of the first meiotic division (Table VI), one can state that in the latter case the error is much smaller. In spite of that, even here the differences between nuclei are random.

TABLE VII

VARIANCE ANALYSIS OF FEULGEN-PHOTOMETRIC DATA ON SECOND MEIOTIC DIVISIONS IN *Ascaris* SPERMATOGENESIS[a]

Source of variation	Degrees of freedom	Sum of squares	Mean square	F Value
Chromosomes of the same cell	1	46.48	46.48	21.72**
Cells	5	88.25	17.65	8.25*
Repetition	12	6.01	0.50	4.28*
Discrepance	5	10.68	2.14	
Total	23	151.42		

[a] The differences among cells and also between the chromosomes of one cell are caused by different chromosomal DNA amounts (**significant at the 1% level; *significant at the 5% level). The mean extinction integral is $M = 25.5$ ∫E. Race univalens.

For the chromosomes of the second meiotic division significant differences were found both (Table VII) between different cells and between the two anaphase chromosomes of the same cell. This last-mentioned difference, together with the difference between dyads within first meiotic division, can be explained—again excluding the possibility of meiotic exchange—on the basis of pre- and postreduction side by side. This corresponds with observations reported in older papers in which the meiotic behavior of supernumerary univalent elements of germline-limited chromatin has been described (3, 6, 8, 19). Furthermore, it should be pointed out that the means for integrals of chromosomes in second meiotic division, first meiotic division, and of whole tetrads show a relation in DNA content of 1:2:4, or, respectively, $\int E = 25.5$, $\int E = 52.6$, and $\int E = 106.5$. This fact is quite comforting as far as the reliability of the method is concerned.

V. Conclusion

As shown by integrating measurements of Feulgen-stained nuclei the DNA content in the germline cells is different from individual to individual in *Ascaris*. There is evidence that in other animal species with germline chromatin the DNA content in germline cells is also not constant. For *Cyclops strenuus* it could be shown with the technique used in this study (2). In insects with germline-limited chromosomes the number of the so-called E chromosomes varies among the individuals of one species (21), which means that the DNA content of germline nuclei cannot be constant. In *Ascaris* the variation of the DNA content is caused by different amounts of germline chromatin. Within one male the DNA content per nucleus is constant, as shown by measuring different primary spermatocytes. No significant differences could be found.

In growing spermatocytes, however, and in the pronucleus stage there is a strong variation of total extinction values after Feulgen-staining. The result is interpreted as a consequence of total despiralization of different chromosomal segments, so that their DNA concentration lies below the threshold of detection with this technique. From the wide variation of DNA content in the sperm of one male (race univalens) one has to conclude that there are a great number of plurivalent chromosomes with different amounts of germline chromatin. The stage during germline development at which the DNA exchange takes place is still conjectural. There are two main possibilities:

(1) The variation in DNA content in spermatids is caused by spermatogonial, that is, mitotic exchange. The differences within pairs of integrals in the first and in the second meiotic division can be explained exclusively on the basis of pre- and postreduction occurring side by side.

(2) Meiotic exchange together with prereductional segregation explains the data as well. Meiotic exchange together with postreduction must be excluded, since in this case an identity of DNA content within pairs of dyads in first meiotic divisions should have been observed, which, however, has not been found.

There is no cytologic evidence for meiotic exchange in spermatogenesis. While oocytes show clear chiasmatic configurations (Fig. 11), in spermatocytes the chromatids are found lying adjacent to one another but strongly condensed. Fine threads, containing DNA, stretch out only towards the spindle pole.

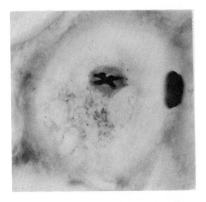

FIG. 11. Hematoxylin-stained section of an *Ascaris* oocyte in first meiotic division. The chromosomes show a chiasmatic configuration. Race univalens.

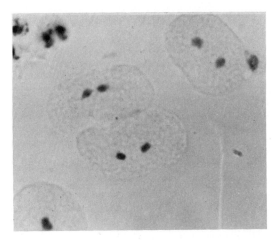

FIG. 12. Tetrad of chromosomes in spermatogenesis. Feulgen-stained squash preparation. Race univalens.

The question of whether there is mitotic, that is spermatogonial, and/or meiotic exchange can be answered unequivocally by cytophotometric analysis of tetrads (Fig. 12), which is presently under way. In the case of exclusively mitotic exchange two chromosomes within each tetrad, namely, those derived from the division of one homolog, should be equal in DNA content. The difference between the unlike chromosomes, however, should be different when several tetrads are compared. In the case of meiotic exchange all four chromosomes of a tetrad should occasionally be different from one another, provided that the chromosomes of the spermatocytes were structurally heterozygous. This actually has been found in most cases.

ACKNOWLEDGMENT

The author wishes to thank Miss Hella Röper for careful technical assistance.

REFERENCES

1. Beermann, S., Chromatin-Diminution bei Copepoden. *Chromosoma* **10**, 504–514 (1959).
2. Beermann, S., A quantitative study of chromatin diminution in embryonic mitosis of Cyclops furcifer. *Genetics* **54**, 567–576 (1966).
3. Bonnevie, K., Über Chromatindiminution bei Nematoden. *Z. Naturw.* (*Jena*) **36**, 275–288 (1902).
4. Boveri, T., Über die Differenzierung der Zellkerne während der Furchung des Eies von Ascaris megalocephala. *Anat. Anz.* **2**, 288–293 (1887).
5. Boveri, T., Die Entwicklung von Ascaris megalocephala mit besonderer Rücksicht auf die Kernverhältnisse. *Festschr. C. von Kupffer* (*Jena*) pp. 383–430 (1899).
6. Boveri, T., Über Geschlechtschromosomen bei Nematoden. *Arch. Zellforsch.* **4**, 132–141 (1909).
7. Boveri, T., Die Potenzen der Ascaris-Blastomeren bei abgeänderter Furchung. Zugleich ein Beitrag zur Frage qualitativ ungleicher Chromosomenteilung. *Festschr. R. Hertwig* (*Jena*) **3**, 131–214 (1910).
8. Geinitz, B. Über Abweichungen bei der Eireifung von Ascaris. *Arch. Zellforsch.* **13**, 588–633 (1915).
9. Guerrier, P., Fixation de la polarité chez Parascaris equorum. *Compt. Rend.* **258**, 3566–3568 (1964).
10. Hörstadius, S., Josefsson, L., and Runnström, J., Morphogenetic agents from unfertilized eggs of the sea urchin Paracentrotus lividus. *Develop. Biol.* **16**, 189–202 (1967).
11. James, J., Constancy of nuclear DNA and accuracy of cytophotometric measurement. *Cytogenetics* (*Basel*) **4**, 19–27 (1965).
12. King, T. J., and Briggs, R., *Cold Spring Harbor Symp. Quant. Biol.* **21**, 271 (1956).
13. Lin, T. P., The chromosomal cycle in Parascaris equorum: Oogenesis and diminution. *Chromosoma* **6**, 175–198 (1954).
14. Moritz, K. B., Die Blastomerendifferenzierung für Soma und Keimbahn bei Parascaris equorum. I. Cytochemische und photometrische Untersuchungen. *Arch. Entwicklungs mech. Organ.* **159**, 31–88 (1967).

15. Moritz, K. B., Die Blastomerendifferenzierung für Soma und Keimbahn bei P. e. Untersuchungen mittels UV-Bestrahlung und Zentrifugierung. *Arch. Entwicklungs mech. Organ.* **159**, 203–266 (1967).
16. Pasteels, J., Recherches sur le cycle germinal chez l'Ascaris. Etude cytochímique des acides nucleiques dans l'oogénèse, la spermatogénèse et le developpement chez Parascaris equorum. *Arch. Biol. (Liege)* **59**, 420–445 (1948).
17. Schrader, F., Notes on the mitotic behavior of long chromosomes. *Cytologia (Tokyo)* **6**, 422–430 (1935).
18. Seidel, F., Entwicklungsphysiologie des Insektenkeimes. *Verhandl. Deut. Zool. Ges. (Freiburg)* **38**, 291–336 (1936).
19. Walton, A. C., Studies on Nematode gametogenesis. *Z. Zellen-Gewebelehre* **1**, 167–239 (1924).
20. White, M. J. D., The chromosome cycle of Ascaris megalocephala. *Nature* **137**, 783 (1936).
21. White, M. J. D., "Animal Cytology and Evolution," 2nd ed., pp. 185–214. Cambridge Univ. Press, London and New York, 1954.

CYTOPHOTOMETRIC DETERMINATION OF THE NUCLEAR DNA AND RNA CONTENT DURING DEDIFFERENTIATION OF CERTAIN PLANT CELLS

*Luise Stange**

INSTITUTE OF BOTANY, TECHNICAL UNIVERSITY, HANNOVER, GERMANY

I. Introduction

In many specialized cells, which do not divide anymore, the cell reduplication cycle is only reversibly blocked. This fact is demonstrated when a mature cell gives up its functional specialization and again enters the reduplication cycle. In plants this activation of divisional functions can be induced in many cases by isolation of specialized cells from meristematic cells, thus interrupting physiologic correlations which had evolved during cell differentiation. Activation of nuclear metabolism in isolated cells has been reported to precede the first nuclear and cell division (*1, 3, 4, 6–11, 15–22*). It can be assumed that those processes which had been blocked during differentiation are activated at first. Thus, knowledge of the sequence of events in the activation process may allow conclusions about the point of attack of the correlations effective in differentiation.

In a number of investigations on cell dedifferentiation in plants, we have chosen the liverwort *Riella* as the experimental plant (*17–20*). This plant shows a relatively simple organization. The adult thallus consists of a multicellular rib or axis and a broad unistratose wing. The meristem is situated at the apex of the plant and, proceeding in the wing from the apex to the base, cells of increasing age are found. Rhizoids are present at the base of the rib. Subdivision of the wing into regions of different age is represented in Fig. 1. By cutting fragments out of the wing dedifferentiation is easily induced

* This work was supported by the Deutsche Forschungsgemeinschaft.

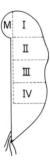

FIG. 1. *Riella helicophylla*. Subdivision of the wing into regions of different age. M: meristem.

in the cells of the isolated tissue. The first change one can observe in all isolated cells is an increase in nucleolar size, which is followed later by nuclear and cell division in many cells and finally by the regeneration of new plants. Under the usual culture conditions, the nucleoli of all cells in the original adaxial zone of the fragment show a significant increase in size 18 hr after isolation, independent of the age of the isolated cells. However, the rate of increase in nucleolar size is much higher in younger cells (from the sub-apical region, FI) than in older cells (from the basal region, FIII). Consequently, the maximal size of nucleoli observed is reached much earlier in younger than in older cells (Fig. 2). Likewise, the nuclei in fragments from the subapical region divide much earlier than those in fragments from the basal region. These results are summarized in Table I.

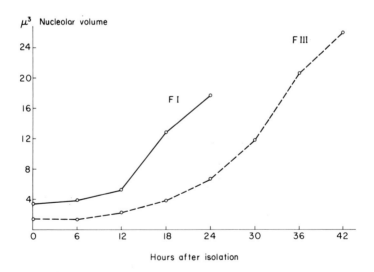

FIG. 2. Nucleolar volume of young (FI) and older (FIII) cells in dependence on time after isolation ($n = 200$). The curves break off with the onset of nuclear divisions.

TABLE I

SUMMARY OF RESULTS ON THE COURSE OF NUCLEAR ACTIVATION IN CELLS OF *Riella* ISOLATED
FROM DIFFERENT WING REGIONS[a]

	Subapical region of the wing (FI)	Basal region of the wing (FIII)
Onset of nucleolar enlargement	18	18
Maximum size of nucleolus	24	42
Onset of nuclear division	30	48
DNA content 1C	18	36
DNA content 2C	30	48

[a] The time after isolation of tissue fragments is given in hours.

II. Determination of Nuclear DNA Content

As in other cases (*7, 9–11, 16*) the cell cycle in the specializing wing cells of *Riella* is blocked in G1. Therefore, DNA must be synthesized before the nuclei can divide. The time of DNA synthesis during dedifferentiation has been determined by microspectrophotometric measurements of nuclei stained with the Feulgen reagent in fragments from the subapical and the basal region of the plant at different times after isolation.

Plants of a female clone of *Riella helicophylla* were grown under constant conditions as described previously (*17*). Out of the subapical region and the basal region of the unistratose wing of plants 6 weeks old, rectangular fragments were cut and kept under the usual culture conditions. 0, 12, 18, 24, 30, 36, 48, 72, and 96 hr after isolation fragments were fixed in ethanol–formol–acetic acid (6:3:1), passed through ethanol 96%, 70%, 40%, and water, and dried onto slides. The fragments were then hydrolyzed in 1 N HCl at 60°C for 10 min and stained with the Feulgen reagent (Pararosaniline, Gurr) for 3 hr. After washing in five changes of SO_2–water and rinsing in distilled water, the slides were passed through ethanol 40%, 70%, 96%, absolute, and were made permanent with Canada Balsam (Harleco).

Measurements of the DNA content of individual nuclei were carried out with the Zeiss universal microspectrophotometer (UMSP). A field $10 \times 10\ \mu$ was scanned using a diaphragm of 0.5 μ diameter and a height of scanning line of 0.5 μ. The extinction at the wavelength 550 nm was measured. Since the walls between cells cause absorption or stray light, it was necessary to select nuclei lying away from the intercellular cell walls. The total extinction of a nucleus is given in arbitrary units.

As reference values for the 1C and 2C content (the thallus of *Riella* is

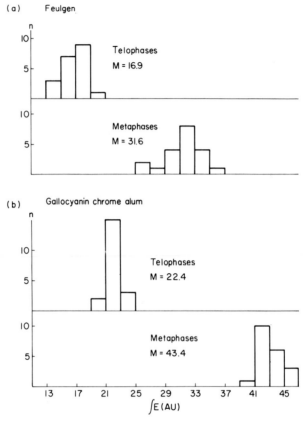

FIG. 3. Frequency distribution diagrams for total extinction (AU, arbitrary units) of telophase and metaphase nuclei. M: mean.

haploid) the extinction of metaphase and telophase nuclei was measured. The results are summarized in Fig. 3a. In this figure frequency distribution diagrams are given for each type of cell. Considering the relatively low number of observations, the ratio between the mean values for telophases (16.9) and metaphases (31.6) fits the expectation quite well.

From the numbers in Fig. 4 the time of DNA synthesis in fragments from the subapical and the basal wing region can be deduced. In comparison with the reference values (Fig. 3), the values for the subapical region 0 and 18 hr after isolation are in accordance with the 1C value. Thus, 18 hr after isolation of these fragments, DNA has not yet been synthesized. But, 30 hr after isolation, the range of the distribution diagram spreads from the 1C to the 2C value. Among the 40 measurements upon which this diagram is based, some nuclei have the clear 2C and 1C values, while others have intermediate values.

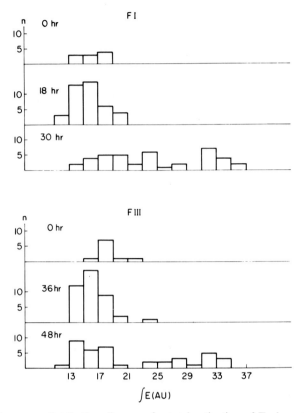

FIG. 4. Frequency distribution diagrams for total extinction of Feulgen-stained nuclei in young (FI) and older (FIII) cells at different time intervals after isolation.

The diagram shows clearly that some cells in fragments from the subapical wing region have doubled their DNA content 30 hr after isolation. This is not the case for fragments isolated from the basal region of the wing. Nuclei in these fragments still have the 1C content 36 hr after isolation. Clearly no nucleus with a 2C content was found. Only 48 hr after isolation, nuclei with a 2C content appear. Although nuclei from intermediate time intervals were not measured, it is evident that in fragments from the subapical region nuclei synthesize DNA much earlier (about 18 hr) than nuclei from the basal region.

Another question is whether or not those nuclei which still have the 1C content at the time when DNA has doubled in some cells of the fragment, do synthesize DNA later on. To answer this question nuclei in fragments from the subapical region were measured 72 hr after isolation. In these fragments, cells with two and even four nuclei were found, indicating that DNA synthesis must have taken place. In numerous cases, directly next to cells with two nuclei, cells with only one nucleus which has the 1C content occur, while in other cases a single nucleus in a cell next to a multinucleate cell has the 2C content. The existence of cells with a single 1C nucleus next to cells with two or four nuclei suggests that differentiation between the cells takes place with respect to the event of DNA synthesis. This statement is strongly supported by measurement of nuclei in fragments 96 hr after isolation in which cells with a single 1C nucleus are found next to cells with 8 or 10 nuclei.

III. Determination of Nuclear RNA Content

By a comparison of the results for the nucleolar enlargement with those for the time of DNA synthesis in Table I it is suggested that RNA synthesis precedes DNA synthesis in nuclear activation during dedifferentiation. But, since the size of the nucleoli does not necessarily reflect their RNA content, we have still determined microspectrophotometrically the RNA content of the nuclei at different intervals after isolation of the cells. The staining techniques with azure B (2) and with cresyl violet (12) could not be used with our material since both dyes were also bound to the cell walls. But gallocyanine chrome alum (5, 13, 14), which we had already used in preparations for the measurements of nucleolar size, proved to be a very useful dye for the staining of nucleic acids in the cells of *Riella*. The staining procedure given by Sandritter, Kiefer, and Rick (13, 14) was also suitable for this plant material.

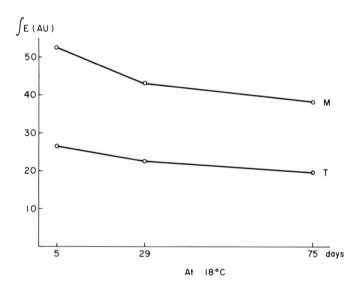

FIG. 5. Total extinction of telophase (T) and metaphase (M) nuclei stained with gallocyanine chrome alum and measured after different time intervals of storage at room temperature.

Using a dye solution at pH 1.64 only nucleic acids are stained in the cells of *Riella*, as could be shown by pretreatment with RNase and DNase.

At first we determined the stoichiometric relationship between bound gallocyanine and DNA by measuring a number of telophase and metaphase nuclei in the meristems of the plant. The tissue has been pretreated by RNase. The results for measurement of total extinction are shown in Fig. 3b. Taking into account the low number of measurements, a rather good ratio of 1C:2C was found. In evaluating these measurements, we observed that some fading of bound dye occurs with the time of storage at room temperature. In Fig. 5 the values for 5 and 75 days after staining belong to the same nuclei which have been measured twice and showed without exception a decrease in total extinction.

For the determination of nuclear RNA synthesis during dedifferentiation, nuclei in tissue fragments from the basal region (FIII) were measured at

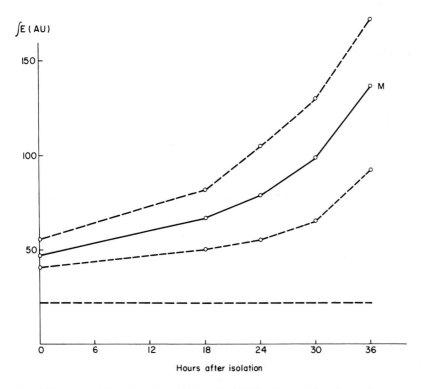

FIG. 6. Total extinction of nuclei of older cells (FIII) stained with gallocyanine chrome alum in dependence on time after isolation ($n = 40$). M, mean; the highest and lowest values of each distribution are also indicated. The horizontal line represents the level of mean DNA content.

different time intervals after isolation. Until now only preparations without enzymic digestion were used. This is possible, since DNA is synthesized in the cells from the basal region later than 36 hr after isolation (Fig. 4, Table I). Therefore, till this time a constant value for the DNA content can be subtracted from the total extinction to find the relative RNA content. The DNA value was determined by measuring nuclei of the same wing region after pretreatment with RNase ($n = 10$, $M = 21.2$; telophases in Fig. 3b: $M = 22.4$). In Fig. 6 it is shown that the RNA content of the nuclei in FIII increases with the time after isolation in almost the same way as the nucleolar size in Fig. 2. The increase is exponential in both cases. It can also be seen from Fig. 6 that the range of the values is increasing with the time after isolation.

IV. Discussion

By comparing the results of the cytophotometric determinations of nuclear DNA and RNA content it is established that RNA synthesis precedes DNA synthesis during nuclear activation in dedifferentiation. Coming back to the initial problem, one can, therefore, assume that the correlations which lead to blockage of the cell reduplication cycle in cell differentiation affect RNA synthesis. A rise in the rate of RNA synthesis preceding the synthesis of DNA has been reported for rabbit kidney cortex cells cultured from the animal (6) and especially for nucleolar RNA, in the liver of rats after partial hepatectomy (21).

While all isolated cells synthesize RNA, only some of them synthesize DNA. The question is: What causes a nucleus to synthesize or not to synthesize DNA? The fact that DNA is synthesized in cells of different age at different time intervals after isolation and only after the mean value for the nucleolar size has reached a maximum, suggests that this critical nucleolar size may be a prerequisite for DNA synthesis. From a statistical analysis of the results for the nucleolar size (20) and from Fig. 6 it can be seen that the variance expressed as percentage and the range of distribution, respectively, increases with time after isolation. This means that although all isolated cells start to synthesize RNA, not all continue this activity. One may assume that among the cells of a fragment correlations come into existence preventing some cells from continuing the reentrance into the cell reduplication cycle. These correlations should have the same nature as the correlations between meristematic and specializing cells in cell differentiation.

In rabbit kidney cortex cells cultured directly from the animal, on the basis of rates of RNA synthesis, three stages of the lag period of about 32 hr can be distinguished before DNA synthesis begins; only during the third stage some of the cells become competent to form DNA (6). The type of RNA synthesized during dedifferentiation in *Riella* is now being investigated in our laboratory.

REFERENCES

1. Bucher, N. L. R., Regeneration in mammalian liver. *Intern. Rev. Cytol.* **15**, 245–300 (1963).
2. Flax, M. H., and Himes, M., Microspectrophotometric analysis of metachromatic staining of nucleic acids. *Physiol. Zool.* **25**, 297–311 (1952).
3. Flickinger, R. A., Biochemical aspects of regeneration. *In* "The Biochemistry of Animal Development" (R. Weber, ed.), pp. 303–337. Vol. 2, Academic Press, New York, 1967.
4. Heitz, E., Das Verhalten von Kern und Chloroplasten bei der Regeneration. *Z. Zellforsch. Mikroskop. Anat.* **2**, 69–86 (1925).
5. Kiefer, G., Kiefer, R., and Sandritter, W., Cytophotometric determination of nucleic acids in UV light and after gallocyanine chromalum staining. *Exptl. Cell Res.* **45**, 247–249 (1966).
6. Lieberman, I., Abrams, R., and Ove, P., Changes in the metabolism of ribonucleic acid preceding the synthesis of deoxyribonucleic acid in mammalian cells cultured from the animal. *J. Biol. Chem.* **238**, 2141–2149 (1963).
7. Lieberman, I., and Ove, P., Deoxyribonucleic acid synthesis and its inhibition in mammalian cells cultured from the animal. *J. Biol. Chem.* **237**, 1634–1642 (1962).
8. MacNutt, M. M., and von Maltzahn, K. E., Cellular differentiation and redifferentiation in Splachnum ampullaceum (L). Hedw. *Can. J. Botany* **38**, 895–908 (1960).
9. Mitchell, J. P., DNA synthesis during the early division cycles of Jerusalem Artichoke callus cultures. *Ann. Botany (London)* [N.S.] **31**, 427–435 (1967).
10. Pätau, K., and Das, N. K., The relation of DNA synthesis and mitosis in tobacco pith tissue cultured *in vitro*. *Chromosoma* **11**, 553–572 (1961).
11. Pätau, K., Das, N. K., and Skoog, F., Induction of DNA synthesis by kinetin and indoleacetic acid in excised tobacco pith tissue. *Physiol. Plantarum* **10**, 949–966 (1957).
12. Ritter, C., Di Stefano, H. S., and Farah, A., A method for the cytophotometric estimation of ribonucleic acid. *J. Histochem. Cytochem.* **9**, 97–102 (1961).
13. Sandritter, W., Kiefer, G., and Rick, W., Über die Stöchiometrie von Gallocyaninchromalaun mit Desoxyribonukleinsäure. *Histochemie* **3**, 315–340 (1963).
14. Sandritter, W., Kiefer, G., and Rick, W., Gallocyanin chrome alum. *In* "Introduction to Quantitative Cytochemistry "(G. L. Wied, ed.), pp. 295–326. Academic Press, New York,1966.
15. Schoser, G., Über die Regeneration bei den Cladophoraceen. *Protoplasma* **47**, 103–134 (1956).
16. Setterfield, G., Growth regulation in excised slices of Jerusalem Artichoke tuber tissue. *Symp. Soc. Exptl. Biol.* **17**, 98–126 (1963).
17. Stange, L. Untersuchungen über Umstimmungs- und Differenzierungsvorgänge in regenerierenden Zellen des Lebermooses Riella. *Z. Botan.* **45**, 197–244 (1957).
18. Stange, L., Regeneration in lower plants. *Advan. Morphogenesis* **4**, 111–153 (1964).
19. Stange, L., Zelldifferenzierung und Embryonalisierung bei dem Lebermoos Riella. *Ber. Deut. Botan. Ges.* **78**, 411–417 (1965).
20. Stange, L., and Kleinkauf, H., Der Zeitpunkt der DNS-Synthese bei der Embryonalisierung von Zellen des Lebermooses Riella in bezug auf RNS-Synthese und Kernteilung. *Planta* **80**, 280–287 (1968).
21. Tsukada, K., and Lieberman, I., Metabolism of nucleolar ribonucleic acid after partial hepatectomy. *J. Biol. Chem.* **239**, 1564–1568 (1964).
22. Zepf, E., Über die Differenzierung des Sphagnumblattes. *Z. Botan.* **40**, 87–118 (1952).

CYTOCHEMICAL ANALYSIS OF THE KARYOTYPE IN THE AMPHIBIAN URODELE PLEURODELES WALTLII MICHAH USING THE USMP-I

S. Bailly and L. Gallien

LABORATORY OF EMBRYOLOGY, FACULTY OF SCIENCES, UNIVERSITY OF PARIS,
PARIS, FRANCE

The work which is presented here is the application of ultramicrospectro-photometric techniques to the study of metaphase chromosomes of the Amphibian *Pleurodeles waltlii* Michah. The chromosomes of this Urodele are relatively few ($2n = 24$) and can be easily identified (9). In their metaphase form they are quite long (8–20 μ). Exposure of newly laid eggs to certain irradiations, especially to γ-rays, can give a variety of chromosomic break-ages, followed or not by rejoining (10). The consequences of such anomalies are shown during the later development of the embryos by serious structural malformations and an important rate of lethality (10). We have undertaken to study, with the maximum available accuracy, the different chromosomic alterations which have been produced. The usual morphologic analysis is not always precise enough and is sometimes unable to render the information sought. Since the studies of Caspersson *et al.*, ultramicrospectrophotometric techniques are now applicable for the quantitative study of chromosomes (3) and particularly for the study of the distribution of their chemical components along the chromosome axis (5–7). Consequently, it seemed useful to apply them to the analysis of abnormal chromosomes.

Before a general study of irradiated chromosomes was undertaken, two preliminary investigations were carried out and published elsewhere: (a) a global spectrophotometric analysis of each chromosome giving an evaluation of their fractional contents of nucleic acids with respect to the total amount carried by the whole complement (1), and (b) an individual study of each of them, for determining the linear distribution of their chemical components (2). Knowing the results of these first investigations, which will be briefly reported here, it is possible to begin the analysis of abnormal chromosomes.

I. Common Materials and Methods for the Three Series of Studies

The metaphase plates were obtained in every case by the squash method from tailbuds of embryos at an age corresponding to stages 30–33 in the chronologic table of development (8). These embryos were submersed for 15–18 hr in colchicine (0.5%), then fixed in acetic acid (50%). According to the type of measurements to be done later, some of the preparations were stained by the Feulgen technique, others prepared on quartz slides without staining.

The microspectrophotometric techniques used here were described in detail elsewhere (1, 2). They are practically the same as those invented and described by Caspersson et al. (5–7). The ultramicrospectrophotometer is the UMSP-I built by Zeiss. The total extinction is calculated, at 546 nm for the Feulgen-stained preparations and at 265 nm for the unstained slides. Total extinction is defined as the product: decadic extinction × area and is expressed in square microns. The absorption for the first wavelength is proper to the Feulgen coloration, and implicitly to DNA, selectively stained by this

TABLE I

EXTINCTION AT 265 NM OF THE METAPHASE
CHROMOSOMES OF *Pleurodeles waltlii*[a]

Pairs of chromosomes	Mean percent of total extinction of entire metaphase plate
I	6.22 ± 0.12
II	5.78 ± 0.12
III	5.50 ± 0.13
IV	5.30 ± 0.16
V	4.77 ± 0.16
VI	4.36 ± 0.17
VII	4.08 ± 0.13
VIII	3.82 ± 0.18
IX	3.03 ± 0.12
X	2.71 ± 0.13
XI	2.55 ± 0.16
XII	1.76 ± 0.06
Total 49.94	

[a] For every pair, the mean value of the extinction (and the standard deviation) of one chromosome per pair is indicated; 15 plates were studied.

technique. The absorption for the second wavelength is proper to both nucleic acids, DNA and RNA, since the absorption of the ultraviolet light at 265 nm by these two acids is much higher than that of other nucleic constituents (4). Nevertheless, because of the fixation used (acetic acid 50%) a certain amount of the RNA may have been lost. All these questions have been discussed elsewhere with more details.

II. Summary of the Results Obtained in the First Two Series

The purpose of the first step (1) was to evaluate the fraction of nucleic acids carried by each chromosome, with respect to the total amount of nucleic acids carried by the whole chromosome complement. To this end, the total extinction at 265 nm for each of the 24 chromosomes of 15 metaphase plates was calculated. The same measurements were carried out at 546 nm on 15 other metaphase plates stained by the Feulgen method.

A mean value was established on the basis of 30 different experiments corresponding to every type of chromosome (Tables 1 and II). The extinction

TABLE II

EXTINCTION AT 546 NM OF THE METAPHASE
CHROMOSOMES OF *Pleurodeles waltlii*[a]

Pairs of chromosomes	Mean percent of total extinction of entire metaphase plate
I	6.41 ± 0.45
II	5.96 ± 0.18
III	5.70 ± 0.15
IV	5.44 ± 0.14
V	4.79 ± 0.25
IV	4.40 ± 0.19
VII	4.05 ± 0.17
VIII	3.78 ± 0.19
IX	2.90 ± 0.17
X	2.56 ± 0.20
XI	2.38 ± 0.11
XII	1.62 ± 0.13

Total 49.99

[a] For every pair, the mean value of the extinction (and the standard deviation) of one chromosome per pair is indicated; 15 plates were studied.

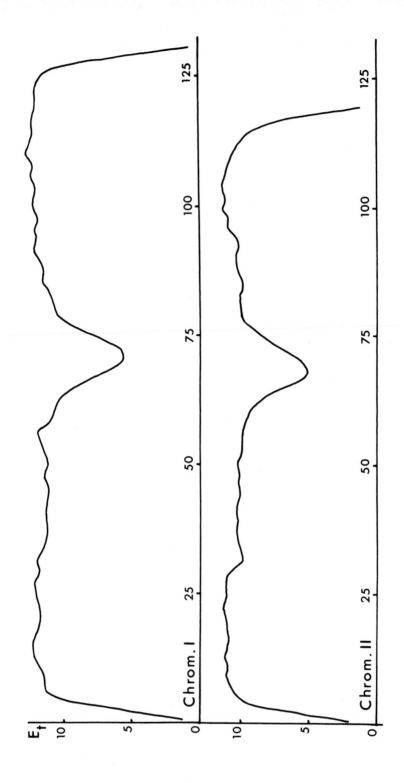

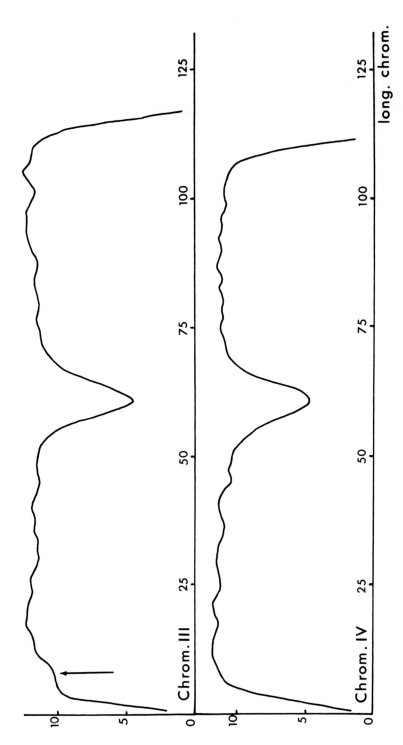

FIG. 1. Chromosomes I, II, III, and IV (first group): profile patterns. The total extinction at 265 nm is indicated on the ordinate ($\times 0.026 \times 10^{-8}$ cm²). These ordinates are averages established, for each chromosome, from 8 to 14 separate experimental results. The length of the chromosome is indicated on the abscissa in arbitrary units. On the III chromosome profile, the arrow indicates the position of the satellite.

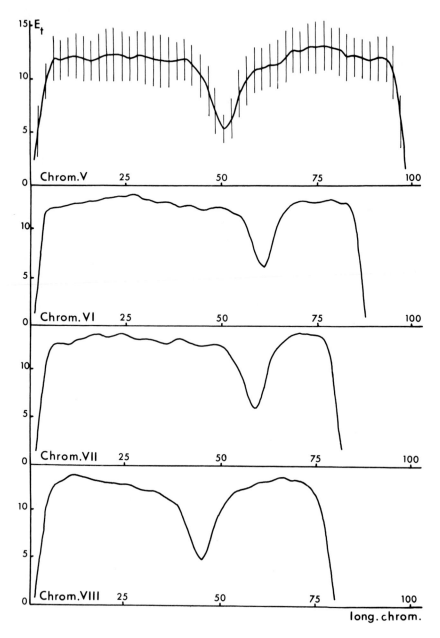

FIG. 2. Chromosomes V, VI, VII, and VIII (second group): profile patterns, (for details see Fig. 1 caption). As an example, the standard deviation is given for every point of the V chromosome.

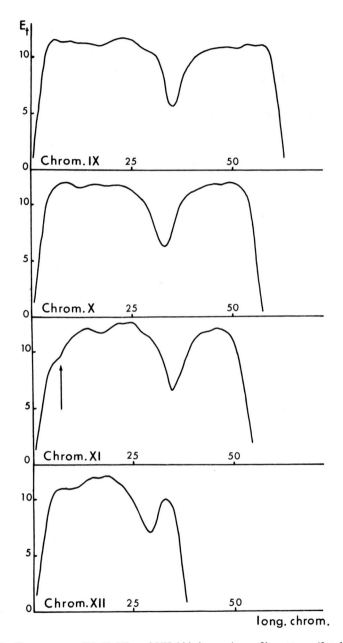

FIG. 3. Chromosomes IX, X, XI, and XII (third group): profile patterns (for details see Fig. 1 caption). On the XI chromosome profile, the arrow indicates the position of the satellite.

values at 265 nm and at 546 nm of the individual chromosomes are presented as percentages of the total extinction value of the whole chromosome complement. They have been found proportional to the size of the chromosomes.

In a second step (2), a profile was established for each chromosome clearly showing the linear distribution of the absorbing components. This study was done in green light on Feulgen-stained chromosomes as well as in UV light on unstained chromosomes. But only the results obtained with the latter wavelength are noted here as being important for the last step of this work.

The basic techniques are essentially the same as in the first step. However, it was only possible to obtain measurements from chromosomes which were sufficiently separated from others, and had a very straight form. The optical scanning was done transversely along their axis and at an interval of 0.3 μ. The diameter of the measuring light beam was also 0.3 μ. The degree of reproducibility of each measurement was 1.5%.

TABLE III

Total Extinction for Each Arm of Every Chromosome

Pairs of chromosomes	Long arm[a]	Short arm[a]
I	53.7±0.8	46.2±0.8
II	58.9±0.1	41.0±0.1
III	51.5±1.2	48.5±1.2
IV	54.1±1.0	45.9±1.0
V	50.8±0.7	49.2±0.7
VI	70.3±1.7	30.0±1.7
VII	74.2±1.3	25.8±1.3
VIII	54.7±1.4	45.3±1.4
XI	58.0±1.5	42.0±1.5
X	59.8±2.3	40.2±2.3
XI	65.2±1.0	34.8±1.0
XII	81.6±1.9	18.3±1.9

[a] For every chromosome, the mean value of the extinction at 265 nm corresponding to each of the arms is given in percentage of the total extinction of the whole chromosome.

To begin with, 325 chromosomes were selected for measurement. The profiles of 206 of them were drawn and finally 138 were used for the construction of a profile pattern (Figs. 1, 2, and 3). For every curve, the value of total extinction at 265 nm was plotted on the ordinate axis. The length of the chromosome is indicated on the abscissa, in arbitrary units (9). The ordinates of every point are, in fact, averages established, according to each case, from 8 to 14 experimental results. For each case, the standard deviation was calculated. For example, these standard deviations are plotted on the profile of the V chromosome. A mathematic study was also made for determining the recorded variations which were really significant.

The general aspect of these curves established from measurements done at 265 nm should, theoretically, show the distribution of the nucleic acids DNA and RNA, though a certain amount of RNA may have been lost because of the fixative used (2). Table III indicates the total extinction of each arm for every chromosome, in percentage of the total extinction of the whole chromosome. The curves show a paramedial depression corresponding to the position of the centromeres.

III. Study of Abnormal Chromosomes

These results being known, it was now possible to study abnormal chromosomes. These were obtained by exposing newly laid eggs to γ-rays, according to a technique devised in our institute (10). The rate of irradiation used was 280 R. The squash preparation was made as formerly indicated, and quartz slides were employed, since the absorption measurements were to be done in ultraviolet light (265 nm). Optical scanning was carried out under the same conditions as those used for profile patterns (2), the scanning interval and diameter of the measuring light beam both being 0.3 μ. Up to now, a few abnormal cases have been studied, and four of these are described here, as being the most typical cases.

The most frequently observed chromosomic anomalies are dicentric chromosomes, obtained by the joining together of two chromosomes after having lost a smaller or larger part of one of their arms. Telocentric chromosomes can also be observed after a nearly total loss of one of their arms (10). In both cases the portions thus separated may or may not be found again in the metaphase plate as acentric fragments. Therefore, the first problem is to be able to identify the two partners of a dicentric chromosome, or determine to which pair the telocentric or acentric portion of a plate belongs. The second one is to know at what point on the dicentric chromosome the joining of its two chromosomes occurs, in order to determine the eventual losses of genetic material.

To this end the following method is used: First, the profile of the studied abnormal chromosome is established by the techniques described above and, at the same time, the absorption of one or two especially well identified chromosomes of the same plate are measured and used as control samples (generally a XI or a XII chromosome). The comparison between the profile of the abnormal chromosome, for instance a dicentric one, and the profile patterns, gives an initial idea as to which kind of chromosome its two distal arms belong. Second, the first hypothesis may be checked by calculation: Using Tables I and III and the value of total extinction found for the control sample chromosome, the values of total extinction that the arms of the dicentric chromosome should have, if the hypothesis is correct, are calculated. These values are then compared to those given by the experimental method. If these two calculated results are close, the hypothesis is thus confirmed and the same methods (comparison of profiles, comparison of extinction values) are then used for the analysis of the medial portion of the dicentric chromosome. In this manner, it becomes known what fraction belongs to which chromosome. Using these comparisons of profiles and results of calculations, it is then finally possible to appreciate what amount of genetic material is lost.

A first example, treated with particular detail, will concretely illustrate this method. A metaphase plate with only 23 chromosomes, one of them being dicentric, is studied (Fig. 4). The analysis of the profile of this dicentric (Fig. 5) shows that it seems to be constituted by the joining of a V and a I chromosome. The two centromeres are in C_1 and C_2. The profile of the AC_1 portion

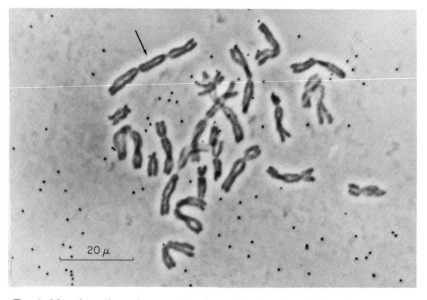

FIG. 4. Metaphase plate, phase contrast, from tailbud of *Pleurodeles* embryo. The arrow shows the dicentric chromosome.

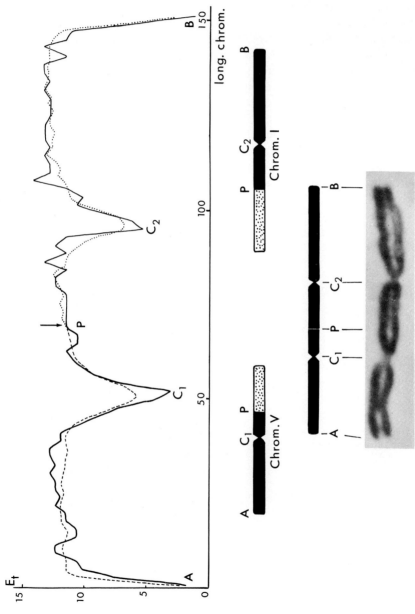

Fig. 5. Profile of the dicentric chromosome noted in Fig. 4. (———) Profile of the abnormal chromosome; (— — —) fragment of the V chromosome profile pattern; (·······) fragment of the I chromosome profile pattern.

looks very much like the V long arm. The control sample chromosome is the XII and its total extinction is 5.35 μ^2. Using Tables I and III, we calculate that the total extinction of the V long arm should theoretically be 7.37 μ^2. The corresponding value given by the measurements is 7.44 μ^2. In the same way, we calculate the total extinction of the BC_2 portion, the profile of which resembles very much the I short arm. Its theoretical value is 8.78 μ^2, against 8.81 μ^2 for experimental result. The total extinction of the C_1P portion, which looks like the proximal part of the V short arm, is 2.36 μ^2 as against 2.28 μ^2 for the same part of the profile pattern. Last, the total extinction of the BC_2 portion, the profile of which seems the same as that of the proximal part of the I long arm, is 4.00 μ^2 as against 3.97 μ^2 (theoretical value). Since the differences between these values are compatible with the margin of error indicated on the Tables I and III, P could be considered as the joint between the V and I chromosomes. These chromosomes lost, respectively, 65% of their short arm and 60% of their long arm; these two parts are illustrated in gray on the diagram of each chromosome (Fig. 5).

The second example to be discussed here is another dicentric chromosome (Fig. 6). As before, the shape of its profile is studied (Fig. 7) and a comparison is made between the total extinction of its different parts and the total extinction of patterns presumed to correspond to them. We conclude to a joining-up

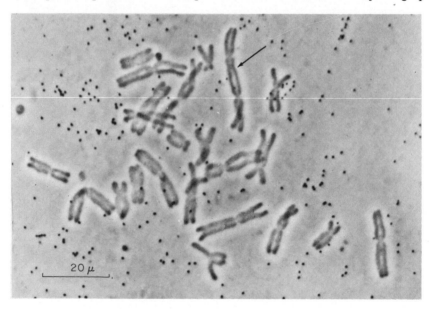

FIG. 6. Metaphase plate, phase contrast. The arrow indicates the dicentric chromosome studied.

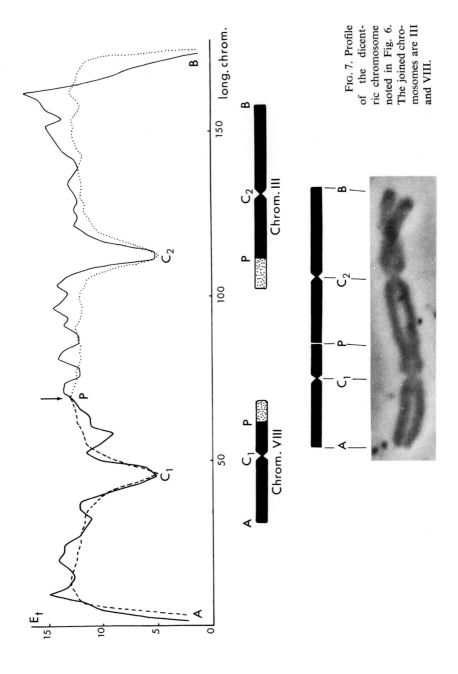

FIG. 7. Profile of the dicentric chromosome noted in Fig. 6. The joined chromosomes are III and VIII.

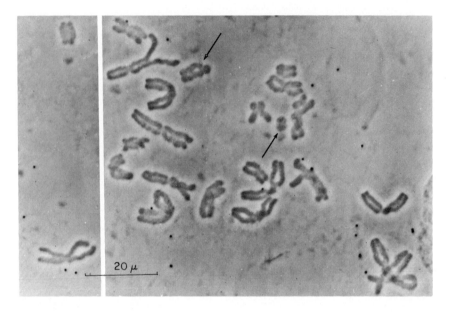

Fig. 8. Metaphase plate, phase contrast. The arrows indicate the different portions of chromosomes studied.

in the P region of the VIII and the III chromosomes by their short and long arms, respectively. There is a 40% loss of material for the VIII short arm and a 33% loss for the III long arm. It is to be noted that the crossing of one chromatid over the other in the B region explains why the profile of this portion is not exactly the same as that of the profile type.

The third example is taken from a metaphase plate having a telocentric chromosome and an acentric fragment (Fig. 8). The analysis by the preceding method shows that the telocentric portion is a fraction of a VIII chromosome (Fig. 9): the long arm, the centromeric part (C), and a part of the short arm CB (21% of it). The acentric portion is the distal part of this same short arm (66%). There is, therefore, a loss of 13% of the short arm in a region close to the point of breakage.

The last example is a rather particular dicentric chromosome (Fig. 10). As is easy to see on the diagram (Fig. 11), the centromeric parts C_1 and C_2 are, respectively, those of the IX and the VI chromosomes. The AP and RB fragments belong to VI, and the whole PR part is formed by the central region of the IX chromosome. There was rupture of VI at P and at R, and then loss of the material which was between these two points, that is to say, about 25% of the long arm. There was then interposition between P and R of the medial part of the IX. This last chromosome has lost about 40% of its

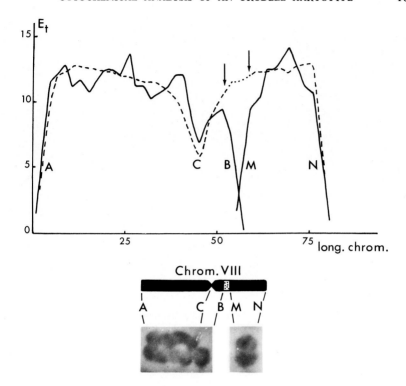

FIG. 9. Profiles of the two chromosomic segments noted in Fig. 8.

long arm and 40% of its short arm. So, in this case the dicentric chromosome is formed by the translocation of a chromosome inside another one.

The examples reported here show very clearly the importance of ultra-microspectrophotometric techniques in the study of abnormal chromosomes. Using them, it is relatively easy to identify the arms of an abnormal chromosome. Of course, with the methods used here (especially the use of comparisons between different values of total extinctions, that is to say, the comparison of different curve shapes), it is not possible to ascertain that the real joining point of two chromosomes is exactly the point mentioned on the diagram. A margin of error exists but it is possible to determine precisely the region of the joint. The identification of every chromosome segment is also very easy. Such accuracy is impossible to obtain with conventional morphologic methods. The last example studied is particularly significant with respect to this point of view.

These first results indicate that it is now possible to successfully apply ultramicrospectrophotometric techniques to the study of chromosomal anomalies, induced or spontaneous, of the karyotype of the Urodeles.

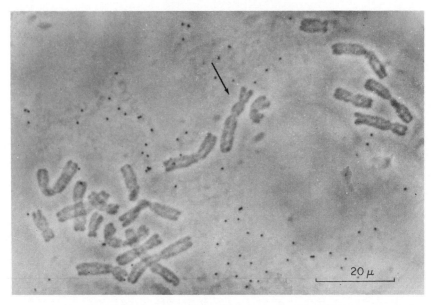

FIG. 10. Metaphase plate, phase contrast. The arrow shows the dicentric chromosome studied. The plate has only 21 chromosomes.

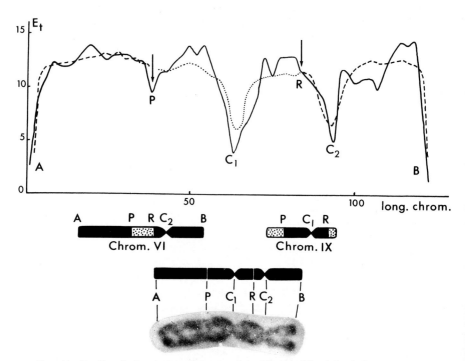

FIG. 11. Profile of the chromosome noted in Fig. 10. The joined chromosomes are VI and IX.

REFERENCES

1. Bailly, S., Etude cytophotométrique de la teneur en acides nucléiques des chromosomes métaphasiques de l'Amphibien Urodèle *Pleurodeles waltlii* Michah. *Exptl. Cell Res.* **48**, 459–556 (1967).
2. Bailly, S., Etude cytophotométrique de la distribution des acides nucléiques le long des chromosomes métaphasiques de l'Amphibien Urodèle *Pleurodeles waltlii* Michah. *Ann. Génét.* **12**, 184–190 (1969).
3. Carlson, L., Caspersson, T., Foley, G. E., Kudynowski, J., Lomakka, G., Simonsson, E., and Sören, L., The application of quantitative cytochemical techniques to the study of individual mammalian chromosomes. *Exptl. Cell Res.* **31**, 589–594 (1963).
4. Caspersson, T., "Cell Growth and Cell Function." New York, Norton, 1950.
5. Caspersson, T., To be published.
6. Caspersson, T., Erhardt, L.-R., Heneen, W., Kudynowski, J., and Simonsson, E., In preparation.
7. Caspersson, T., Farber, S., Foley, G. E., Kudynowski, J., Modest, E. J., Simonsson, E., Wagh, U., and Zech, L., Chemical differentiation along metaphase chromosomes. *Exptl. Cell. Res.* **49**, 219–222 (1968).
8. Gallien, L., and Durocher, M., Table chronologique du développement chez *Pleurodeles waltlii* Michah. *Bull. Biol. France Belg.* **91**, 97–114, (1957).
9. Gallien, L., Labrousse, M., Picheral, B., and Lacroix, J. C., Modifications expérimentales du caryotype chez un Amphibien Urodèle (*Pleurodeles waltlii* Michah.) par irradiation de l'oeuf et la greffe nucléaire. *Rev. Suisse Zool.* **72**, 59–85, (1965).
10. Labrousse, L., Analyse des effets des rayonnements appliqués à l'oeuf sur les structures caryologiques et sur le développement embryonnaire de l'Amphibien Urodèle *Pleurodeles waltlii* Michah. *Bull. Soc. Zool. France* **91**, 491–590 (1966).

THE ROLE OF THE MOUNTING MEDIUM IN UV MICROPHOTOMETRY

Günter Kiefer

DEPARTMENT OF PATHOLOGY, UNIVERSITY OF FREIBURG, FREIBURG, GERMANY

It is now 30 years ago that Caspersson worked out the basis for UV microscopy and microspectrophotometry. He pointed out the limitations for this method and the conditions necessary to assure sufficient exactness in measurement.

Since then this method has been improved considerably—mainly with respect to the instrumentation. Today we have at our disposal sensitive photomultipliers and electronic parts that assure fairly exact measurements in accordance with the conditions set by Caspersson. Also the development of better and more achromatic optical elements has played a considerable role in this progress. One of the final products of this development is the universal microspectrophotometer of Zeiss. It seems that further principal optical improvements of microphotometers are neither possible nor necessary. The situation today is that the prime errors occurring in microphotometry and especially in UV microphotometry are caused by the object to be measured. The most serious difficulties are introduced by the inhomogeneity of the object with respect to its refractive index. When looking at an absorption spectrum of a cell, we find a maximum at 260 nm and a shoulder at 280 nm. This corresponds to nucleic acid and protein absorption. Besides this, we find a loss of light intensity above 300 nm, i.e., in a region in which the aromatic and heterocyclic ring systems no longer absorb.

We therefore may differentiate two different types of light loss: (a) actual specific absorption or aromatic ring systems and (b) light loss caused by scattering of cellular structures. We can call these two components a consumptive and a conservative absorption. Both components add to each other.

The same can be observed in the absorption spectrum of a protein solution. In this simple system the conservative or scattering absorption increases with a shortening of the wavelength, and this occurs in an exponential fashion. Rayleigh formulated this function as follows:

$$E = K \cdot \lambda^{-m}$$

Accordingly, the extinction or absorbance caused by scattering is inversely proportional to a power of the wavelength. The exponent of the wavelength depends largely on the size of the particles in solution and is found to lie between 0 and 4,

$$0 \leqq m \leqq 4$$

If the particle size is one-tenth of the wavelength used, the exponent is approximately 4; if it is in the order of the wavelength, then m is approximately 2. If we find a logarithm for the Rayleigh formula, the function becomes linear, i.e.,

$$\log E = \log K - m \log \lambda$$

In a bilogarithmic coordinate system the scattering curve appears as a straight line, the slope of which corresponds to the wavelength exponent m in the Rayleigh formula.

It is possible to extrapolate this scattering curve and to calculate the pure, i.e., the consumptive, absorption at 280 or 265 nm. This correction is required when absorption is evaluated for absolute determinations and this correction can be accomplished easily for a protein solution.

In a fixed microscopic preparation the conditions are much more complicated. We must note that we deal with particles varying considerably in size. In addition to very small particles which behave according to the Rayleigh formula, larger conglomerates are also found, for which the laws of geometrical optics are to be considered. We therefore cannot assume that a microscopic preparation contains pure Rayleigh conditions. This makes it impossible to calculate the amount of light loss by scattering. Several years ago Caspersson (2) had attempted, therefore, to measure the angular distribution of scattered light behind the preparation. For this he made use of the following apparatus: The specimen and the illuminating system could be rotated around an axis through the object. The light energy leaving the specimen was measured with a fixed photomultiplier in relation to the angle formed by the photomultiplier axis and the optical axis of the illuminating system. The object of this investigation was to determine the amount of light which does not leave the object in the direction of the optical axis. Results of these measurements made Caspersson demand a high numeric aperture for the objective. These measurements also had further significance, namely, the amount of light scattered sideways depends on the medium surrounding the specimen. If the object is surrounded by air, a rather great amount of light is scattered. If alcohol or paraffin is used much less light is scattered. This showed that the intensity of scattered light not only depends on the shape of the object but also on the difference of refractive index of object and mounting material. The amount of stray light is small if the refractive indices are closely matched, i.e., the decrease in light intensity is mainly due to consumptive absorption of cellular material and hence our measurements can be interpreted with the greatest degree of reliability. Caspersson's apparatus could have been used for selecting an embedding medium which would give the smallest amount of scattered light. But it has never been used for this purpose. Instead, one has more often chosen an easier method. The Rayleigh relation is used with microscopic preparations and one can calculate that part of absorbance at 265 nm belonging to the scattered light. However, as the scattering light exponent m is unknown, one must assume some value between 1 and 4. Moberger (4) always calculated two values for the nucleic acid concentration—a maximum and a minimum value, by assuming an $m = 1$ exponent and then an exponent $m = 4$. Walker (8) published a diagram from which one can read off the amount of scattered light at 265 nm. To use this diagram one has to measure the absorbance at 265 nm and outside the region of consumptive absorption at 313 nm. The amount of scattered light loss can be calculated from the diagram if a certain Rayleigh exponent is assumed. A similar method was chosen by Sandritter (6). He set up a formula for the calculation of nucleic acid content of cells, starting with the following values: The ratio of extinction E_{280} to E_{265} is 0.56 for nucleic acids and for proteins a ratio of 0.82 was assumed. The extinction coefficient is taken as 20,000. Finally it was assumed that the scattering light decreases with increasing wavelength with an exponent of 1.

Thus, the whole situation seems to be very unsatisfying. We own complicated microphotometers that allow us to take very precise measurements, but we cannot make use of the preciseness since our specimen is so badly defined. One way to improve this situation is to prepare the object in such a way that light scattering is minimal and thus the error is small. The mode of fixation of the material plays an important role. One should choose a fixation method that causes a minimum of inhomogeneity with respect to variation of refractive index in the cell. For instance, freeze substitution seems to be very suitable for most materials.

In addition to this the conditions can be improved if it is possible to find an embedding medium having a suitable index or refraction. Quite a number of embedding materials have been mentioned in the literature, especially glycerin (1, 3). However, its index of refraction is too low. The index can be increased by dissolving $ZnCl_2$ or chloral hydrate in it. With these solutions one can obtain indices of refraction as high as 1.6.

It is not necessary to know the absolute value of the refractive index. What

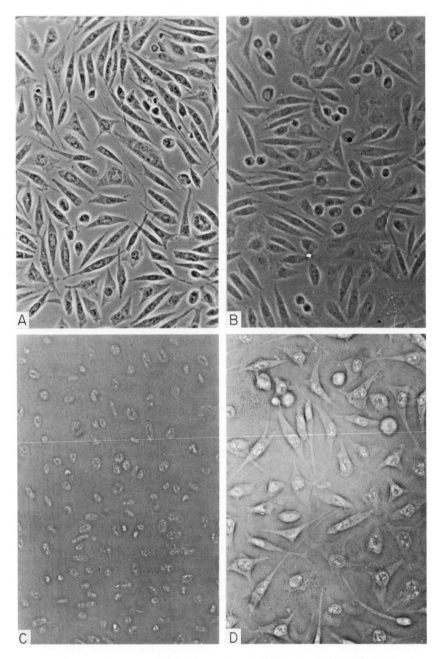

FIG. 1. Phase contrast images (at $\lambda = 546$ nm) of fixed mouse fibroblasts in culture (L strain), mounted in media of different refractive indexes: (A) $n = 1.52$, (B) $n = 1.54$, (C) $n = 1.56$, (D) $n = 1.58$.

is wanted is that the mounting medium should have a refractive index close to that of the specimen. In our laboratory we have asked ourselves the following question: Is there a simple test with the aid of which we can detect the matching of the index of refraction? This test must be carried out at the wavelength to be used for the absorption measurments, i.e., in our special case in UV. There exist four possibilities for such a test.

(1) *Phase Contrast.* For microphotometry in the visible region of the spectrum there are a great number of mounting media available having different indices of refraction and we can match the refraction index by the following method: The object is embedded in different media having different indices of refraction and it is then observed whether phase effects occur (Fig. 1). Very low and very high refraction indices give high contrasts. These contrasts become less intensive, the less difference there is in the refraction indices of the object and the medium. If the object becomes invisible, the two refraction indices are identical. We have tried the same procedure in UV light. But, as may have been expected, the results were poor. The object must be observed with an image converter or with a UV television system or a photograph must be made of it. This, however, would make it impossible to say whether the contrast at 265 nm is caused by phase effects or by the real absorption. The procedure can only be used above 320 nm but would then not show matching of the index of refraction at 265 or 280 nm.

(2) *Height of Absorbance.* One can assume that the measured extinction value or its surface integral is equivalent to the sum of the pure (consumptive) absorbance plus the scattering light loss. So, if we were to measure the same cell embedded in different media, it should be possible to select an embedding medium that leads to a minimum in the scattered light intensity. We did this experiment using smears of rat thymus lymphocytes (Table I). It was found that the total extinction of these cells is least if they are mounted in water-free glycerin. In this case we used different preparations because only small differences between different cells of this type exist. If we deal with other cells where this is not the case, we have to measure the same cell many times and change the mounting medium between measurements. It is obvious that the

TABLE I

TOTAL ABSORBANCE OF THYMUS LYMPHOCYTES ($\lambda = 265$ NM; FIXATION: FREEZE SUBSTITUTION)

Embedding medium	Integrated extinction
Glycerin	51.60 ± 2.81
Glycerin + $ZnCl_2$	57.74 ± 2.78
Glycerin–chloral hydrate	58.96 ± 3.32

method using the relative height of the total extinction as a criterion for the quality of the refractive index matching is very laborious.

(3) *Absorption Spectrum.* It would be an advantage if one knew the wavelength exponent of the scattering light curve. Then the absorption values could be corrected more exactly by the use of the Rayleigh formula. In accordance with the method of Schauenstein for protein solutions (7), one must know the absorption spectrum of the object above 300 nm. As the extinction values in this region are very small, relatively high accuracy is required. The first question is: Does light scattering in a microscopic object follow the Rayleigh relation? The absorbance spectrum of thymus lymphocytes drawn in the bilogarithmic coordinate system behaves as a straight line between 500 and 300 nm (Fig. 2). This means that there exists a relationship between the absorbance and the wavelength that can be expressed in terms of the Rayleigh formula. There is a difference in the slope of the light-scattering curves depending on the mounting medium used. The Rayleigh exponent m for glycerin was found to be 2.37, for glycerin plus $ZnCl_2$ it was 1.19, and for glycerin plus chloral hydrate it was 3.25. The lowest value is obtained when using a concentrated solution of $ZnCl_2$ in glycerin. One may therefore presume that in this case the refraction index difference at 265 nm will be especially small. With this method we therefore can estimate matching of the the refraction index of the embedding medium.

Moreover, we can make use of the Rayleigh exponent to correct the absorption measurements at lower wavelengths, similar to what Walker and Moberger recommended. In general, we will have to measure more than only one absorption spectrum because of the inhomogenity of the tissue.

(4) *Dark Field.* As the fourth method, mention is made of observations of the specimen in dark field illumination. The dark field image of the cell

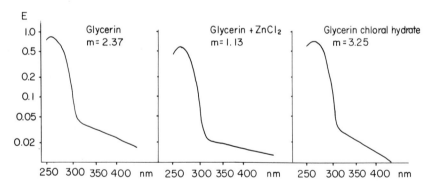

FIG. 2. Absorbance spectra of rat thymus lymphocytes mounted in different media. Bilogarithmic coordinates; m, slope of the straight part of the curve above 320 nm, is the Rayleigh exponent.

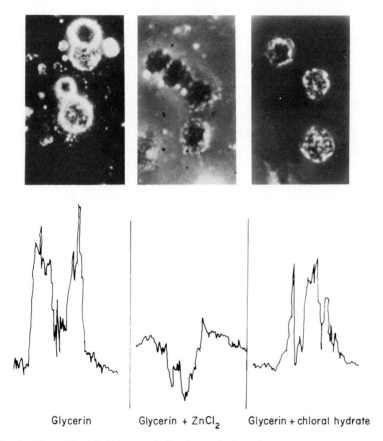

Glycerin   Glycerin + ZnCl$_2$ Glycerin + chloral hydrate

FIG. 3. Above: Dark field images (at $\lambda = 265$ nm) of rat thymus lymphocytes mounted in different media. Below: Corresponding densitometer tracings across the photographic images.

mainly depends on its refractive properties. Contrary to the situation in phase-contrast, absorption of the object influences its dark field image only to a small extent. Therefore, this method can also be applied at wavelengths of 265 and 280 nm. Rudkin and Corlette (5) used the dark field observations in UV with giant chromosomes in order to choose a suitable embedding material. Also Jansen (3) finds that the dark field observation is very sensitive for a demonstration of differences in the refractive index. For our investigations we made use of a UV microscope, produced by Messrs. Leitz, Wetzlar. Illumination is achieved by a xenon lamp. The main advantage of the apparatus is that one can watch the object directly with an image converter. Therefore, it is not necessary to make a photograph. The microscope has reflecting optics, i.e., the central mirror causes a central shading in the objective.

We now have chosen a condenser whose exterior aperture was smaller than the interior one of the objective. Thus no light falling directly through the object reached the objective and therefore we obtained a dark field image. When looking at thymus lymphocytes in this microscope at 265 nm the contrast changes with the mounting medium (Fig. 3). It is smallest in glycerin saturated with $ZnCl_2$. We also assess the image in an objective way by taking a microphotograph of the cell and trace a scanning line through the photograph of the cells' image. When using $ZnCl_2$ solutions the densitometer curve shows only very small deflections. Yet in the case of dark field illumination it is not necessary to make measurements, since looking at the cell by an image converter, or UV–television system, or a photograph of the cell, is enough for deciding on the most suitable mounting medium.

This method is therefore recommended for the control of refractive index matching in the UV region. But other methods should also be applicable, especially the use of absorption spectra at wavelengths greater than 300 nm. One of the advantages of this latter method is that one finds a number with which one is able to correct the extinction values at 265 nm for the scattered light loss.

A number of experiments aiming at further improvements in the preparation of tissues and cells for UV microspectrophotometry can be thought of, and it is indeed necessary to continue these efforts because UV microspectrophotometry is a very potent tool in cell physiology.

REFERENCES

1. Boguth, W., and Pracht, I., Vergleich von UV-Absorptionsspektren des Schilddrüsenkolloids normaler und Vitamin-E-verarmter Ratten. *Z. Wiss. Mikroskop.* **67**, 240–243 (1966).
2. Caspersson, T., "Cell Growth and Cell Function." Norton, New York, (1950).
3. Jansen, M. T., On the refractive index of histological tissue sections in visible and ultraviolet light. *Exptl. Cell. Res.* **15**, 239–242 (1958).
4. Moberger, G., Malignant transformation of squamous epithelium. A cytochemical study with special reference to cytoplasmic nucleic acid and proteins. *Acta Radiol.* **112**, Suppl., 1–108 (1954).
5. Rudkin, G. T., and Corlette, S. L., Refractive index matching in the UV using dark field photomicrography. *J. Histochem. Cytochem.* **4**, 438 (1956).
6. Sandritter, W., Ultraviolettmikrospektrophotometrie. "Handbuch der Histochemie" (W. Graumann and K. Neumann, eds.), Vol. 1, part 1, pp. 220–338. Fischer, Stuttgart, 1958.
7. Schauenstein, E., and Bayzer, H., Über die quantitative Berücksichtigung der Tyndall-Absorption im UV-Absorptionsspektrum von Proteinen. *J. Polymer Sci.* **16**, 45–51 (1955).
8. Walker, P. M. B., Ultraviolet microspectrophotometry. "General Cytochemical Methods" (J. F. Danielli, ed.), Vol. 1, pp. 163–217. Academic Press, New York, 1958.

A PRECISION STAGE FOR TWO-DIMENSIONAL SCANNING MOVEMENT

Leon Carlson

INSTITUTE FOR MEDICAL CELL RESEARCH AND GENETICS, MEDICAL NOBEL INSTITUTE, KAROLINSKA INSTITUTE, STOCKHOLM, SWEDEN

The principle to be described in this chapter is generally applicable when precision movements of restricted length in two dimensions are needed. One important use is in scanning stages for cytophotometric instruments, such as microspectrophotometers, microinterferometers, microdensitometers, and microfluorometers.

In a cytophotometer of the object-scanning type, the specimen is placed on a stage that can be moved with high precision in two perpendicular directions. The optical analysis can thus be performed simply by a stationary beam of light passing symmetrically around the optical axis, an arrangement that has many advantages (2).

Some types of scanning stages have been described earlier in the literature (2, 3). The one described here is a novel construction and has the special qualities of being very stable, lightweight, and versatile. This special design is only one of many possible applications of the basic principle.

I. General Description of Stage

Figure 1 is a schematic perspective diagram of the scanning stage. The center line (Z) represents the optical axis of the instrument to which the stage is fitted. The specimen is supported by a special holder placed in the circular hole of the mobile plate A. This single plate can be moved independently in two directions $(X$ and $Y)$ perpendicular to each other and to the optical axis.

II. Support of Mobile Plate

Plate A is supported by three springs (B, C, and D). These are shaped with a rigid middle portion and short flexible portions at both ends which act as joints without play. The springs are fixed at one end to plate A and at the other end to the base plate (E), the first two being situated below and the third above plate A.

When plate A is moved out from its central position the points fixed to the

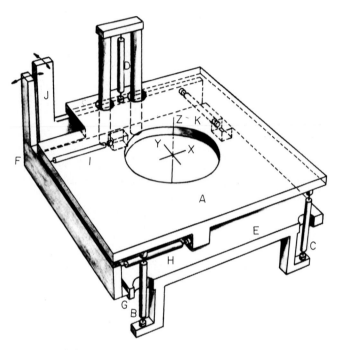

Fig. 1. Schematic perspective diagram of the scanning stage.

springs will move in circular arcs. Provided plate A is moved without twisting around the Z axis, the ends of B and C will move downward just the same distance as the end of D will move upward. Consequently, the vertical shift produced at the central axis by the small vertical shifts at B and C will be, to a great extent, neutralized by the shift at D. For example, a stage having a spring length of 40 mm and the springs situated at a distance of 50 mm from the X axis will allow the scanning of an area 1.2 mm × 1.2 mm with a vertical shift at the optic axis of less than 0.1 μm. Thus, defocusing of the objective will not take place for ordinary microscopic scanning applications.

III. Guided Movement in Two Perpendicular Directions

The movement in the X direction is produced by tilting a lever (F). This is fixed by a long leaf spring (G) to the base plate (E). Two springs (H and I), of construction similar to B, C, and D, transmit the movement to plate A and, because they are fixed at identical distances from G, they force plate A to move parallel to its original direction.

Fig. 2. The scanning stage fitted to an ordinary microscope stand in a microinterfero-meter.

Similarly, the movement in the Y direction is produced by tilting the lever J, the movement of which is transmitted by the spring K. In this case only one spring is necessary, because plate A is already restricted to parallel movement by H and I.

When plate A is moved in the Y direction and the X lever is kept in a fixed position, the springs H and I will bend sideways and every point on plate A will move in a circular arc. However, the radius of curvature of this arc is very great compared to the distances traversed in ordinary scanning applications. Thus, the deviation from a straight line will be only 0.03 μm for a 100-μm scan with springs of 40-mm length. The same argument is valid when the X lever is moved and the Y lever is fixed.

The movement of the levers F and J can be actuated in different ways. Precision screws driven by synchronous or stepping motors have been used for moving the upper ends of the levers. Hydraulic systems have also been applied in a similar way. Figure 2 shows the stage fitted to a scanning micro-interferometer (1). Due to the compactness and minimal weight of the scanning stage, it has been possible to mount it on an ordinary microscope stand even though the focusing must be done by raising and lowering the stage. The movement is transmitted by vertical shafts actuated at their upper ends by stepping motors, giving them small rotary movements. The lower ends of the shafts are shaped as levers with one end attached to the stand and the other to one of the levers F and J.

IV. Summary

The stage for two-dimensional movement consists of a single mobile plate capable of movement in two perpendicular directions with one movement being independent of the other. The plate is supported by three springs limiting the movement to one plane, and two levers connected to the plate by other springs guide the movement within this plane. An outstanding feature is the good definition of position because all joints are free of play.

REFERENCES

1. Carlson, L., Caspersson, T., Lomakka, G., and Silverbåge, S., A rapid scanning and integrating microinterferometer for large-scale population work. *In* "Introduction to Quantitative Cytochemistry-II" (G. L. Wied and G. F. Bahr, eds.), Academic Press, New York, 1970.
2. Caspersson, T. O., and Lomakka, G. M., Scanning microscopy techniques for high resolution quantitative cytochemistry. *Ann. N. Y. Acad. Sci.* **97**, 449–463 (1962).
3. Trapp, L., Instrumentation for recording microspectrophotometry. *In* "Introduction to Quantitative Cytochemistry" (G. L. Wied, ed.), pp. 427–435. Academic Press, New York, 1965.

A RAPID SCANNING AND INTEGRATING MICROINTERFEROMETER FOR LARGE-SCALE POPULATION WORK*

L. Carlson, T. Caspersson, G. Lomakka, and S. Silverbåge

INSTITUTE FOR MEDICAL CELL RESEARCH AND GENETICS, MEDICAL NOBEL INSTITUTE,
KAROLINSKA INSTITUTE, STOCKHOLM, SWEDEN

The interference microscope is one of the most useful tools available for the determination of dry weight in biological specimens. The optical path difference (OPD), the physical quantity measured by the interference microscope, is directly proportional to the dry weight per unit area (1, 2, 4). To obtain accurately the total dry weight of a cell, or part of a cell, it is necessary to integrate a great number of measurements, taken as small points, distributed over the entire object area (scanning measurements). The ordinary instrument for such measurements is a recording, scanning, and integrating microinterferometer (7).

For measurement of the total dry weight of single cells lying well separated from each other, a very common situation in biomedical "population work," it is not necessary to record every point measurement but only the final value obtained by integration of these point values over the whole cell. An instrument for such nonrecording measurements can be made for rapid operation (6) and the one to be described is of this type.

I. General Description of the Instrument

Figure 1 shows a simplified block diagram of the instrument. The interference microscope is of the shearing type using polarized light (5) and is made by the Zeiss Company. The light is split into two beams, one passing

* Supported by grants from the Swedish Board for Technical Development and from Incentive Research and Development AB, 161 30 Bromma 11, Sweden.

117

through the object and the other through an adjacent object-free region. The difference in optical path between these two beams is the quantity measured by the instrument. This is accomplished by using a Senarmont compensator consisting of a quarterwave plate and an analyzer. The light passing through the quarterwave plate is linearly polarized and the angle of the plane of polarization is a linear function of the optical path difference. Measurement

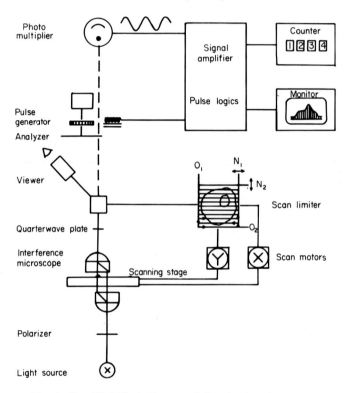

FIG. 1. Simplified block diagram of the microinterferometer.

of this angle is performed 120 times per second with the aid of a rotating analyzer. While being measured, the cell is scanned at a speed of 60 μm/sec along lines spaced at even distances. When the whole cell is covered, the scanning stops automatically and the integrated result of the measurement is displayed on the digital readout unit.

A. SCANNING MECHANISM

The object-scanning principle is applied. This means that the specimen is placed on some type of a scanning precision stage that is moved in a well-defined pattern while the measuring beam remains stationary. The principle of the scanning stage is described in a separate chapter in this volume (*3*). Thus only the motor arrangement, which has been newly developed for this instrument, will be briefly described here. The scanning pattern consists of a set of parallel, equidistant lines (*X* scan). Either 0.5, 1, 2, or 4 μm can be chosen as the interline distance. Movement along the *X* scan is produced by a synchronous motor via a ball-bearing screw and a system of levers. The length of the scan lines is determined by adjustable contacts which also reverse the direction of movement. At every switching point there is a stepwise movement in the perpendicular direction (*Y* axis). This is accomplished with a stepping motor via a screw and a system of levers as described for the *X*-scan. Also this movement is limited by contacts which determine the extention of the scan area in the *Y* direction. The setting of the scan-limiting contacts is easily done because they are coupled to a set of luminous, adjustable index lines which indicate the scan areas. They are projected into the field of view and are superimposed on the image of the specimen. The rectangular area to be scanned is thus shown enclosed between four index lines.

B. Analyzing System

The analyzer is a circular piece of Polaroid foil placed on the shaft of a synchronous motor and rotated at 3600 rpm. The intensity of the light after it passes through the analyzer varies sinusoidally as a function of the angle of rotation (curve A of Fig. 2). When the optical path difference is changed,

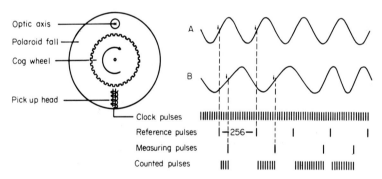

Fig. 2. Analyzer with signal diagrams.

the phase angle of this curve is changed proportionally. The function of the analyzer is to measure this phase angle. To do so, only one characteristic point need be measured. In this instance, the inflection point with a positive slope (*i*) has been chosen. The analyzing system gives a measuring pulse during every cycle at this point. This has been accomplished by strong amplification of the photomultiplier signal and subsequent detection of the zero crossing point by a Schmitt trigger.

C. PULSE GENERATOR

In order to measure the angular position of the analyzer, it has been fitted with an angular scale that consists of a cogwheel of magnetic material fixed to the analyzer and rotating with it. A magnetic tape recorder pick-up head is placed in a fixed position close to the cogwheel and gives an electric pulse as each cog passes. In this instrument one wavelength of optical path difference is represented by 256 such clock pulses. Because of this digital characterization of the analyzer angle, it has become possible to fully automate the measuring procedure.

D. ZERO REFERENCE MEMORY

The optical path difference should always be measured relative to the value in free space just outside the cell. Hence, it is important that the instrument has a memory to remember the phase in free space. This is realized by a reference scaler which functions as follows: When a new X scan is about to start, the reference scaler is set to zero and is waiting for the first measuring pulse. When this pulse comes, it opens the gate to the reference scaler and the clock pulses stream in. The reference scaler counts the pulses up to 256. It then gives out one reference pulse and starts on a new cycle to 256. Thus, reference pulses will be delivered every time the analyzer passes the position it had when the measuring pulse in free space was given and also at 180° apart from this. This information is kept in the memory until a new reference position is fed into it, that is, when free space is reached after the completion of one scan line and the next is about to be started.

E. OPTICAL PATH DIFFERENCE INTEGRATION

The measuring pulses are produced by the analyzer during the whole measuring procedure and show the momentary phase position (curve B of Fig. 2) once for every half-revolution of the analyzer. As the specimen is moved at a speed of 60 μm/sec and the analyzer takes 120 readings/sec, the mean spacing between the measuring points will be 0.5 μm. The OPD value is read for every point, which is done as follows: The reference pulse opens a gate letting clock pulses pass into the main counter. The gate will stay open until it is closed by the measuring signal. The number of pulses collected by the counter is then proportional to the phase difference between reference and measuring signal. The next reference signal will open the gate again and another train of pulses will be accumulated in the counter. When the whole object has been scanned the sum of pulses corresponds to the surface integral

FIG. 3. Appearance of the microinterferometer as commercially produced by Incentive Research and Development AB.

of the OPD. It is immediately displayed on the counter tubes and can easily be transferred to, for example, cards, punch tape, or magnetic tape for further data treatment.

F. CELL MONITOR

Every pulse train is shown on an oscilloscope screen as a vertical line. The horizontal sweep is synchronized to the X movement. In this way "cross sections" through the objects are shown, a representation that has been familiar from old recording microinterferometers. Even if no measurements are made on the screen images they are very useful to check that all settings are correctly done. The appearance of one version of the microinterferometer is shown in Fig. 3.

II. Standard Measuring Routine

Due to the highly automated function of the instrument, the measuring routine is very simple.

(1) The cell is placed in the angle between the two fixed index lines O_1 and O_2 (Fig. 1) with a suitable margin in between.

(2) The index lines N_1 and N_2 are moved to enclose the cell by manipulation of knobs on the front panel.

(3) The prism which deflects the microscope light into the ocular is pulled to the side, thus giving free passage of the beam to the photomultiplier. The scanning is then automatically started and continues until the whole area between the index lines is covered. Scanning is then automatically stopped and the result can be read from the counter.

(4) When the prism is pushed back to the inspection position, the scanning stage automatically returns to the zero position and the counter is automatically set to zero.

III. Manual Zero Setting

The automatic zero setting might be impossible to apply if, for example, the background is not free along the vertical index lines to the left and to the right of the cell. If one of these lines is free but the other is not, it is possible to choose zero setting on only one line. If neither of the lines is free, the measuring point can be moved manually to any suitable point in the field. At this point the instrument can be ordered to note the phase, and this value is then remembered and used as zero in the subsequent scanning measurement.

IV. Summary

The instrument herein described is specially designed to make possible rapid and accurate measurements of dry weight on large cell populations. It utilizes a continuously rotating analyzer with a digitized angle transmitter. The scanning is performed by a precision stage moving the object in a pattern of parallel lines. Special features to facilitate high working speed are rapid scanning, automatic zero setting, a convenient limiting system for the measuring field, automatic integration, and digital readout.

REFERENCES

1. Barer, R., Interference microscopy and mass determination. *Nature* **169**, 366 (1952).
2. Carlson, L., Constants necessary for the interferometric mass measurement of fixed material, determined by microradiography. *Acta Histochem.* **6**, Suppl., 397–402 (1965).
3. Carlson, L., A precision stage for two-dimensional scanning movement. *In* "Introduction to Quantitative Cytochemistry-II" (G. L. Wied and G. F. Bahr, eds.), Academic Press, New York, 1970.
4. Davies, H. G., and Wilkins, M. H. F., Interference microscopy and mass determination. *Nature* **169**, 541 (1952).

5. Lebedeff, A., L'interféromètre à polarisation et ses applications. *Rev. Opt.* **9**, 385–413 (1930).
6. Lomakka, G., A rapid scanning and integrating microinterferometer. *Acta Histochem.* **6**, Suppl., 393–396 (1965).
7. Svensson, G., A scanning interference microphotometer. *Exptl. Cell Res.* **12**, 406–409 (1957).

CHANGES IN NUCLEAR STAINABILITY ASSOCIATED WITH SPERMATELIOSIS, SPERMATOZOAL MATURATION, AND MALE INFERTILITY

*Barton L. Gledhill**

DEPARTMENT OF CLINICAL STUDIES, SCHOOL OF VETERINARY MEDICINE, UNIVERSITY OF PENNSYLVANIA, NEW BOLTON CENTER, KENNETT SQUARE, PENNSYLVANIA

I. Introduction

The genetic substance in spermatozoal heads is known as deoxyribonucleic acid (DNA) and is chemically bound to certain basic nuclear proteins such as histones and protamines to form a deoxyribonucleoprotein (DNP) complex. According to classic genetic principles, the number of chromosomes and therefore the amount of DNA is not altered during spermiogenesis in the process of spermateliosis, i.e., the metamorphosis of spermatids to spermatozoa. Each spermatozoon should contain what might be termed a haploid amount of DNA. Since within any given species, the haploid chromosome number in essence is constant, the haploid amount of DNA should likewise remain constant.

There are, however, a number of reports which seem to challenge this concept of a constant haploid amount of DNA within a species. Some of these reports deal with studies made on male infertility and therefore are of prime interest for this chapter.

Little more than a decade ago a group of investigators (*86*), suggested that one form of human male infertility might be correlated to a deficient and grossly fluctuating amount of DNA in spermatozoa. Shortly thereafter a similar suggestion was advanced (*85*) to explain certain cases of infertility

* This review was written during the progress of research supported in part by funds from The National Institutes of Health (1-R01-HD-03577-01 and 1-F3-HD-14,733-01) and in part by a grant from the Pennsylvania Department of Agriculture (ME-95).

125

in bulls. The existence of an inconsistency in the haploid amount of DNA, again associated with male infertility, seemed to gain further support when another group of workers (105) found that spermatozoa from infertile bulls contained more DNA than did spermatozoa from normal bulls. In contrast to these investigations, Welch and Resch (136) found a statistically nonsignificant difference between arithmetic means for the DNA content of ejaculated spermatozoa from fertile and infertile bulls. Paufler and Foote (106) studied 12 bulls in regular artificial breeding use. Based upon 27,383 first inseminations, the fertility of the bulls ranged from 69 to 76% and the among-bull correlation between relative values for spermatozoal Feulgen-DNA and percent 60- to 90-day nonreturn to estrus was nonsignificant.

The striking fluctuation between normal and abnormal DNA amounts observed with repeated sampling of the same infertile individuals (87) does not point to aberrations in chromosome number as the causative factor of DNA deviations. In a later publication, Leuchtenberger (83) stated clearly that the variability and deficiency of DNA could not be ascribed to loss or irregular distribution of chromosomes. Bahr and Wied (7) have postulated the existence of a characteristic distribution of redundant chromosomal DNA, but this small amount of DNA, if it exists, would not be nearly sufficient to account for the differences described by the Leuchtenberger group (85, 86).

The effects of in vitro storage on the Feulgen-positive material (DNA) in ejaculated bull spermatozoa were studied by Salisbury et al. (115). These investigators observed that spermatozoa, after 5 days' storage at 5°C in egg-yolk citrate diluent, showed a pronounced reduction in DNA content and suggested that this reduction in DNA upon storage may explain the approximately parallel decrease in conception rate and increase in embryonic mortality of cattle known to occur after artificial insemination with similarly aged semen. Later results by different groups of workers (70, 71, 78) substantiated the observation of a decrease in Feulgen-DNA during in vitro storage.

Chang (42) compared the DNA content of rabbit spermatozoa recovered from the uterus and Fallopian tubes to the DNA content of spermatozoa in an ejaculate sample. He found that the spermatozoa recovered from the female reproductive tract contained more DNA than did the spermatozoa in the ejaculate sample. This finding has been substantiated recently by Esnault et al. (50) who found an average increase of 25% in the Feulgen-DNA content of rabbit spermatozoa, incubated in vivo 9 hr in the uterus of a doe.

The main common denominator in the reports cited above seems to be the method used to estimate the DNA content of spermatozoa. These reports make one wonder if the decreased or increased values obtained with this method, i.e., Feulgen microspectrophotometry of spermatozoa, in fact, do represent true differences in DNA. There are suggestions in the literature that the stoichiometry of the Feulgen reaction, when applied to cells, might be modified by the protein bound to the DNA (6, 35, 134) or the cellular environment (54, 56).

Concomitant with the characteristic morphologic changes of spermateliosis, there is, in several species of animals, an alteration of the typical somatic type of histone into a more basic and arginine-rich histone (protamine). This phenomenon has been studied in detail in salmon with biochemical methods (80, 94). Cytochemical methods also have been used for studies on salmon (2) as well as snails (26), squid (21, 22), grasshopper (23), Drosophila (45), house cricket (75), rats (90, 123, 130), mice (96, 97), the live-bearing fish, Poecilia (109), and grass-snake (124). The results obtained in these investigations suggest that a change in basic nuclear protein occurs not only in fish but also in other species of animals.

In the bull, the existence of a histone to protamine "transition" of the type which has been demonstrated in salmon and other fish had not been conclusively shown prior to the present studies. Vendrely et al. (132) found that the DNP biochemically extractable from bull testicular homogenates exclusively contained histone of the somatic type. Berry and Mayer (18) biochemically examined ejaculated, caput epididymidal and testicular bull spermatozoa and found that the basic protein in ejaculated spermatozoa did not differ from calf thymus histone with respect to mole percent of basic amino acids. These authors thought that the basic proteins of both caput epididymidal and testicular spermatozoa were "histone-like."

The possibility of an association between a form of male infertility and spermatozoa with an atypical DNA content is biologically, and for most domestic animals economically, of such importance as to warrant further investigation. This review summarizes the results of studies (61, 62, 66) initiated to investigate such a possibility. In this chapter, evidence is collated and offered to suggest the existence of infertility in the bull due to a biochemical immaturity of the protein associated with spermatozoal DNA, rather than a defect in the DNA per se. Apparently an arrest in protein alteration during spermateliosis is responsible and is still evident during subsequent maturation in the epididymis.

II. Materials and Methods

A total of 18 Swedish Red and White breed bulls was used over a period of 3 years and between 40,000 and 50,000 individual gametes were analyzed. Nine bulls comprised the normal group and were considered as control bulls.

Of the remaining nine bulls that had lowered fertility, seven had no or only slight deviations from accepted standards of normal semen quality. Despite extensive clinical examinations, including, in one case, somatic chromosome analysis, no explanation of the infertility could be found. The other two animals showed clinical signs of testicular degeneration which was later histopathologically confirmed. One of the infertile bulls eventually became the subject of an extensive investigation (67) into the spermiogenic changes in DNP associated with infertility of clinically undiagnosed origin. The age range of the 18 bulls was 1.5 to 15 years.

The manner in which germinal cells from the testicle, caput, corpus, and cauda epididymidis and ejaculate were obtained and prepared for cytochemical analyses is described in detail elsewhere (34, 60, 66). Only morphologically normal spermatozoal heads were selected for measurement although limited studies have shown that total DNA content does not differ between the more common types of misshapen heads and normally shaped heads (60, 69). Diploid spermatozoa (14, 51, 58, 59, 105, 113–115) are, of course, an exception and were carefully excluded from all measurements. Terminology descriptive of spermatogenesis is used in accordance with Bishop and Walton (20).

When the core-sampling technique of measurement in a fixed light-beam cytophotometer is used for cytochemical analysis of, for example, spermatozoal DNA content, special attention must be paid to the problems of variability in geometry of the nuclei and variability in the degree of homogeneity of distribution of chromatin (10, 31, 82, 116, 129, 135). Problems of this sort have been largely circumvented in the present series of studies by

TABLE I

METHODS USED FOR CYTOPHOTOMETRY OF BULL SPERMATIDS AND SPERMATOZOA

DNA Determinations	
1. Ultraviolet light absorption	(265 nm)
2. Feulgen reaction	(546 nm)
Nuclear Protein Analyses	
1. Bromphenol blue reaction	(592 nm)
2. Sakaguchi reaction	(480 nm)
DNP-Phosphate Group Determinations	
1. Acridine orange fluorescence	(530 nm)
2. Methyl green reaction	(590 nm)
Dry Mass Determinations	
1. Interferometry	
2. X-Ray microradiography	

using recently developed rapid-scanning instruments (37–39) especially suited for use in large-scale cytochemical investigations. Total quantities of cellular substances are determined with these high-resolution instruments by automatic integration of numerous spot measurements in each individual cell or cell part.

Table I summarizes the techniques used quantitatively and qualitatively to study the DNA, the DNP complex, and the nuclear protein as well as the methods for determination of the dry mass (weight) of bull spermatids and spermatozoa [see Sandritter and Grosser (118) for additional methods used by other authors].

A. DNA DETERMINATIONS

The DNA content of spermatids and spermatozoa was determined by Feulgen and ultraviolet (UV) microspectrophotometry. Total extinction of individual Feulgen stained cells was measured (37) at 546 nm (Fig. 1). With

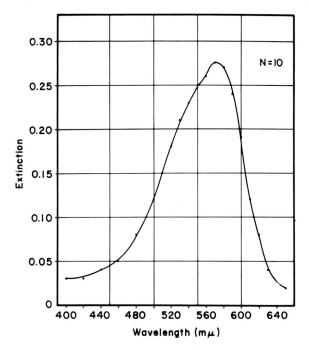

FIG. 1. The absorption spectrum of ejaculated bull spermatozoal heads stained by the Feulgen reaction as outlined by Leuchtenberger (82). This extinction curve is notably similar to those published by Garcia and Iorio (56) for Feulgen-stained human leukocytes and by Bahr and Wied (7) for Feulgen-stained ejaculated bull spermatozoa. Ten cells were measured in a UMSP-I and each point on the curve is the mean of ten duplicated measurements.

minor modifications, Feulgen staining was performed in accordance with
the technique outlined by Leuchtenberger (*82*). After hydrolysis in 1 *N* HCl
for 9 min at 60°C, cells were stained with a fuchsin solution (pH 1.5) (*43*)
at room temperature for 60 min. An additional method (*24*) in which 1 *N*
TCA (trichloroacetic acid) was substituted for 1 *N* HCl was also used. The
optimum hydrolysis time (9 min) for maximum color development with
both the HCl- and the TCA-Feulgen methods (*68*) agrees well with previously
published data (*7*, *116*, *135*). Smears of fresh fixed human blood from the
same individual were included as a stain reference system each time spermatozoal preparations were stained. The same factor by which the mean of
the stained polymorphonuclear leukocytes had to be adjusted so as to be
equal to a predetermined and arbitrary 100% value was applied to the means
of the spermatozoal preparations.

Total extinction at 265 nm of individual unstained spermatozoal heads

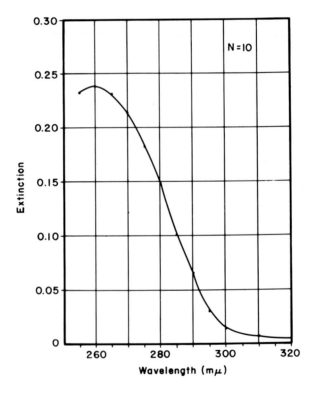

FIG. 2. The absorption spectrum of washed, fixed, and unstained ejaculated bull
spermatozoal heads immersed in glycerol. This extinction curve is very similar to that
published by Bahr and Wied (*7*). Ten cells were measured in a UMSP-I and each point on
this extinction curve is the mean of ten duplicated measurements.

(Fig. 2) and spermatid nuclei (after removal of RNA) has been measured in a scanning cytophotometer (91) and has been taken as an estimate of the amount of DNA present in each cell.

B. Nuclear Protein Analyses

The proteins conjugated with DNA in spermatozoal heads are of the basic type (93). The alkaline bromphenol blue (BPB) binding reaction for basic proteins (26) as modified by Ringertz and Zetterberg (111) has been used as a measure of total basicity (BPB binding capacity). The Sakaguchi reaction for protein-bound arginine (88) was also used for the analysis of basic nuclear protein composition. Microspectrophotometry (91) of BPB stained cells and cells stained by the Sakaguchi reaction has been performed at 592 nm and 480 nm, respectively.

C. DNP-Phosphate Group Determinations

The strength with which the nuclear protein is bound to the DNA molecule is also of interest, particularly in respect to the condensation of chromatin which takes place during spermiogenesis. One way to investigate this binding is to determine the number of negatively charged phosphate groups within the deoxyribonucleoprotein complex available for binding by cationic dyes. Two such basic dyes, i.e., the fluorescence dye acridine orange (AO) and methyl green (MG), have been utilized for this purpose. The AO fluorescence intensity of individual spermatids and spermatozoa was measured at 530 and 590 nm in the microspectrofluorometer (40). Fluorescence intensity at 530 nm, after correction for intensity contributed by AO-binding $RNP-PO_4^-$ groups (590 nm total fluorescence), was taken to represent the number of AO-binding, $DNP-PO_4^-$ groups (110). The MG binding capacity of individual cells was microspectrophotometrically (91) estimated at 590 nm.

D. Dry Mass Determinations

When quantitatively studying spermatozoal DNA and nuclear protein with light absorption methods, dry mass determinations are valuable as a point of reference. A scanning microinterferometer (92) has been used to determine the surface integral of the optical path difference (OPD) between the spermatozoal head and its background. The optical path difference values are proportional to the total dry mass but can be converted to grams only when the specific proportionality factor for the particular cell type and immersion medium is known. Determination of this factor for spermatozoal heads by the combination of soft X-ray microradiography (89) and microinterferometry (13, 49) was done.

III. Results

A. ALTERATIONS IN DEOXYRIBONUCLEOPROTEIN DURING SPERMATELIOSIS ASSOCIATED WITH NORMAL FERTILITY (66)

1. DNA Determinations

Prior to removal of RNA, round spermatids, but not elongated spermatids and testicular spermatozoa, exhibited higher 265 nm total extinction values (total extinction = surface integral of the optical density denoting absorption) than could be solely attributed to the DNA content of their nuclei. However, after removal of RNA by RNase digestion and cold trichloroacetic acid extraction, the UV absorption (265 nm) of round spermatids was essentially the same as that of elongated spermatids and testicular spermatozoa and was slightly less than one-quarter that of RNA extracted primary spermatocytes. An equal amount of DNA was present in spermatids in all stages of spermiogenesis (Fig. 3). This constant amount of DNA was in agreement with the fact that all of these cells were haploid.

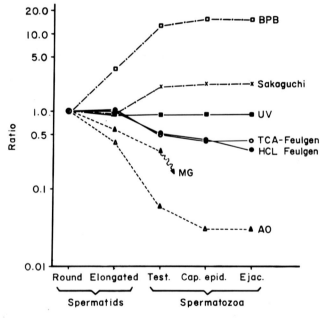

FIG. 3. A summary of the alterations in DNP seen during spermateliosis associated with normal fertility in the bull. For each cytochemical parameter the mean values for the different cell types are related to the corresponding mean values for round spermatids to yield a ratio which is plotted on a logarithmic scale (ordinate). Reproduced from Gledhill, Gledhill, Rigler, and Ringertz (66) with permission of Academic Press.

Feulgen stained round spermatids exhibited 546 nm total extinction values which were approximately one-quarter of those for primary spermatocytes. Total extinction values of Feulgen stained cells showed a pronounced decrease during spermateliosis between elongated spermatids and testicular spermatozoa in a ratio of approximately 2:1 (Fig. 3).

2. Nuclear Protein Analyses

As spermatids transformed into spermatozoa, their content of protein-bound arginine and their total basicity increased (Fig. 3). Testicular spermatozoa stained with the Sakaguchi reaction exhibited more than doubled absorption values in comparison to those for spermatids. Elongated spermatids bound slightly more than 3 times and testicular spermatozoa slightly more than 13 times more bromophenol blue than did round spermatids.

3. DNP-Phosphate Group Determinations

The number of $DNP-PO_4^-$ groups available for binding by cationic dyes was reduced (Fig. 3). The fluorescence intensity of AO stained elongated spermatids and testicular spermatozoa, in comparison to round spermatids, was markedly diminished. It was reduced by factors of approximately 3 and 18, respectively. A decrease was also seen in MG binding during spermateliosis. Elongated spermatids, when compared to round spermatids, had absorption values which were reduced by a factor of slightly less than 2. Nearly always, testicular spermatozoa bound so little methyl green as to be unmeasurable.

B. ATYPICAL ALTERATIONS IN DEOXYRIBONUCLEOPROTEIN DURING SPERMATELIOSIS ASSOCIATED WITH INFERTILITY (67)

1. DNA Determinations

UV microspectrophotometry at 265 nm of round spermatids after removal of RNA showed that the DNA content of these cells was approximately one-quarter that of primary spermatocytes. Spermatid and spermatozoal DNA content was the same as found in the normally fertile bulls. The cellular content of DNA determined by UV techniques was unchanged during spermiogenesis and the ensuing period of epididymal passage (Fig. 4a).

Round and elongated spermatids from this infertile bull exhibited a normal Feulgen reaction in that their absorption values were also approximately one-quarter those of primary spermatocytes. Testicular as well as caput epididymidal and ejaculated spermatozoa all showed a reduction in Feulgen stainability. However, this reduction was not nearly as marked as that seen in similar cells from the fertile bulls (Fig. 4a).

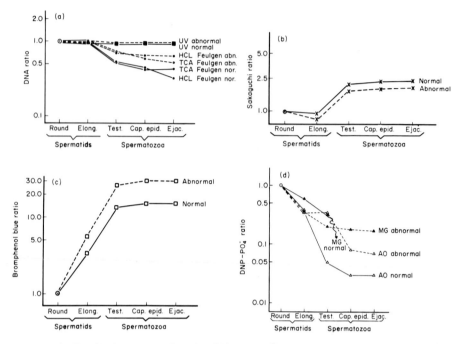

FIG. 4. Graphical summary of results of (a) UV-DNA and Feulgen-DNA determinations, (b) Sakaguchi protein-bound arginine analyses, (c) alkaline bromphenol blue analyses, and (d) DNP-phosphate group determinations with acridine orange and methyl green. In each graph, the mean values for the different cell types from the infertile bull (————, abnormal) and control bulls (————) (see Fig. 3) are related to the corresponding mean values for round spermatids to yield a ratio which is plotted on a logarithmic scale (ordinate). Redrawn from Gledhill, Gledhill, Rigler, and Ringertz (67).

2. Nuclear Protein Analyses

The Sakaguchi reaction values for protein-bound arginine were slightly depressed in comparison with control values (Fig. 4b). The bromphenol blue binding capacity was somewhat elevated (Fig. 4c). These deviations in Sakaguchi and bromphenol blue values, however, were not so clear-cut as those seen with Feulgen, AO, and MG.

3. DNP-Phosphate Group Determinations

The capacity of spermatozoal DNP to bind the cationic dyes, acridine orange and methyl green, also differed from that seen in the control bulls (Fig. 4d). A decrease in AO binding did occur but was not so profound as that seen in cellular counterparts from the fertile animals. Normally the

AO binding capacity of ejaculated spermatozoa, when compared to round spermatids, was reduced by a factor of 37. In this instance, the decrease was only 16-fold.

The MG binding capacity of male gametes as they matured normally decreased to such an extent that absorption measurements on caput epididymidal and ejaculated spermatozoa were impossible. In the present case of infertility, however, caput epididymidal and ejaculated spermatozoa bound measurable amounts of MG.

C. CYTOCHEMICAL REACTIVITY OF SPERMATOZOAL DEOXYRIBONUCLEOPROTEIN IN RELATION TO MATURATION AND TO FERTILITY STATUS (61, 62)

One way to estimate the reliability of experimental measurements is to determine, assuming the absence of a systematic error, the magnitude of the random error on the basis of duplicate determinations. The random error (s) for a single determination can be calculated from the formula:

$$s = \sqrt{\frac{\sum(a - b)^2}{2n}}$$

where a and b denote the separate values in a duplicate determination and n denotes the number of duplicate determinations. The relative random error for a single determination (random error expressed as percent of the mean value) was calculated from 11 duplicate determinations, i.e., duplicate preparations from each of 11 ejaculates from seven normally fertile bulls. This error (Table II) is greater for Feulgen measurements than for UV or dry mass determinations; nonetheless, it is not unreasonably large.

TABLE II

RELATIVE RANDOM ERROR OF A SINGLE DETERMINATION[a]

Parameter	Relative random error %
Total extinction at 546 nm (Feulgen)	3.4
Total extinction at 265 nm (UV)	0.7
Integrated OPD (dry mass)	0.2

[a] Taken from Gledhill (63).

1. *DNA Determinations*

There was no statistic difference in mean values for the spermatozoa in either fertility group during epididymal maturation with respect to UV absorption or total dry mass. Furthermore, there was no difference in mean values between fertility groups in either of these parameters for spermatozoa collected from the caput, corpus, or cauda epididymidis or for ejaculated spermatozoa (*61, 62*).

"Relative Feulgen reactivity" is defined herein as the ratio of the mean total extinction value of Feulgen-stained spermatozoa to that of round spermatids from the same individual. The relative Feulgen reactivity of caput epididymidal and ejaculated spermatozoa from the bulls with lowered fertility was significantly higher ($p < 0.001$) than that of spermatozoa from the bulls with normal fertility.

A decrease in mean Feulgen absorption values as spermatozoa passed through the epididymis, albeit nonsignificant ($p > 0.05$), appeared to be less accentuated in the infertile bulls than in the fertile bulls. The mean absorption value of Feulgen stained caput epididymidal spermatozoa was significantly higher ($0.01 > p > 0.001$) in the bulls of lowered fertility than in the fertile bulls. Likewise, there was a significant ($0.05 > p > 0.01$) difference in 546 nm absorption of ejaculated spermatozoa between the two fertility groups, with the infertile bulls higher (*61, 62*).

2. Nuclear Protein Analyses and DNP-Phosphate Group Determinations

The changes in nuclear protein and $DNP-PO_4^-$ which normally occur during spermateliosis continued at a reduced rate as testicular spermatozoa migrated into and partially through the caput epididymides of the control bulls and were no longer apparent during further epididymal passage (Figs. 3 and 4b, c, d).

D. DRY MASS DETERMINATIONS ON SPERMATOZOAL HEADS FROM BULLS WITH NORMAL FERTILITY (*34*)

The mean total dry mass of morphologically normal, formaldehyde fixed, bull spermatozoal heads was determined by soft X-ray microradiography and was found to be 13.2×10^{-12} gm per head (*34*). This is substantially higher than the estimate of approximately 7.9×10^{-12} gm obtained by Müller *et al.* (*99*) using a similar method. However, it is very similar to the estimates of 13.0 and 13.35×10^{-12} gm obtained in two experiments using quantitative electron microscopy (*8*). Biochemically determined estimates of the dry weight of unextracted bull spermatozoa, ranging from 21 to 27×10^{-12} gm, have been reviewed by Salisbury and VanDemark (*117*). If, as these authors state, the spermatozoal head accounts for 49% of the total weight of the nonextracted cell, the head would then weigh approximately 12×10^{-12} gm. On the other hand, using the value of 64% reported elsewhere (*8*), one obtains the somewhat higher estimate of 15×10^{-12} gm.

In the present study, the microradiograms on which the dry mass determinations were done had been exposed before the spermatozoal preparations were immersed in water for microinterferometry. Subsequently, the integrated optical path difference (OPD) of each of the heads measured by microradiography was determined microinterferometrically. The mean integrated OPD per head was 1.54×10^{-12} cm^3. Before accepting the validity of this value, it is of prime importance to know if the immersion in water resulted in extraction of measurable amounts of substances from the fixed heads. No evidence of extraction by water was found (34); thus, one may assume that 1.54×10^{-12} cm^3 is the integrated OPD which corresponds to the microradiographically obtained spermatozoal head total dry mass of 13.2×10^{-12} gm.

A proportionality factor which has been termed *specific optical path difference* (ψ) by Carlson (33) is necessary for the mathematic transformation of the microinterferometrically obtained integrated OPD into grams. This factor is obtained by dividing the integrated OPD by the total dry mass and, in the case of bull spermatozoal heads in water, is very nearly 0.12 cm^3/gm ($\psi_w = 1.54 \times 10^{-12}$ $cm^3/13.2 \times 10^{-12}$ gm).

Obviously, the validity of this calculated value is dependent upon the validity of the two experimentally determined values from which it is derived. Both of these experimentally determined values are apparently correct as one of them (total dry mass) agrees well with several reports in the literature (8, 117) and extraction of cellular substances by water is not a problem in the other (integrated OPD). Additionally, both of them are combined results of three separate experiments between which there was no significant difference ($p > 0.05$). Therefore, it follows that 0.12 cm^3/gm is the proportionality factor specific for bull spermatozoal heads in water.

IV. Discussion

A. SPERMATELIOSIS AND SPERMATOZOAL MATURATION

The current results obtained with both the Feulgen and the UV microspectrophotometric methods for determination of DNA are in agreement with many earlier reports (84, 126–128) which state that the amount of DNA in primary spermatocytes is four times greater than that in spermatids. From this point onward, however, the results of DNA estimations obtained by the two microspectrophotometric methods differ. The greatest reduction in Feulgen stainability occurred at the same time in spermateliosis as the most marked alterations in nuclear protein. This is, of course, not proof of a cause-and-effect relationship. Nonetheless, the present observations, coupled with the reports cited below, are strong circumstantial evidence that the Feulgen stainability of bull spermatozoa is influenced not only by the DNA content but also by the composition and binding of the protein component of the DNP.

A precise physicochemical explanation for the reduction in Feulgen re-activity of male germ cells as they mature, despite constant amounts of DNA, is not yet available. This is not at all surprising since all factors determining the behavior of this reaction when applied to cells are not fully understood (36, 54, 55, 74, 134).

The observed decrease in Feulgen stainability cannot be explained by a difference in optimum hydrolysis time for each of the various types of spermatids and spermatozoa since the time used was that which resulted in the greatest amount of color development in the ejaculated spermatozoa. Obviously, any alteration in the hydrolysis time used would have only resulted in greater differences between the cell types.

The well-recognized existence of differing optimum hydrolysis times and differing shapes of hydrolysis curves for various tissues (1, 82, 119) implies that the reactivity of a given amount of DNA toward the Feulgen reaction is influenced by its intracellular environment. Lison (90) suggested that a decrease in basophilia which occurred at the same time as an increase in arginine content during the maturation phase of rat spermateliosis was not due to a decrease in DNA content but was the consequence of an increase in positively charged guanidine groups of arginine side chains. Data presented by Bloch and Hew (27) on histone transition in the snail, Helix aspersa, shows, between two stages of spermatid maturation, a conspicuous decrease in DNA content which was determined with Feulgen microphotometry. Between the same stages, the basic nuclear protein changed from a somatic type of histone to an arginine-rich protein and the fine structure of the nucleus changed from that of a typical interphase nucleus to one showing filament formation. These authors did not comment upon the possibility of these latter two changes being related to the discrepancy in Feulgen-DNA values but rather attributed the discrepancy to an increasing error of measurement.

The results of the present studies indicate that an alteration of the sperma-tid histone into a more basic arginine-rich protein occurs during spermatel-iosis in the bull. Although a similar alteration has been found in several species, its biologic implications are largely unknown. Histones have been implicated in the regulation of gene activity (28, 72, 121) by virtue of their ability to inhibit DNA-dependent RNA synthesis. Vaughn (131) contends that, based on his observations of mature spermatozoa from the decapodal crustacean, Emerita analoga, the hypothesis which holds that arginine-rich histones and protamines occur in association with spermatozoal DNA in order to completely inhibit genotypic expression must be reexamined. How-ever, the cessation of RNA synthesis early in spermiogenesis in the grass-hopper (98) and the mouse (97) does seem to occur at approximately the same time as the nuclear protein alterations found in the bull.

The observed reduction in the number of negatively charged phosphate groups and the increase in basicity of the DNP complex suggest a strengthening of the electrostatic binding between the nuclear protein and the DNA molecule and hence an increase in the stability of the complex. Murray and Peacocke (*100*) have found that the double helical structure of DNA is stabilized against the action of thermal denaturation by the binding of histone to DNA. Relatively high temperatures, in fact, are necessary to induce thermal denaturation of nucleohistone in intact boar spermatozoal heads (*41*).

The structural changes in chromatin known to occur during spermateliosis (*52, 112*) may be associated with these observations. Changes in the macromolecular structure of DNA result in a decreased methyl green binding capacity (*133*). Thus, the double helical structure of DNA, upon becoming more neutralized, may decrease in rigidity and become more tightly spiralled, thereby allowing consolidation of the chromatin. The resistance of spermatozoal DNA to extraction by DNase, observed with conventional procedures (*44*), might also be a reflection of the dense packing of spermatozoal DNP inhibiting the enzyme from reaching at least part of its substrate.

The unusual value for the specific optical path difference of ejaculated bull spermatozoal heads (*34*) may be associated with the dense packing of the DNP complex. Estimates of the dry mass of bull spermatozoal heads (7 to 9×10^{-12} gm) obtained with interferometric procedures have been reported by some workers (*48, 85, 120*). These estimates are only in apparent contradiction to that presented in this report since the former authors utilized 0.18 cm^3/gm to calculate the total dry mass instead of 0.12 cm^3/gm.

The biologic implications of this value for ψ are at present unknown. The value is substantially lower than values characteristic (*47*) for most chemical substances commonly found in cells. Perhaps it indicates that the spermatozoal head has hydrophobic or unusual electrostatic properties not found in other biologic material which may be associated with the densely packed DNP. One definite conclusion may be drawn. It is necessary to determine the numeric value of the specific optical path difference for each cell type and immersion medium in order to properly convert microinterferometrically obtained data into grams.

Brief mention should be made of the oft examined problem of identifying X and Y chromosome-bearing spermatozoa (*16, 19*). Jensen and Leeflang (*73*) report that they might have found two populations, *vis.*, of X and Y spermatozoa, whose Feulgen-DNA contents differed by approximately 3%, the theoretic difference between X and Y chromosomal DNA content when based on the total haploid amount (*11*).

Although Stolla (*122*) thought that UV cytophotometry was a suitable means to differentiate the DNA content of male- and female-determining spermatozoa from the boar, no statistically valid difference in DNA content was found. Extensive analysis of the present data (UV, dry mass and Feulgen) has failed to reveal a bimodal distribution (64). An attempt has been made (65) to measure a somewhat more gross chromatin distributive imbalance in ejaculated spermatozoa from a subfertile boar known to be a translocational heterozygote as revealed by cultures of peripheral blood. Approximately three-quarters of one of the big acro- or telocentric chromosomes had been translocated to one of the small metacentric chromosomes. Both UV-DNA content (265 nm) and interferometrically determined total dry mass of spermatozoa from the translocation boar were significantly lower than corresponding values from a normally fertile control boar. Additionally, the variation in DNA content of individual spermatozoa from the affected boar was greater than in the control boar. Perhaps future refinement of instrumentation and use of more sensitive fluorometric techniques will allow the differentiation of male- and female-determining spermatozoa.

B. Male Infertility

Underdevelopment or injury of the spermatogenic epithelium nearly always results in lowered fertility and usually can be recognized by the presence of excessive numbers of morphologically abnormal spermatozoa in the semen as well as deviations in other semen characteristics (12, 76, 81). The infertile bulls investigated in the present studies were judged to be free from diseases known to cause infertility and, with two exceptions, did not exhibit high percentages of morphologically abnormal spermatozoa. Their infertility, in fact, could not be explained on the basis of semen analyses and other clinical examinations. This not infrequent situation implies that factors must exist which can lower the fertilizing potential of spermatozoa and yet remain visually unrecognizable. Defects in the spermiogenic biochemical alteration might be one such factor.

The decrease in stainability of more advanced cells, as shown by relative Feulgen reactivity, was less marked in the infertile bulls than in the fertile ones. However, UV values for caput epididymidal spermatozoa did not differ between the two fertility groups, which indicated that the DNA content did not differ. The less extensive reduction, i.e., the higher Feulgen absorption values for caput epididymidal spermatozoa from infertile bulls, could be a reflection of defects in the nuclear protein alteration of spermateliosis since a reduction in Feulgen stainability normally occurs concomitantly with this alteration.

Because the amount of DNA and dry mass was unchanged during epididymal maturation, by analogy, the amount of protein can be expected to be constant. Therefore, the trend towards lowered Feulgen values in the more distal portions of the epididymis in normally fertile bulls probably reflects the final stages of changes in nuclear protein which originate in the testicle. This phenomenon has also been noted in rabbit spermatozoa (29) where the Feulgen values decreased gradually during epididymal passage while, as in the present investigation, UV values were unchanged. In another publication, Bouters et al. (30) attributed the difference in quantitative response of rabbit spermatozoa to the Feulgen reaction to an aging process in the cells. Interestingly, Paufler and Foote (107) noted no difference in Feulgen-DNA content between epididymal and ejaculated rabbit spermatozoa. Significantly lower Feulgen-DNA values for bull spermatozoa obtained from the ampullae of the ducti deferenti when compared to those values for epididymal spermatozoa have been interpreted as suggesting that the ampullar environment is unfavorable for *in vivo* storage of bull spermatozoa (32).

Such a progressive reduction in Feulgen stainability of spermatozoa as they pass through the excurrent duct system should be considered in respect to the acquisition of fertility. Normally, caput epididymidal spermatozoa are less capable of fertilization than spermatozoa from the lower portion of the epididymis (15, 137) and embryos developing following insemination of lower corpus or distal cauda epididymidal spermatozoa tend to die or disappear before implantation (103). The observed reduction in Feulgen stainability may be one indication that the cell has matured and can realize its full potential in regard to fertility. In the present material, the Feulgen reduction trend during epididymal passage was not so apparent in spermatozoa from the infertile bulls, again suggesting a defect originating in the gonadal nuclear protein alteration.

It is tempting to speculate that one form of male infertility may be associated with an arrested or impeded DNP alteration during spermateliosis. Results of the detailed examination of one case of infertility in which atypical alterations were found support this speculation (67). Spermatozoa formed under such conditions would be likely to contain a type of immature DNP complex. Such an immature DNP complex might be caused by a defect in the binding of nuclear protein to the DNA-PO$_4^-$ groups. If spermatozoal nuclear protein were not as tightly bound to DNA due to atypical basic amino acid units, the entire DNP complex would be chemically less stable than normal. Although the DNA would be identical in both instances, the DNP complex with lesser stability can be considered immature because of its nuclear protein.

To do more than briefly theorize as to why individuals ejaculating spermatozoa with a defective DNP complex might be infertile is premature. Perhaps the genetic material is more susceptible to damage during the interval between ejaculation and fertilization since the DNP is apparently stabilized to a lesser extent. Another possibility is that the immature DNP complex of the spermatozoal head cannot decondense following penetration of an ovum (fertilization) and therefore interference with fusion and syngamy of the pronuclei might occur. Or, perhaps embryonic development does start but the paternal protein is so atypical that death of the zygote occurs at an early stage. Alfert (3), Bloch and Hew (27), Das et al. (46), and Vaughn (131) have presented cytochemical studies relative to fertilizing spermatozoa and probable nuclear histone transitions which seem to be particular to the various embryonic systems.

The exciting data of Esnault et al. (50) showing an increased Feulgen reactivity of ejaculated and a decreased reactivity of epididymal spermatozoa following identical conditions of in vivo incubation in the uterus, suggest that spermatozoal capacitation may be related in some way to changes in the DNP (104) since only the ejaculated spermatozoa revert to the more stainable form following incubation. Although this may well prove to be true, an explanation of how nuclear material can influence the ability of spermatozoa to penetrate ova, which is the basic tenet underlying the concept of capacitation, is difficult to envision. Of prime importance to this communication, however, is the fact that the difference in Feulgen stainability of the spermatozoa, which were fundamentally dissimilar in relative age only, occurred with no alteration of the nuclear 265 nm absorption. The failure of incubated epididymal spermatozoa to increase in tinctorial affinity is also being advanced as of significance in regard to abnormalities observed following fertilization with epididymal spermatozoa.

The results of the present studies at first seem to agree with some reports in the literature in which atypical Feulgen values have been found in cases of male infertility. Yet it is apparent from the present studies that the Feulgen reaction, although admittedly specific for DNA, when applied quantitatively to bull spermatozoa seems to be influenced by factors other than the DNA amount. Berchtold (17) and Bähr (9) have also come to this conclusion. On the other hand, quantitative UV microspectrophotometry and dry mass determinations by microinterferometry are unaffected by these factors and do not show atypical spermatozoal DNA content in the cases of infertility investigated. Biochemical determination of DNA content (125) likewise has not demonstrated fertility-related DNA differences. Therefore, the relationship between decreased fertility and atypical amounts of spermatozoal DNA suggested by earlier investigators and based on Feulgen microspectrophotometry must be questioned. It seems that Feulgen microspectrophotometry alone is not adequate for the estimation of DNA content in spermatozoa.

Possibly, what has been reported (71, 115) as a decrease in Feulgen-positive material during in vitro storage of bull spermatozoa is a reflection of changes in the stainability of the DNP complex occasioned by increasing senility with its concomitant alterations of nuclear proteins. Investigations (78) on the changes in DNA content of rabbit spermatozoa during storage *in vitro* indicate that, despite a significant decrease in Feulgen-DNA with increasing storage time, no quantitative loss of DNA occurs when estimated by methods involving tritiated-thymidine labeling of the DNA. However, qualitative changes in the DNP complex may occur during these circumstances and such changes may be associated with the reduced fertilizing ability of stored rabbit spermatozoa (79). The increased proportion of denatured DNA in ram spermatozoa reported by Quinn and White (108) upon incubation, subsequent to deep-freezing, could be such a qualitative change. Reasonably, one might speculate that aging of spermatozoal DNP may inactivate part of the genome and that transmission of the resultant incomplete genetic message by fertilization causes death of the embryo. Anand (4) carefully draws attention to this speculation while reporting that no DNA loss was detected in aged live boar spermatozoa. Anand and First (5), again using chemical extraction and colorimetric identification of DNA with diphenyl-amine, have shown that loss of DNA upon storage is only from dead spermatozoa. This latter fact was earlier demonstrated in bull spermatozoa (71) with Feulgen absorption techniques. Miller and Blackshaw (95), contrary to what has generally been reported, have found no significant differences in the Feulgen absorption and acridine orange fluorescence of 1- and 9-day-old *in vitro*-stored rabbit spermatozoa.

Similar interpretations may be appropriate to the relationship between a decrease in Feulgen-DNA and male infertility as reported by the Leuchten-berger group (83, 85–87). Normally, during epididymal passage, a thorough mixing of variously aged spermatozoa occurs (102). If, as has been indicated, *in vitro* and *in vivo* (male) aging of spermatozoa are associated with a non-DNA decrease in Feulgen stainability, then perhaps the infertility observed by Leuchtenberger and her colleagues (85, 86) was more related to some defect resulting in the ejaculation of an entire population of abnormally old spermatozoa which gave lower than normal Feulgen-DNA values. In a manner of speaking, this would be a concept of biochemically over-mature spermatozoa and should be differentiated from the hypothesis presented earlier which proposes the existence of biochemically immature spermatozoa, although the two ideas are mutually compatible.

Frequent ejaculation in the bull may or may not influence epididymal transit time of spermatozoa (77, 101). Daily versus weekly ejaculation of boars seemed to have no effect on the DNA content of spermatozoa as identified and measured with diphenylamine (53).

The Feulgen findings in the present studies are in agreement with those of Parez *et al.* (105) who also observed increased Feulgen values for spermatozoa from infertile bulls. However, unlike these authors, the increase herein is not interpreted as a true increase in the amount of DNA but is thought to reflect a defective nuclear protein component in the DNP complex.

V. Conclusions and Summary

There is normally a biochemical maturation of DNP during spermateliosis in the bull which is characterized by an alteration of the spermatid histone into a more basic, arginine-rich protein and by an increase in strength of electrostatic binding of nuclear protein to the DNA molecule. Concomitantly, there is a decrease in the Feulgen reactivity despite the fact that there is no quantitative change in DNA. There are strong indications that the binding and composition of the basic nuclear protein linked to the DNA molecule effect the Feulgen stainability of spermatids and spermatozoa.

During epididymal passage, there is no quantitative change in DNA or total dry mass of bull spermatozoal heads irrespective of the fertility of the animal. Caput, corpus, cauda epididymidal, and ejaculated spermatozoal heads from infertile bulls contain the same mean amount of DNA and have the same mean total dry mass as do their counterparts from normally fertile bulls. The reduction in Feulgen stainability which occurs during spermateliosis is significantly less pronounced in spermatozoa from certain bulls with well documented but unexplainable infertility than in spermatozoa from fertile ones. The occurrence of a higher mean value for Feulgen-stained heads from the epididymides and the ejaculates of such infertile bulls in comparison to similar heads from bulls with normal fertility is probably due to a spermatozoal DNP complex in which the nuclear protein component is in some way defective. Thereby, morphologically normal yet biochemically immature spermatozoa are produced and presumably form the basis for the observed infertility.

Interpretation of quantitative estimations of DNA in spermatozoa by Feulgen microspectrophotometry must be made with caution. Whenever possible, additional cytophotometric techniques should be utilized.

Acknowledgments

The original data summarized herein represent work performed and published by the author and several colleagues. Public acknowledgment of the valuable contributions made throughout the entire study by these colleagues, Dr. Nils R. Ringertz, Dr. Rudolf Rigler, Jr., Mr. Leon Carlson, and Mrs. Marianne P. Gledhill, is made with great pleasure. This research was done at the Department of Obstetrics and Gynaecology, Royal Veterinary College (Head: Professor Allan Bane) and the Institute for Medical Cell Research and Genetics, Nobel Medical Institute, Karolinska Institutet (Head: Professor Torbjörn Caspersson), Stockholm, Sweden. Financial support granted by the Swedish Agricultural

Research Council, the Wallenberg Foundation, and the Damon Runyon Memorial Fund for various aspects of these studies is gratefully acknowledged. The author received personal and research funds from the American Veterinary Medical Research Trust for the early parts of these studies and later from the National Institute of General Medical Sciences, U.S. Public Health Service (Grants 1-F2-GM-14, 733-01A2 and 5-F2-GM-14, 733-02).

REFERENCES

1. Agrell, I., and Bergqvist, H., Cytochemical evidence for varied DNA complexes in the nuclei of undifferentiated cells. *J. Cell Biol.* **15**, 604–606 (1962).
2. Alfert, M., Chemical differentiation of nuclear proteins during spermatogenesis in the salmon. *J. Biophys. Biochem. Cytol.* **2**, 109–114 (1956).
3. Alfert, M., Cytochemische Untersuchungen an basischen Kernproteinen während der Gametenbildung, Befruchtung und Entwicklung. *Coll. Ges. Physiol. Chem.* **9**, 73–84 (1958).
4. Anand, A. S., Factors affecting the loss of DNA from spermatozoa. Ph.D. Thesis, University of Wisconsin (1967).
5. Anand, A. S., and First, N. L., Effect of aging of boar semen on the DNA content of live and dead spermatozoa. *Abstr. 6th Congr. Intern. Reprod. Animale, Paris*, 1968, p. 217.
6. Atkin, N. B., Mattinson, G., Beçak, W., and Ohno, S., The comparative DNA content of 19 species of placental mammals, reptiles and birds. *Chromosoma* **17**, 1–10 (1965).
7. Bahr, G. F., and Wied, G. L., Cytochemical determinations of DNA and basic protein in bull spermatozoa. Ultraviolet spectrophotometry, cytophotometry and microfluorometry. *Acta Cytol.* **10**, 393–412 (1966).
8. Bahr, G. F., and Zeitler, E., Study of bull spermatozoa. Quantitative electron microscopy. *J. Cell Biol.* **21**, 175–189 (1964).
9. Bähr, H., Mikrospektrophotometrische Untersuchungen zur Bestimmung der Desoxyribonukleinsäure (DNS) in Samenzellen von Bullen. Dissertation, Vet. Med., Munich (1965).
10. Baker, F. N., Bouters, R., and Salisbury, G. W., Optical density profiles in Feulgen-stained bovine spermatozoan nuclei. *J. Dairy Sci.* **47**, 1104–1105 (1964).
11. Baker, F. N., Salisbury, G. W., and Fechheimer, N. S., Measurements of bovine chromosomes. *Nature* **208**, 97 (1965).
12. Bane, A., and Nicander, L., Electron and light microscopical studies on spermateliosis in a boar with acrosome abnormalities. *J. Reprod. Fertility* **11**, 133–138 (1966).
13. Barer, R., Interference microscopy and mass determination. *Nature* **169**, 366–367 (1952).
14. Beatty, R. A., Genetics of gametes. V. The frequency distribution of the head length of rabbit spermatozoa and a search for dimorphism. *Proc. Roy. Soc. Edinburgh* **B68**, 72–82 (1961).
15. Bedford, J. M., Development of the fertilizing ability of spermatozoa in the rabbit epididymis. *J. Reprod. Fertility* **10**, 286–287 (1965).
16. Benedict, R. C., Schumaker, V. N., and Davies, R. E., The buoyant density of bovine and rabbit spermatozoa. *J. Reprod. Fertility* **13**, 237–249 (1967).
17. Berchtold, M., Absorptionsmessungen an Bullenspermien in ultravioletten Licht. *Zuchthyg.* **1**, 22–27 (1966).
18. Berry, R. E., and Mayer, D. T., The histone-like basic protein of bovine spermatozoa. *Exptl. Cell Res.* **20**, 116–126 (1960).

19. Bhattacharya, B. C., Bangham, A. D., Cro, R. J., Keynes, R. D., and Rowson, L. E. A., An attempt to predetermine the sex of calves by artificial insemination with spermatozoa separated by sedimentation. *Nature* **211**, 863 (1966).

20. Bishop, M. W. H., and Walton, A., Spermatogenesis and the structure of mammalian spermatatozoa. *In* " Marshall's Physiology of Reproduction" (A. S. Parkes, ed.), 3rd ed., pp. 1–129, Longmans, Green, New York, 1960.

21. Bloch, D. P., Histone synthesis in non-replication chromosomes. *J. Histochem. Cytochem.* **10**, 137–144 (1962).

22. Bloch, D. P., Cytochemistry of the histones. *Protoplasmatologia* **5**, 1–56 (1966).

23. Bloch, D. P., and Brack, S. D., Evidence for the cytoplasmic synthesis of nuclear histone during spermiogenesis in the grasshopper *Chortophaga viridifasciata* (*de Geer*). *J. Cell Biol.* **22**, 327–340 (1964).

24. Bloch, D. P., and Godman, G. C., A microphotometric study of the synthesis of desoxy-ribonucleic acid and nuclear histone. *J. Biophys. Biochem. Cytol.* **1**, 17–28 (1955).

25. Bloch, D. P., and Godman, G. C., Evidence of differences in the desoxyribonucleo-protein complex of rapidly proliferating and non-dividing cells. *J. Biophys. Biochem. Cytol.* **1**, 531–550 (1955).

26. Bloch, D. P., and Hew, H. Y. C., Schedule of spermatogenesis in the pulmonate snail, *Helix aspersa,* with special reference to histone transition. *J. Biophys. Biochem. Cytol.* **7**, 515–532 (1960).

27. Bloch, D. P., and Hew, H. Y. C., Changes in nuclear histones during fertilization, and early embryonic development in the pulmonate snail, *Helix aspersa. J. Biophys. Biochem. Cytol.* **8**, 69–81 (1960).

28. Bonner, J., and Huang, R. C., Properties of chromosomal nucleohistone. *J. Mol. Biol.* **6**, 169–174 (1963).

29. Bouters, R., Esnault, C., Ortavant, R., and Salisbury, G. W., Comparison of DNA revealed by Feulgen and by ultraviolet light in rabbit spermatozoa after storage in the male efferent ducts. *Nature* **213**, 181–182 (1967).

30. Bouters, R., Esnault, C., Salisbury, G. W., and Ortavant, R., Discrepancies in analyses of deoxyribonucleic acid in rabbit spermatozoa, involving Feulgen staining (Feulgen-DNA) and ultraviolet light absorption (UV-DNA) measurements. *J. Reprod. Fertility* **14**, 335–363 (1967).

31. Bouters, R., Salisbury, G. W., and Baker, F. N., Feulgen staining of rabbit sperma-tozoa: Hydrolysis time and distribution of chromatin. *Stain Technol.* **40**, 193–197 (1965).

32. Bouters, R., and Vandeplassche, M., Vergleichende mikrospektrophotometrische Bestimmungen von Feulgen-DNS an epididymären, ampullären, ejakulierten und durch Massage gewonnenen Spermien von Bullen. *Zuchthyg.* **3**, 1–5 (1968).

33. Carlson, L., Constants necessary for the interferometric mass measurement of fixed material, determined by microradiography. *Acta Histochem.* Suppl. 6, 396–402 (1965).

34. Carlson, L., and Gledhill, B. L., Studies on the dry mass of bull spermatozoal heads. Comparative measurements using soft X-ray microradiography and microinter-ferometry. *Exptl. Cell Res.* **41**, 376–384 (1966).

35. Caspersson, T. O., Die quantitative Bestimmung von Thymonucleinsäure mittels fuchsinschwefliger Säure. *Biochem. Z.* **253**, 97–110 (1932).

36. Caspersson, T. O., " Cell Growth and Cell Function," pp. 14–77. Norton, New York, 1950.

37. Caspersson, T. O., and Lomakka, G. M., Scanning microscopy techniques for high resolution quantitative cytochemistry. *Ann. N.Y. Acad. Sci.* **97**, 449–463 (1962).

38. Caspersson, T. O., and Lomakka, G. M., Microscale spectroscopy. *In* " Instrumenta-tion in Biochemistry " (T. W. Goodwin, ed.), pp. 25–40. Academic Press, New York, 1966.

39. Caspersson, T. O., Lomakka, G. M., and Caspersson, O., Quantitative cytochemical methods for the study of tumour cell populations. *Biochem. Pharmacol.* **4**, 113–127 (1960).

40. Caspersson, T. O., Lomakka, G. M., and Rigler, R., Jr., Registrierender Fluoreszenzmikrospektrograph zur Bestimmung der Primär-und Sekundärfluoreszenz verschiedener Zellsubstanzan. *Acta Histochem.* Suppl. 6, 123–126 (1965).

41. Chamberlain, P. J., and Walker, P. M. B., The thermal denaturation of nucleoprotein in boar sperm. *J. Mol. Biol.* **11**, 1–11 (1965).

42. Chang, M. C., Fertilizing capacity of spermatozoa. *In* "Recent Progress in the Endocrinology of Reproduction" (C. W. Lloyd, ed.), p. 153. Academic Press, New York, 1959.

43. Coleman, L. C., Preparation of leucobasic fuchsin for use in the Feulgen reaction. *Stain Technol.* **13**, 123–124 (1938).

44. Daoust, R., and Clermont, Y., Distribution of nucleic acids in germ cells during the cycle of the seminiferous epithelium in the rat. *Am. J. Anat.* **96**, 255–283 (1955).

45. Das, C. C., Kaufmann, B. P., and Gay, H., Histone-protein transition in *Drosophila melanogaster*. I. Changes during spermatogenesis. *Exptl. Cell Res.* **35**, 507–514 (1964).

46. Das, C. C., Kaufmann, B. P., and Gay, H., Histone-protein transition in *Drosophila melanogaster*. II. Changes during early embryonic development. *J. Cell Biol.* **23**, 423–430 (1964).

47. Davies, H. G., The determination of mass and concentration by microscope interferometry. *Gen. Cytochem. Methods* **1**, 55–161 (1958).

48. Davies, H. G., Deeley, E. M., and Denby, E. F., Attempts at measurement of lipid, nucleic acid and protein content of cell nuclei by microscope-interferometry. *Exptl. Cell Res.* **4**, 136–149 (1957).

49. Davies, H. G., and Wilkins, M. H. F., Interference microscopy and mass determination. *Nature* **169**, 541 (1952).

50. Esnault, C., Orgebin-Crist, M. -C., and Ortavant, R., Unpublished data (1968); cited in Orbegin-Crist (104).

51. Esnault, C., and Ortavant, R., Origine des spermatozoïdes diploïdes présents dans l'éjaculat d'un taureau Charolais. *Ann. Biol. Animale, Biochim., Biophys.* **7**, 25–28 (1967).

52. Fawcett, D., The structure of the mammalian spermatozoon. *Intern. Rev. Cytol.* **7**, 195–234 (1958).

53. First, N. L., Personal communication (1968).

54. Garcia, A. M., Cytophotometric studies on haploid and diploid cells with different degrees of chromatin coiling. *Ann. N.Y. Acad. Sci.* **157**, 237–249 (1969).

55. Garcia, A. M., Stoichiometry of dye binding versus degree of chromatin coiling. *In* "Introduction to Quantitative Cytochemistry-II" (G. L. Wied and G. F. Bahr, eds.), pp. 153–170. Academic Press, New York, 1970.

56. Garcia, A. M., and Iorio, R., Potential sources of error in two-wavelength cytophotometry. *In* "Introduction to Quantitative Cytochemistry" (G. L. Wied, ed.), pp. 220–221. Academic Press, New York, 1966.

57. Garcia, A. M., and Iorio, R., Studies on DNA in leukocytes and related cells of mammals. V. The fast green-histone and the Feulgen-DNA content of rat leukocytes. *Acta Cytol.* **12**, 46–51 (1968).

58. Gledhill, B. L., Quantitative ultramicrospectrophotometry of presumed diploid bovine spermatozoa. *Proc. 5th Congr. Intern. Reprod. Animale, Trento,* 1964. Vol. 3, pp. 489–494.

59. Gledhill, B. L., Cytophotometry of presumed diploid bull spermatozoa. *Nord. Veterinär med.* **17**, 328–335 (1965).

60. Gledhill, B. L., Studies on the DNA content, dry mass and optical area of morphologically normal and abnormal bull spermatozoal heads. *Acta Vet. Scand.* **7**, 1–20 (1966).
61. Gledhill, B. L., Studies on the DNA content, dry mass and optical area of bull spermatozoal heads during epididymal maturation. *Acta Vet. Scand.* **7**, 131–142 (1966).
62. Gledhill, B. L., Studies on the DNA content, dry mass and optical area of ejaculated spermatozoal heads from bulls with normal and lowered fertility. *Acta Vet. Scand.* **7**, 166–174 (1966).
63. Gledhill, B. L., "Studies on the DNA content, Nuclear Protein, and Dry Mass of Bull Spermatids and Spermatozoal Heads with Aspects on Fertility," p. 12. Balder, Stockholm, 1966.
64. Gledhill, B. L., Unpublished observations (1967).
65. Gledhill, B. L., Gledhill, M. P., and Henricson, B., An attempt to measure a chromatin distributive imbalance in spermatozoa. *Hereditas* **60**, 407–409 (1968).
66. Gledhill, B. L., Gledhill, M. P., Rigler, R., Jr., and Ringertz N. R., Changes in deoxyribonucleoprotein during spermiogenesis in the bull. *Exptl. Cell Res.* **41**, 652–665 (1966).
67. Gledhill, B. L., Gledhill, M. P., Rigler, R., Jr., and Ringertz, N. R., Atypical changes of deoxyribonucleoprotein during spermiogenesis associated with a case of infertility in the bull. *J. Reprod. Fertility* **12**, 575–578 (1966).
68. Gledhill, B. L., and Ringertz, N. R., Unpublished observations (1965).
69. Godowicz, B., and Krzanowska, H., DNA content of mouse spermatozoa from inbred strain KE of low male fertility. *Folia Biol.* (*Warsaw*) **14**, 235–242 (1966).
70. Handa, A., Hiroe, K., and Tomizuka, T., DNA content in bull spermatozoa during storage in yolk-citrate diluent at 4°C. (Trans. title). *Japan. J. Animal Reprod.* **10**, 109–113 (1965).
71. Handa, A., Hiroe, K., and Tomizuka, T., Effects of *in vitro* aging on the DNA content of bull spermatozoa. (Trans. title.) *Japan. J. Animal Reprod.* **11**, 91–94 (1965).
72. Huang, R. C., and Bonner, J., Histone, a suppressor of chromosomal RNA synthesis. *Proc. Natl. Acad. Sci. U.S.* **48**, 1216–1222 (1962).
73. Jensen, M. T., and Leeflang, C. W., Feulgen-DNA content of individual bull spermatozoa. *Exptl. Cell Res.* **44**, 614–616 (1966).
74. Kasten, F. H., The Feulgen reaction—an enigma in cytochemistry. *Acta Histochem.* **17**, 88–99 (1964).
75. Kaye, J. S., and McMaster-Kaye, R., The fine structure and chemical composition of nuclei during spermiogenesis in the house cricket. I. Initial stages of differentiation and the loss of nonhistone protein. *J. Cell Biol.* **31**, 159–179 (1966).
76. Knudsen, O., Cytomorphological investigations into the spermiocytogenesis of bulls with normal fertility and bulls with acquired disturbances in spermiogenesis. *Acta Pathol. Microbiol. Scand.* Suppl. 101, 79 (1954).
77. Koefoed-Johnsen, H. H., Influence of ejaculation frequency on the time required for sperm formation and epididymal passage in the bull. *Nature* **185**, 49–50 (1959).
78. Koefoed-Johnsen, H. H., Fulka, J., and Kopecný, V., Undersøgelser over spermiekernens DNA indhold under opbevaring of saed. *Aarsberetn. Inst. Sterilitetsforskn.* pp. 9–18 (1967).
79. Koefoed-Johnsen, H. H., Fulka, J., and Kopecný, V., Stability of thymine in spermatozoal DNA during storage in vitro. *Abstr. 6th Congr. Intern. Reprod. Animale, Paris,* 1968 p. 230.
80. Kossel, A., "The Protamines and Histones." Longmans, Green, New York, 1928.

81. Lagerlöf, N., Morphologische Untersuchungen über Veränderungen im Spermabild und in den Hoden bei Bullen mit Verminderter oder Aufgehobener Fertilität. *Acta Pathol. Microbiol. Scand.* Suppl. 19, 254 (1934).
82. Leuchtenberger, C., Quantitative determination of DNA in cells by Feulgen microspectrophotometry. *Gen. Cytochem. Methods* 1, 219–278 (1958).
83. Leuchtenberger, C., The relation of the deoxyribosenucleic acid (DNA) of sperm cells to fertility. *J. Dairy Sci.* 43, Suppl. 1, 31–50 (1960).
84. Leuchtenberger, C., Leuchtenberger, R., Schrader, F., and Weir, D. R., Reduced amounts of deoxyribose nucleic acid in testicular germ cells of infertile men with active spermatogenesis. *Lab. Invest.* 5, 422–440 (1956).
85. Leuchtenberger, C., Murmanis, I., Murmanis, L., Ito, S., and Weir, D. R., Interferometric dry mass and microspectrophotometric arginine determinations on bull sperm nuclei with normal and abnormal DNA content. *Chromosoma* 8, 73–86 (1956).
86. Leuchtenberger, C., Schrader, F., Weir, D. R., and Gentile, D. P., The deoxyribose nucleic acid (DNA) content in spermatozoa of fertile and infertile human males. *Chromosoma* 6, 61–78 (1953).
87. Leuchtenberger, C., Weir, D. R., Schrader, F., and Murmanis, L., The deoxyribose nucleic acid (DNA) content in spermatozoa of repeated seminal fluids from fertile and infertile men. *J. Lab. Clin. Med.* 45, 851–864 (1955).
88. Liebman, E., Permanent preparations with the Thomas arginine histochemical test. *Stain Technol.* 26, 261–263 (1951).
89. Lindström, B., Roentgen absorption spectrophotometry in quantitative cytochemistry. *Acta Radiol.* Suppl. 125, 206 (1955).
90. Lison, L., Variation de la basophilic pendant la maturation du spermatozoïde chez la rat et sa significantion histochimique. *Acta Histochem.* 2, 47–67 (1955).
91. Lomakka, G. M., A rapid scanning and integrating cytophotometer. *Acta Histochem.* Suppl. 6, 47–54 (1965).
92. Lomakka, G. M., A rapid scanning and integrating microinterferometer. *Acta Histochem.* Suppl. 6, 393–396 (1965).
93. Mann, T., "The Biochemistry of Semen and of the Male Reproductive Tract," pp. 139–160. Methuen, London, 1964.
94. Miescher, F., "Die histochemischen und physiologischen Arbeiten," Vogel, Leipzig, 1897.
95. Miller, O. C., and Blackshaw, A. W., The DNA of rabbit spermatozoa aged *in vitro* and its relation to fertilization and embryo survival. *Abstr. 6th Congr. Intern. Reprod. Animals, Paris*, 1968, p. 233.
96. Monesi, V., Autoradiographic evidence of a nuclear histone synthesis during mouse spermiogenesis in the absence of detectable quantities of nuclear ribonucleic acid. *Exptl. Cell Res.* 36, 683–688 (1964).
97. Monesi, V., Synthetic activities during spermatogenesis in the mouse. RNA and protein. *Exptl. Cell Res.* 39, 197–224 (1965).
98. Muckenthaler, F. A., Autoradiographic study of nucleic acid synthesis during spermatogenesis in the grasshopper, *Melanoplus differentialis. Exptl. Cell Res.* 35, 531–547 (1964).
99. Müller, D., Sandritter, W., Schiemer, H. G., and Endres, K., Röntgenhistoradiographische und interferenzmikroskopische Trockengewichtsbestimmungen an Zellausstrichen. *Histochemie.* 1, 438–444 (1959).
100. Murray, K., and Peacocke, A. R., Thymus deoxyribonucleoprotein. I. Preparation and the thermal denaturation. *Biochim. Biophys. Acta* 55, 935–942 (1962).

101. Orgebin-Crist, M. -C., Recherches expérimentales sur la durée de passage des spermatozoïdes dans l'épididyme du taureau. *Ann. Biol. Animale. Biochim., Biophys.* **2**, 51–108 (1962).
102. Orgebin-Crist, M. -C., Passage of spermatozoa labelled with thymidine-^{3}H through the ductus epididymidis of the rabbit. *J. Reprod. Fertility* **10**, 241–251 (1965).
103. Orgebin-Crist, M. -C., Maturation of spermatozoa in the rabbit epididymis: Fertilizing ability and embryonic mortality in does inseminated with epididymal spermatozoa. *Ann. Biol. Animale, Biochim., Biophys.* **7**, 373–389 (1967).
104. Orgebin-Crist, M. -C., Studies on the function of the epididymis. *Biol. Reprod.* **1**, 155–175 (1969).
105. Parez, M., Petel, J. -P., and Vendrely, C., Sur la teneur en acide désoxyribonucléique des spermatozoïdes de taureaux présentant différents degrés de fécondité. *Compt. Rend.* **251** 2581–2583 (1960).
106. Paufler, S. K., and Foote, R. H., Influence of light on nuclear size and deoxyribonucleic acid content of stored bovine spermatozoa. *J. Dairy Sci.* **50**, 1475–1480 (1967).
107. Paufler, S. K., and Foote, R. H., Nucleus size and Feulgen-DNA content of epididymal and ejaculated rabbit spermatozoa. *Abstr. 6th Congr. Intern. Reprod. Animale. Paris,* 1968, p. 239.
108. Quinn, P. J., and White, I. G., A spectral analysis of ram sperm DNA after cold shock and deep freezing. *Abstr. 6th Congr. Intern. Reprod. Animale, Paris,* 1968 p. 242.
109. Rasch, E. M., Darnell, R. M., and Abramoff, P., Differentiation of nuclear proteins during spermatogenesis in *Poecilia. J. Cell Biol.* **31**, 156A (1966).
110. Rigler, R., Jr., Microfluorometric characterization of intracellular nucleic acids and nucleoproteins by Acridine Orange. *Acta Physiol. Scand.* **67**, Suppl. 267, 122 (1966).
111. Ringertz, N. R., and Zetterberg, A., Cytochemical demonstration of histones and protamines. Mechanism and specificity of the alkaline bromphenol blue binding reaction. *Exptl. Cell Res.* **41**, 243–259 (1966).
112. Ris, H., Ultrastructure and molecular organization of genetic systems. *Can. J. Genet. Cytol.* **3**, 95–120 (1961).
113. Salisbury, G. W., and Baker, F. N., Nuclear morphology of spermatozoa from inbred and linecross Hereford bulls. *J. Animal Sci.* **25**, 476–479 (1966).
114. Salisbury, G. W., and Baker, F. N., Frequency of occurrence of diploid bovine spermatozoa. *J. Reprod. Fertility* **11**, 477–480 (1966).
115. Salisbury, G. W., Birge, W. J., de la Torre, L., and Lodge, J. R., Decrease in nuclear Feulgen-positive material (DNA) upon aging in *in vitro* storage of bovine spermatozoa. *J. Biophys. Biochem. Cytol.* **10**, 353–359 (1961).
116. Salisbury, G. W., Lodge, J. R., and Baker, F. N., Effects of age of stain, hydrolysis time, and freezing of the cells on the Feulgen-DNA content of bovine spermatozoa. *J. Dairy Sci.* **47**, 165–168 (1964).
117. Salisbury, G. W., and Van Demark, N. L., "Physiology of Reproduction and Artificial Insemination of Cattle," pp. 221–224. Freeman, San Francisco, California, 1961.
118. Sandritter, W., and Grosser, K. D., Quantitative histochemische Untersuchungen an Spermien. *Symp. Biol. Hung.* **4**, 63–77 (1964).
119. Sandritter, W., Jobst, K., Rakow, L., and Bosselmann, K., Zur Kinetic der Feulgenreaktion bei Verlängerter Hydrolysezeit. Cytophotometrische Messungen im sichtbaren und ultravioletten Licht. *Histochemie* **4**, 420–437 (1965).

120. Sandritter, W., Schiemer, H. G., and Uhlig, H., Interferenzmikroskopische Trocken-gewichtsbestimmungen an Zellen mit haploidem und diploidem Chromosomensatz. *Acta Histochem.* **10**, 155–173 (1960).
121. Stedman, E., and Stedman, E., Cell specificity of histones. *Nature* **166**, 780–781 (1950).
122. Stolla, R., Ultraviolettmikrospektrophotometrische Untersuchungen an Samenzellen unter besonderer Berücksichtigung des DNA-Gehaltes der X- und Y- Spermien. Dissertation, München, (1968).
123. Sud, B. N., Morphological and histochemical studies of the chromatoid body and related elements in the spermatogenesis of the rat. *Quart. J. Microscop. Sci.* **102**, 495–505 (1961).
124. Sud, B. N., Morphological and histochemical studies of the chromatoid body in the grass-snake. Natrix natrix. *Quart. J. Microscop. Sci.* **102**, 51–58 (1961).
125. Summerhill, W. R., and Olds, D., Levels of deoxyribonucleic acid in bovine sperma-tozoa and their relationship to fertility. *J. Dairy Sci.* **44**, 548–551 (1961).
126. Swierstra, E. E., The cytology and kinetics of spermatogenesis in the rabbit, and the desoxyribonucleic acid content of spermatogenic cells. Ph.D. Thesis, Cornell University (1962).
127. Swift, H., The desoxyribose nucleic acid content of animal nuclei. *Physiol. Zool.* **23**, 169–198 (1950).
128. Swift, H., and Kleinfeld, R., DNA in grasshopper spermatogenesis, oögenesis, and cleavage. *Physiol. Zool.* **26**, 301–311 (1953).
129. Torre, L. de la, and Salisbury, G. W., Feulgen-DNA cytophotometry of bovine spermatozoa. *J. Dairy Sci.* **47**, 284–292 (1964).
130. Vaughn, J. C., The relationship of the *sphere chromatophile* to the fate of displaced histones following histone transition in rat spermiogenesis. *J. Cell Biol.* **31**, 257–278 (1966).
131. Vaughn, J. C., Changing nuclear histone patterns during development. I. Fertilization and early cleavage in the crab, *Emerita analoga. J. Histochem. Cytochem.* **16**, 473–479 (1968).
132. Vendrely, R., Knobloch, A., and Vendrely, C., An attempt of using biochemical methods for cytochemical problems. The desoxyribonucleoprotein of spermatogenetic cells of bull testis. *Exptl. Cell Res.* **4**, 279–283 (1957).
133. Vercauteren, R., The structure of desoxyribose nucleic acid in relation to the cyto-chemical significance of the methylgreen-pyronin staining. *Enzymologia* **14**, 134–140 (1950).
134. Walker, P. M. B., and Richards, B. M., Quantitative microscopical techniques for single cells. *In* "The Cell" (J. Brachet and A. E. Mirsky, eds.), Vol. 1, pp. 91–138. Academic Press, New York, 1959.
135. Welch, R. M., and Hanley, E. W., The experimental error of Feulgen cytophoto-metry in the analysis of bull spermatozoa over an extended period of time. *Texas Univ., Publ.* **6014**, 7–30 (1960).
136. Welch, R. M., and Resch, K., The deoxyribonucleic acid (DNA) content of semen spermatozoa and testis germinal cells from Santa Gertrudis bulls of known fertility and infertility. *Texas Univ. Publ.* **6014**, 31–37 (1960).
137. Young, W. C., A study of the function of the epididymis. III. Functional changes undergone by spermatozoa during their passage through the epididymis and vas deferens in the guinea pig. *J. Exptl. Biol.* **8**, 151–162 (1931).

STOICHIOMETRY OF DYE BINDING VERSUS DEGREE OF CROMATIN COILING*

Alfredo Mariano Garcia†

DEPARTMENT OF ANATOMY, STATE UNIVERSITY OF NEW YORK, UPSTATE MEDICAL CENTER, SYRACUSE, NEW YORK

In the first volume of this series we stated as a concluding remark "With up to 90% accuracy we can analyze DNA content in cells. Beyond that 90%, however, errors will make hypotheses risky" (*15*). It was clear at that stage of our work that the spread of values within a class of Feulgen-stained cells could not be solely attributed to instrumental fluctuation since, after subtraction of the latter, one could detect a "true standard deviation among stained nuclei" whose coefficient of variation was about 8–15% of the mean. Whatever the underlying cause, whether cytochemical vagaries or true differences in DNA content, it was, at that time, a moot point and therefore we limited ourselves to a mere mention of "potential sources of error" rather than trying to solve questions which we were unable to elucidate [see for details and pertinent bibliography, Garcia (*11, 12*) and Garcia and Iorio (*15*)].

Before plunging into the analysis of more recent work it may be useful to reexamine some of our early incursions in the field, since those results were seen in a context different from that which seems now a more cogent explanation. Table I summarizes data from Feulgen-stained cells of rabbit bone marrow (*12*) which were measured by the two-wavelength method of Ornstein (*34*) and Patau (*35*) in a cytophotometer designed and constructed by Pollister and Moses (*37*). The cell types represent diploid, nonreplicating stages; of these, the metamyelocytes, with an incipient degree of chromatin

* Work supported by U.S. Public Health Service Grant 5 RO1 AM10016-03 HEM of the National Institute of Arthritis and Metabolic Diseases.

† NIH Development Career Awardee 5 KO3 GM 11,790-03.

153

TABLE I

FEULGEN-DNA VALUES OF RABBIT BONE MARROW

Statistics (two-wavelength method)	Cell type		
	Metamyelo- cytes	Granulo- cytes	Lympho- cytes
Mean Feulgen content (extinction times area in μ^2)	19.17	17.10	16.03
Coefficient of variation of the instrument (%)	19	19	12
Coefficient of variation " between cells " (%)	15	12	6

compaction, show the highest dye uptake; the granulocytes, which can be classified as "medium coiled" yield values which are 10% lower, while a further 10% decrease is observed for the lymphocytes which in this particular set were almost invariably small and compact. The coefficient of variation of this simple apparatus is rather large, but when the instrumental fluctuation is subtracted by the appropriate statistic treatment, a smaller, but nevertheless significant coefficient of variation within populations or "between cells" becomes evident. It can also be noticed that the small lymphocytes are the most homogeneous population of the three sets and correspondingly, they show the smallest spread.

Table II shows some data which were included in the first volume of this

TABLE II

FEULGEN-DNA VALUES OF HUMAN LEUKOCYTES

Statistics (two-wavelength method)	Cell type			
	Granulo- cytes	Lympho- cytes	Mono- cytes	All types
Mean Feulgen content (extinction times area in μ^2)	20.7	20.4	21.2	20.7
Coefficient of variation of the instrument (%)	7	6	10	8
Coefficient of variation " between cells " (%)	7	9	16	11

series (*15*). They were extracted from 392 Feulgen-stained human peripheral leukocytes distributed in 14 slides; three different Schiff batches were used in an attempt to randomize results. The measurements were made in the cytophotometer built by Canal Industrial Corporation by means of the two-wavelength method (*34, 35*). When lymphocytes are taken as a single population (encompassing small, medium, and large cells) they show the same mean dye content as granulocytes. However, monocytes, which as a rule and because of their high degree of metabolic activity (*43*), have a large nucleus and diffuse chromatin, yield values which are 5–10% higher. At that time we traced this " departure from constancy " to an incorrect selection of wavelengths. The other possibility, namely the existence of a real increase of Feulgen stain, escaped us. As before a coefficient of variation " between cells " of near 10% became apparent after statistic analysis. It must be added that analogous conclusions were reached by a different measuring procedure known as the one-wavelength, two-area method (*16*).

The random fluctuation of the cytophotometers, tested by two measuring procedures, did not seem to be the only cause for the spread of the frequency curve. Therefore, we started to wonder about another potential source of variation, i.e., the chemical or physical manipulations to which we submit our material in order to achieve the production of a tangible or measurable image. Even if we admit that progress in the field of quantitative cytochemistry has made the instrumental and data extracting phases of the work relatively safe and reliable, we must reckon with the rather unpleasant fact that our objects are still images obtained by vicarious mechanisms; in other words, they are not absolute biologic entities but mere colored compounds which we optimistically assume to bear a point-to-point correspondence, in quantity and quality, with the substrate they represent.

In an attempt to find some leads into one of the most fundamental and controversial problems of biometrics and with a healthy mistrust of absolute or parametric references, we decided to pursue the search as before, using partially or totally randomized block designs (*3, 11*). In this context it occurred to us that a comparison between staining methods aimed at more than one functional group, studied by means of different cytophotometers and measuring procedures, might provide some clues.

Smears of centrifuged peripheral rat blood were fixed in methanol–formalin and stained for histones with fast green (*1*) and for DNA with the Feulgen reaction (*6, 8, 9, 15, 17, 24, 25, 28, 40, 41*). The color content was assessed by means of two instruments and three methods of quantitation: the two-wavelength (*34, 35*) and the one-wavelength, two-area (*16*) methods were used with the cytophotometer built by Canal Industrial Corporation, while the image-scanning procedure of Deeley (*7*) was employed in connection with the Barr and Stroud Integrating Microdensitometer Model GN 2.

Four cell populations were considered: granulocytes, small lymphocytes (nuclei 5 μ in diameter or less with clumped chromatin), medium lymphocyte (nuclei 5 to 6.7 μ in diameter with thick but discernible chromatin strands) and large lymphocytes and monocytes (nuclei above 6.7 μ in diameter having a delicate, rather diffuse chromatin pattern). In order to normalize the data and cancel out differences due to extinction coefficients and/or instrumental calibrations, rounded spermatids were used as a baseline and assigned a value of 100 in every case.

Table III represents the data per stain, per cell type, per method. In any of the six series the trend is the same: the large lymphocytes and monocytes show the highest dye content, the opposite occurs in small lymphocytes, while medium lymphocytes and granulocytes show intermediate values.

Table IV details the mean dye content per cell type, per instrument, per method, regardless of the cytochemical technique employed. Table V shows the fast green/Feulgen ratios; with all the data normalized, the expected 1:1 ratio is unmistakably upheld, with a maximal deviation of 4.4% which is well within the limits of experimental error, while Table VI confirms the data of Table V showing mean values per stain per cell type, irrespective of measuring or instrumental variables. It is obvious that in any of the tables "cell type" seems to be the only variable prone to exercise a systematic influence on the final outcome.

Some remarks about these results are deemed appropriate.

(1) The differences cannot be accounted for by distributional error. Although the crushing condenser (5) was not used with the scanner, the measurements were taken at off-peak wavelengths with the condenser closed to a numeric aperture of 0.3. Moreover, the differences among cell types were, if anything, accentuated by the two-area and the two-wavelength methods which correct for distributional error in a similar fashion, especially when the plug measurement, for the two-area method, is taken at the wavelength of half-maximal extinction (15–17).

(2) To trace the cause to some uneven response of the integrating micro-densitometer would be even more unlikely. The Canalco cytophotometer, with its simpler design, offers a means for an accurate check: this type of apparatus is used for one- or two-wavelength measurements; they are somehow more laborious and perhaps less stable than scanners, but because of their less complicated construction, the probability of systematic or hidden errors is minimized, since there are easy ways to test their performance (15). In sum, it seems improbable that instruments designed in accordance with different principles (arithmetic response of transmittance versus logarithmic response of extinction integrated over area) may be biased by the same flaws, although, of course, such a possibility cannot be entirely discarded.

TABLE III

FEULGEN-DNA AND FAST GREEN–HISTONE VALUES OF RAT LEUKOCYTES[a]

Instrument	Method of measurement	Cytochemical reaction	Cell types			
			Granulocytes	Small lymphocytes	Medium lymphocytes	Large lymph. and monocytes
Canalco cytophotometer	1-wavelength, 2-area	Fast gr.–hist.	190	161	171	195
		Feulgen-DNA	182	152	179	198
	2-wavelength	Fast gr.–hist.	180	158	173	192
		Feulgen-DNA	186	153	185	203
Barr and Stroud microdensitometer	Scanning	Fast gr.–hist.	181	165	178	184
		Feulgen-DNA	186	165	182	191

[a] Rounded spermatids = 100 extinction units, used as reference; 530 cells; 2120 measurements.

TABLE IV

MEANS PER METHOD OF MEASUREMENT FOR RAT LEUKOCYTES

Instrument	Method of measurement	Cell types			
		Granulocytes	Small lymphocytes	Medium lymphocytes	Large lymph. and monocytes
Canalco cytophotometer	1-wavelength, 2-area	186	156	175	197
	2-wavelength	183	155	178	198
Barr and Stroud microdensitometer	Scanning	183	165	180	188

TABLE V

FAST GREEN/FEULGEN RATIOS OF RAT LEUKOCYTES

Instrument	Method of measurement	Cell types			
		Granulocytes	Small lymphocytes	Medium lymphocytes	Large lymph. and monocytes
Canalco cytophotometer	1-wavelength, 2-area	1.044	1.059	0.955	0.985
	2-wavelength	0.968	1.033	0.935	0.946
Barr and Stroud microdensitometer	Scanning	0.973	1.000	0.978	0.963
Average		0.995	1.031	0.956	0.965

TABLE VI

RAT LEUKOCYTE MEANS PER STAIN

Stain	Cell types			
	Granulocytes	Small lymphocytes	Medium lymphocytes	Large lymph. and monocytes
Fast green-hist.	184	161	174	190
Feulgen-DNA	185	157	182	197
Grand mean per cell type	184	159	178	194

(3) Tables V and VI show that the differential response cannot be explained on the basis of factors affecting either the Feulgen or the fast green reactions separately and thus, it stems neither from differential HCl hydrolysis (39), not from the quality of the dyes used (10). An argument similar to the one discussed with regard to instruments can be brought in here: We are dealing with staining mechanisms based on entirely different principles, and both reacted exactly in the same fashion. Obviously, if we are confronted here with a cytochemical artifact, such an artifact seems inclined to behave in a consistent and almost predictable way regarding both the electropositive groups of the histone and the aldehyde groups of the apurinic acid.

In order to specify more exactly the relation between nuclear size and Feulgen-DNA content, the whole procedure was repeated in smears of human blood using three methods of measurement and two instruments as before. The results are presented in Fig. 1. In every instance the amount of color increases with nuclear size.

From the work so far discussed it appears reasonable to conclude that aging of the lymphocyte leads to a decrease of nonhistone protein, reduction in nuclear size, chromatin shrinkage, and disappearance (or great reduction) of diffuse chromatin. Whether the decrease in dye-binding capacity evinced a real DNA diminution or was the effect of steric hindrances, remained to be seen. In an effort to find the answer to these two alternatives it was decided to take the male haploid cells as a baseline. If we accept the premise that neither chromosomal complement nor DNA amount change from spermatid to sperm (19, 20, 27, 45), then the major difference between these two developmental stages should be a qualitative and not a quantitative one, namely the degree of nuclear compaction which seems similar to the phenomena of lymphocytic senescence.

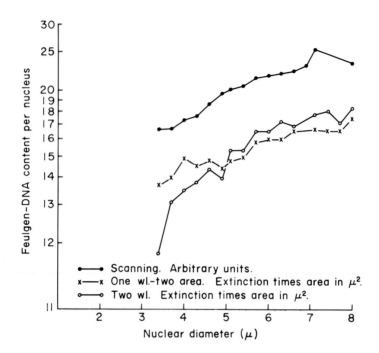

Fig. 1. Human peripheral lymphocytes. Mean Feulgen-DNA content as a function of nuclear size. A comparison between two instruments and three methods of measurement.

Smears and prints of testis and peripheral rat blood were made on cover-slips, air-dried, fixed in 9:1 methanol:formalin, hydrolyzed in 5 N HCl at room temperature (6, 9, 15, 24, 25) and stained by means of the Feulgen reaction (8, 28, 40, 41). For each hydrolysis the slides and the Schiff batch were made anew, so that the results can be considered reasonably random-ized. The nuclear dye content was measured with the Barr and Stroud micro-densitometer (7) equipped with the crushing condenser (5). Eight cell popula-tions were considered: rounded spermatids, semicompact spermatozoa or sperm b, compact testicular spermatozoa or sperm a [see Gledhill et al. (19, 20) for a more detailed classification], diploid spermatogonia, granulo-cytes, small lymphocytes, medium lymphocytes, and large lymphocytes and monocytes. A total of 750 cells were analyzed by quadruplicate measurements before and after cell crushing.

Table VII details the values per cell type per hydrolysis time, after normal-izing the data with respect to spermatids. It must be said, as a passing comment, that the crushing condenser detected no distributional error, i.e., the values were the same before and after squashing the specimen; obviously then the differences in chromatic intensity are due not to distributional error

TABLE VII

RAT CELL FEULGEN-DNA VALUES EXPRESSED AS PERCENTAGE OF DYE CONTENT OF SPERMATIDS[a]

HCl hydrolysis (min)	Relative Feulgen value per cell type							
	Testicular sperm a (compact)	Testicular sperm b (semicoiled)	Rounded spermatids	Diploid spermatogonia	Granulocytes	Small lymphocytes	Medium lymphocytes	Large lymph. and monocytes
45	77	88	100	191	168	144	161	191
60	78	89	100	192	172	149	172	189
90	74	82	100	196	184	177	188	196
120	81	85	100	198	177	146	166	197
150	79	90	100	205	192	156	188	202
180	84	90	100	193	164	153	166	163
240	73	79	100	199	172	163	170	191
Mean	78	86	100	196	176	155	173	190

[a] Barr and Stroud integrating microdensitometer.

but to uneven amounts of dye bound to the substrate. As before, nuclear diffusion (or chromatin uncoiling) and color development seem to be positively correlated. In the haploid series the semicoiled sperm is more than 10% lower than the spermatid while the Feulgen value of the compact, testicular sperm equals 78% of the latter. A similar relation is demonstrated for the diploid series: as expected, diploid spermatogonia and large lymphocytes and monocytes show the highest color content, the opposite is the case for small lymphocytes, while granulocytes and medium lymphocytes are placed midway between the two extremes.

Table VIII affords a better way to objectify the results: Here the three types of haploids—spermatid, sperm *a*, and sperm *b*—are used as numerators while the means of other types are used as denominators. Sperm *a* to spermatid, instead of the theoretic 1.00 equals 0.78, while sperm *b* to spermatid equals 0.86. The significant haploid:diploid ratios are given in boldface in the table and these values per se prove the point better than any discussion. Regarding Feulgen intensity the compact sperm seems the "haploid" of the small lymphocyte, the partially coiled sperm seems the "haploid" of medium lymphocytes and granulocytes, while the rounded spermatid with its characteristically diffuse chromatin seems the "haploid" of diploid spermatogonia, large lymphocytes, and monocytes. We are faced here then with an almost perfect haploid:diploid ratio for three different degrees of chromatin compaction. These data, as well as data from other laboratories (*2, 9, 19–21, 30, 31, 36*), suggest that the physical state of the deoxyribonucleohistone complex influences the dye-binding capacity of its reacting groups, be they amino, guanidyl, aldehyde, or phosphate. A striking proof is afforded by the reactivation undergone by lymphocytes in phytohemagglutinin (*32*). After a certain lapse the small lymphocytes enlarge and the nuclear material becomes more diffuse (*23, 42*). This phenomenon takes place during the first 24 hours and precedes the onset of DNA synthesis (*4, 29*). This loosening of the chromatin without *de novo* synthesis results in an increased Feulgen stainability (*22*), increased acridine orange (*26*) and fast green binding (*44*). Conversely, as shown clearly by Gledhill (*19*) and Gledhill *et al.* (*20*) spermiogenesis in the bull entails decreased Feulgen, acridine orange, and methyl green values, with practically no change in DNA as judged by ultraviolet or dry mass measurements. The close agreement between Gledhill's and Mayall and Mendelsohn's results (*30, 31*) and our own amply reinforces what seems to be the logical conclusion, that is, the existence of a nonstoichiometric dye-binding mechanism which is systematically influenced by the degree of nuclear compaction.

A group of Belgian cytophotometrists with long and active standing in the field has interpreted Feulgen variations in a different context: Their prevailing view has been that these results express true differences in DNA content and

TABLE VIII

RAT CELL FEULGEN-DNA VALUES[a]—MEAN HAPLOID : DIPLOID RATIOS

Numerator of ratio	Denominator of ratio							
	Sperm a	Sperm b	Spermatid	Spermatogonia (diploid)	Granulocyte	Small lymphocytes	Medium lymphocytes	Large lymph. and monocytes
Sperm a (compact)	1.00	0.92	0.78	0.40	0.44	0.50	0.45	0.41
Sperm b (semicoiled)	1.09	1.00	0.86	0.43	0.49	0.56	0.50	0.46
Spermatid	1.28	1.16	1.00	0.51	0.57	0.65	0.58	0.52

[a] Barr and Stroud microdensitometer.

are not artifacts induced by distributional error. In order to get a picture as multilateral and comprehensive as possible, the interested reader should consult a very complete monograph by Roels (38) where these views are deftly presented, together with an exhaustive review of the pertinent literature.

The logical sequence of the studies so far reported was to treat leukocytes with hypotonic solutions to see if such a physical stimulus would induce changes in Feulgen or fast green staining in a fashion comparable to phyto-hemagglutinin.

Peripheral leukocytes of rabbit were incubated in a variety of hypotonic solutions: (a) 3 parts of blood diluted with 2 parts of water; (b) 0.95 % sodium citrate (18); (c) 0.2 M sucrose; and (d) Ohnuki's medium (33). The cells were then smeared, fixed, and stained for DNA with the Feulgen reaction as previously detailed (17) and for histones with the alkaline fast green method of Alfert and Geschwind (1). The Barr and Stroud measuring instrument was used and a total of 2794 cells were studied.

The results of these experiments have been published in full detail elsewhere (13, 14) and only the most important points will be emphasized here. Figure 2

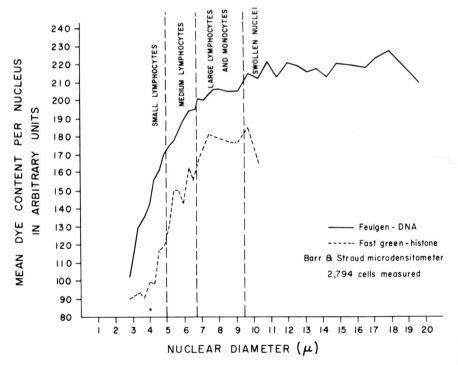

Fig. 2. Rabbit mononuclears after hypotonic treatment. Mean dye content as a function of nuclear size. A comparison between two staining methods.

shows the mean Feulgen-DNA and fast green–histone values plotted against nuclear size; it illustrates the parallel behavior of both stains with respect to the degree of nuclear diffusion. The small lymphocyte (around 3 μ) yields about 50% of the total amount of color shown by large lymphocytes, monocytes, and moderately swollen cells. However, beyond a size of 10–12 μ there seems to be a loss of histone which tends to reduce the fast green content of those nuclei.

Figure 3 shows the frequency polygons of Feulgen-DNA values plotted in

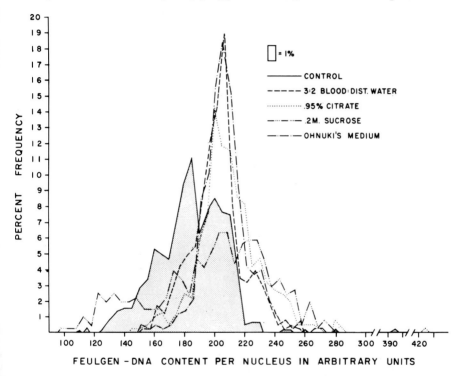

FIG. 3. Rabbit leukocytes after hypotonic treatment. Frequency polygons of Feulgen-DNA values per treatment. Intervals on abscissa chosen for increments of 2.5%. (Reproduced by permission of editors and publishers of the *Journal of Histochemistry and Cytochemistry*.)

2.5% increments. The control shows a tripartite form. The small peak at the left (160 units) corresponds to small lymphocytes; the central and largest mode (180–185 units) corresponds to medium lymphocytes and granulocytes, while the third mode at the right (195–205) reflects the dye content of large lymphocytes and monocytes. Three of the hypotonic treatments tend to induce similar changes, namely, blood:water, citrate, and Ohnuki's medium; the distributional curve is narrower and displaced to the right, between 200 and 210 units. In contrast, leukocytes incubated in sucrose show a protracted distribution: there is a low plateau between 125 and 150, a small fall, and finally a slow rise between 165 and 240 units.

When Feulgen-DNA values are plotted according to their cumulative

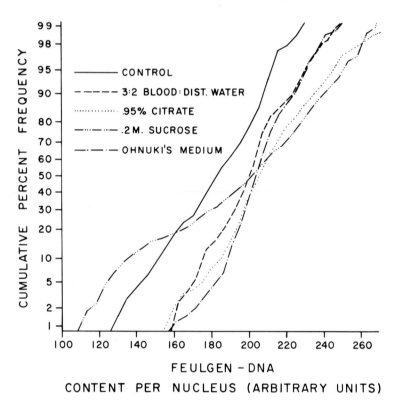

FIG. 4. Rabbit leukocytes after hypotonic treatment. Cumulative frequency distribution of Feulgen-DNA values per treatment. (Reproduced by permission of editors and publishers of the *Journal of Histochemistry and Cytochemistry*.)

frequencies the differences become even sharper, as seen in Fig. 4. The median for the control equals 182; the tracings for blood:water and Ohnuki's mixture are displaced to the right while the curve for sucrose reflects a typical platykurtic distribution. We must also mention here [for details, see Garcia (*13*)] that there is a point-to-point correspondence between the frequency distribution of Feulgen-DNA and that of nuclear sizes.

Figure 5 shows fast green-histone values plotted according to their cumulative frequencies. Sucrose was omitted from this experiment because of its ambiguous response. The values for blood:water and Ohnuki are displaced

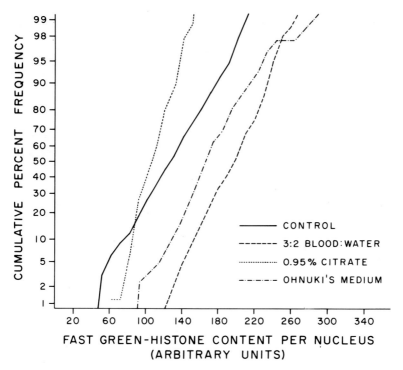

FIG. 5. Rabbit mononuclears after hypotonic treatment. Cumulative frequency distribution of fast green–histone values. (Reproduced by permission of editors and publishers of the *Journal of Histochemistry and Cytochemistry.*)

to the right, while the tracing for citrate is almost entirely at the left, due probably to some histone extraction. As for Feulgen, a positive correlation exists between nuclear size and total amount of color (*14*).

It seems reasonable, therefore, to state that physical changes in the deoxyribonucleoprotein complex may induce concomitant alterations in the reactivity of its dye-binding sites, be they aldehyde, phosphate, or amino groups, and that the differences discussed here are biologic in the qualitative sense, but only cytochemical as far as quantitation is concerned.

While instruments and measuring techniques have rendered the quantitation of a colored end product relatively safe and free of errors, the second stage, the correspondence between the cytochemical reaction and the biologic substrate it represents, is not always complete. At the risk of looking too optimistic it seems, nevertheless, that nonstoichiometric responses of deoxyribonucleoprotein to staining mechanisms can be traced to its physical state (i.e., diffuse or compact), the latter being a reflection of its metabolic activity. Cytophotometry can become in this way not only a measuring tool but, more important, a means of qualitative assay, since it offers the possibility of correlating departures from stoichiometry with nuclear metabolism. This new approach, used in conjunction with pattern recognition and radioauto-graphy may provide, in the future, a formidable addition to the armamentarium of the cell biologist.

ACKNOWLEDGMENTS

The cooperation of Miss Gloria Shetler in the preparation of the material and the processing of the data is greatly appreciated. We are also indebted to Messrs. Leo F. Walsh and Robert G. Jennings of the electronics department for their help in the care and maintenance of the integrating microdensitometer. The fluent style of Prof. Chester L. Yntema exercised a salutary influence upon the manuscript's English. Last but not least, we express our gratitude to editors and publishers of the *Journal of Histochemistry and Cytochemistry* for their kind permission to reproduce Figures 3, 4, and 5.

REFERENCES

1. Alfert, M., and Geschwind, I. I., A selective staining method for the basic proteins of cell nuclei. *Proc. Natl. Acad. Sci. U.S.* **39**, 991–999 (1953).
2. Atkins, N. G., and Richards, B. M., Deoxyribonucleic acid in human tumors as measured by microspectrophotometry of Feulgen stain: A comparison of tumors arising at different sites. *Brit. J. Cancer* **10**, 769–786 (1956).
3. Bartels, P. H., Bahr, G. F., and Wied, G. L., Information theoretic approach to cell identification by computer. *In* "Automated Cell Identification and Cell Sorting" (G. L. Wied and G. F. Bahr, eds.). Academic Press, New York, 1970.
4. Bender, M. A., and Prescott, D. M., DNA synthesis and mitosis in cultures of human peripheral leucocytes. *Exptl. Cell Res.* **27**, 221–229 (1962).
5. Davies, H. G., Wilkins, M. H. F., and Boddy, R. G., Cell crushing: A technique for greatly reducing errors in microspectrophotometry. *Exptl. Cell Res.* **6**, 550–553 (1954).
6. De Cosse, J. J., and Aiello, N., Feulgen hydrolysis: Effect of acid and temperature. *J. Histochem. Cytochem.* **14**, 601–604 (1966).
7. Deeley, E. M., An integrating microdensitometer for biological cells. *J. Sci. Instr.* **32**, 263–267 (1955).
8. Deitch, A. D., Cytophotometry of nucleic acids. *In* "Introduction to Quantitative Cytochemistry" (G. L. Wied, ed.), pp. 327–354. Academic Press, New York, 1966.
9. Deitch, A. D., Wagner, D., and Richart, R. M., Conditions influencing the intensity of the Feulgen reaction. *J. Histochem. Cytochem.* **16**, 371–379 (1968).
10. Duijn, P. van, Remarks in "Panel on DNA constancy" (P. van Duijn, moderator). *2nd Intern. Tutorial Quant. Cytochem. Univ. Chicago, Chicago, Ill.,* 1968.
11. Garcia, A. M., Studies on DNA in leucocytes and related cells of mammals. I. On microspectrophotometric errors and statistical models. *Histochemie* **3**, 170–177 (1962).
12. Garcia, A. M., Studies on DNA in leucocytes and related cells of mammals. IV. The Feulgen-DNA content of peripheral leucocytes, megakaryocytes and other bone marrow cell types of the rabbit. *Acta Histochem.* **17**, 246–258 (1964).
13. Garcia, A. M., Studies on DNA in leukocytes and related cells of mammals. VI. The Feulgen-DNA content of rabbit leukocytes after hypotonic treatment. *J. Histochem. Cytochem.* **17**, 47–55 (1969).

14. Garcia, A. M., Studies on DNA in leukocytes and related cells of mammals. VII. The fast green-histone content of rabbit leukocytes after hypotonic treatment. *J. Histochem. Cytochem.* **17**, 475–481 (1969).
15. Garcia, A. M., and Iorio, R., Potential sources of error in two-wavelength cytophotometry. *In* "Introduction to Quantitative Cytochemistry" (G. L. Wied, ed.), pp. 215–237. Academic Press, New York, 1966.
16. Garcia, A. M., and Iorio, R., A one-wavelength, two-area method in cytophotometry for cells in smears or prints. *In* "Introduction to Quantitative Cytochemistry" (G. L. Wied, ed.), pp. 239–245. Academic Press, New York, 1966.
17. Garcia, A. M., and Iorio, R., Studies on DNA in leukocytes and related cells of mammals. V. The fast green-histone and the Feulgen-DNA content of rat leukocytes. *Acta Cytol.* **12**, 46–51 (1968).
18. German, J., The pattern of DNA synthesis in the chromosomes of human blood cells. *J. Cell Biol.* **20**, 37–55 (1964).
19. Gledhill, B. L., Studies on the DNA content, dry mass and optical area of bull spermatozoal heads during epididymal maturation. *Acta Vet. Scand.* **7**, 131–142 (1966).
20. Gledhill, B. L., Gledhill, M. P., Rigler, R., Jr., and Ringertz, N. R., Changes in deoxyribonucleoprotein during spermiogenesis in the bull. *Exptl. Cell Res.* **41**, 652–665 (1966).
21. Hale, A. J., The leukocyte as a possible exception to the theory of deoxyribonucleic acid constancy. *J. Pathol. Bacteriol.* **85**, 311–326 (1963).
22. Hale, A. J., and Cooper, E. H., Studies on DNA replication in leukaemic and non-leukaemic leucocytes. *In* "Current Research in Leukemia" (F. G. J. Hayoe, ed.), pp. 95–107. Cambridge Univ. Press, London and New York, 1965.
23. Inman, D. R., and Cooper, E. H., Electron microscopy of human lymphocytes stimulated by phytohaemagglutinin. *J. Cell Biol.* **19**, 441–445 (1963).
24. Itikawa, O., and Ogura, Y., The Feulgen reaction after hydrolysis at room temperature. *Stain Technol.* **29**, 13–15 (1954).
25. Jordanov, J., On the transition of desoxyribonucleic acid to apurinic acid and the loss of the latter from tissues during Feulgen reaction hydrolysis. *Acta Histochem.* **15**, 135–152 (1963).
26. Killander, D., and Rigler, R., Jr., Initial changes of deoxyribonucleoprotein and synthesis of nucleic acid in phytohaemagglutinine-stimulated human leucocytes *in vitro. Exptl. Cell Res.* **39**, 701–704 (1965).
27. Leslie, I., The nucleic acid content of tissues and cells. *In* "The Nucleic Acids" (E. Chargaff and J. N. Davidson, eds.), Vol. 2, pp. 1–50. Academic Press, New York, 1955.
28. Leuchtenberger, C., Quantitative determination of DNA in cells by Feulgen microspectrophotometry. *Gen. Cytochem. Methods* **1**, 219–278 (1958).
29. MacKinney, A. S., Stohlman, F., Jr., and Brecher, G., The kinetics of cell proliferation in cultures of human peripheral blood. *Blood* **19**, 349–358 (1962).
30. Mayall, B. H., The detection of small differences in DNA content with microspectrophotometric techniques. *J. Cell Biol.* **31**, 74A (1966) (abstr.).
31. Mayall, B. H., Variability in the stoichiometry of deoxyribonucleic acid stains. *J. Histochem. Cytochem.* **15**, 762–763 (1967) (abstr.).
32. Nowell, P. C., Differentiation of human leukemic leucocytes in tissue culture. *Exptl. Cell Res.* **19**, 267–277 (1960).
33. Ohnuki, Y., Demonstration of the spiral structure of human chromosomes. *Nature* **208**, 916–917 (1965).
34. Ornstein, L., The distributional error in microspectrophotometry. *Lab. Invest.* **1**, 250–265 (1952).

35. Patau, K., Absorption microphotometry of irregular shaped objects. *Chromosoma* **5**, 341–362 (1952).
36. Perugini, S., Torelli, U., and Soldati, M., Ricerche citofotometriche sulla componente proteica delle cellule ematiche. 2) Il contenuto nucleare di proteina istonica e di acido desossiribonucleico nei linfociti e negli epatociti del ratto. *Riv. Istochim. Norm. Patol.* **2**, 449–460 (1956).
37. Pollister, A. W., and Moses, M. J., Simplified apparatus for photometric analysis and photomicrography. *J. Gen. Physiol.* **32**, 567–577 (1949).
38. Roels, H., "Metabolic" DNA: A cytochemical study. *Intern. Rev. Cytol.* **19**, 1–34 (1966).
39. Sandritter, W., Jobst, K., Rakow, L., and Bosselman, K., Zur Kinetik der Feulgen-reaktion bei verlängerter Hydrolysezeit. Cytophotometrische Messungen im sichtbaren und ultravioletten Licht. *Histochemie.* **4**, 420–437 (1965).
40. Stowell, R. E., Feulgen reaction for thymonucleic acid. *Stain Technol.* **20**, 45–58 (1945).
41. Swift, H., Cytochemical techniques for nucleic acids. *In* "The Nucleic Acids" (E. Chargaff and J. N. Davidson, eds.), Vol. 2, pp. 51–92. Academic Press, New York, 1955.
42. Tanaka, Y., Epstein, L. B., Brecher, G., and Stohlman, F., Jr., Transformation of lymphocytes in cultures of human peripheral blood. *Blood* **22**, 614–629 (1963).
43. Torelli, U., Grossi, G., Artusi, T., Emilia, G., Attiya, I. R., and Mauri, C., RNA turnover rates in normal peripheral mononuclear leucocytes. *Exptl. Cell Res.* **36**, 502–509 (1964).
44. Torelli, U., Quaglino, D., Sauli, S., and Mauri, C., A cytochemical study on the DNA and histone changes occurring in phytohaemagglutinin stimulated leucocytes. *Exptl. Cell Res.* **45**, 281–288 (1967).
45. Vendrely, R., The deoxyribonucleic acid content of the nucleus. *In* "The Nucleic Acids" (E. Chargaff and J. N. Davidson, eds.), Vol. 2, pp. 155–180. Academic Press, New York, 1955.

ERRORS IN ABSORPTION CYTOPHOTOMETRY: SOME THEORETICAL AND PRACTICAL CONSIDERATIONS*

Brian H. Mayall and Mortimer L. Mendelsohn

DEPARTMENT OF RADIOLOGY, UNIVERSITY OF PENNSYLVANIA, PHILADELPHIA, PENNSYLVANIA

I. Introduction

The techniques of quantitative cytochemistry allow measurement of cellular components in the context of cell morphology. Thus measurements can be related to single cells or parts of cells in a way that is unmatched by any other approach. As such, these techniques have a unique contribution to make in cell biology. However, enthusiasm for the potentialities of quantitative cytochemistry must be tempered by an appreciation of the difficulties associated with the verification of results obtained with a methodology that has no alternative. Much of the history of this field is replete with circular or near-circular efforts at self-justification, and concern with errors of theory and practice continues to be of fundamental importance. This chapter reviews this problem for absorption cytophotometry.

Anomalies associated with measurements of the DNA content of leukocytes of human peripheral blood provide a good context in which to discuss the errors of absorption cytophotometry. Leukocytes are readily obtained and prepared as flattened spreads on microscope slides, and they are easily classified into types on the basis of their morphological characteristics. They are not synthesizing DNA, and they are a population of postmitotic presumably diploid cells; thus, they would be expected to show DNA constancy. However, cytophotometric estimations of their DNA stain content show

* This work was supported by U.S. Public Health Service Contract PH 43–62–432 and U.S. Public Health Service Grants 5 R01 CA03896 and 5 K06 CA18,540 from the National Cancer Institute, National Institutes of Health, Bethesda, Maryland.

significant differences both among cell types and among cells within types, with monocytes measuring higher than both small lymphocytes and neutrophilic granulocytes (1, 17, 19, 20, 26).

Do these differences represent departures from DNA constancy, or are they due to errors in preparation and measurement affecting cells to differing degrees? It seems important that monocytes are the leukocytes which have the palest nuclei and the highest stain content. Many cytophotometric errors are negative functions of optical density, and so could cause measurements on dark cells to be lower than measurements on light cells. But can the differences among leukocytes be explained by such errors? These are examples of the questions which must be asked in the interpretation of cytophotometric data, and in the discussion that follows an attempt will be made to evaluate errors in terms that are relevant to such questions.

II. General Approaches to Absorption Cytophotometry

The Beer-Lambert absorption law relates the optical density to the amount of chromophore present in a measuring field, and it provides the theoretical basis for both macroscopic and microscopic absorption photometry. The law may be expressed in the following forms:

$$\log \left(\frac{I_0}{I}\right) = -\log T = A = \frac{km}{B} = kcl \tag{1}$$

where I_0 is the flux of light incident on the chromophore, I is the emergent flux, T is the transmission of the field, A is the absorbance or optical density, k is the specific absorptivity of the chromophore at the measuring wavelength, c is the concentration of chromophore, l is the pathlength through the chromophore, m is the mass of the chromophore in the field, and B is the area of the measuring field. This exponential law applies strictly for conditions in which the illumination is monochromatic, the pathlengths are uniform, the chromophore is homogeneously concentrated, and absorption is the only effect of the interaction between light and chromophore.

In practice the actual quantities measured are I and I_0, the light flux with and without the chromophore in the field, and B, the area of the measuring field. These measurements then are used to calculate either the absolute mass of chromophore, m, or the relative mass of chromophore, km. Obviously an error in the measurement of any of these three quantities will cause a corresponding error in the computed values. However, with present-day instrumentation such errors will normally be very small. The errors that cause concern are those due to departures from the conditions necessary for the exact application of the absorption law. These departures are an inevitable consequence of operating in a cytomorphological context. Cells and cell constituents are simply not uniform in thickness, homogeneous in concentration, or free of optical effects other than absorption.

Heterogeneity of chromophore within a measuring field leads to distributional error, a negative error that as Patau has shown is roughly proportional to the product of the mean optical density and the variance of optical density (39). Referring back to Eq. (1), the error occurs whenever the transmission of a field is an average of two or more unequal transmissions. The logarithm of such a mean is always less than the mean of the individual logarithms.

Three fundamentally different approaches have been developed to overcome the effects of specimen heterogeneity. The methods differ in their theory and in the efficiency with which they reduce distributional error (33). The plug method (46) and its variants, the multiple plug method (42) and the two-area one-wavelength method (16, 18), measure the average optical density over a relatively homogeneous portion of the cell and extrapolate this to the entire area. The two-wavelength method (32, 37, 39) allows the integrated optical density of the cell to be measured free from gross distributional error and without having to measure the area of the cell directly. The whole cell is contained within the field and the transmission is measured at two-wavelengths chosen to give specific absorptivities of two-to-one for the chromophore. In integrating scanning cytophotometers (6, 7, 13, 14, 23, 28, 34, 50) the cell is divided, in effect, into a large number of small, regularly spaced, relatively homogeneous samples. The optical density of each sample is measured and summed to give the total optical density for the cell.

The photographic methods of Caspersson (6), Ornstein (37), den Tonkelaar and van Duijn (48), Kelly (24), and Rudkin (41) can all be thought of as variants of the scanning method, and the photographic method of Mendelsohn (31) is a modification of the two-wavelength method. All the photographic methods have the advantages that stem from going from a microscopic to a macroscopic domain, but they are limited in their accuracy by the errors associated with the photographic transformation of the image.

In general, measuring errors fall into two types: stochastic errors and systematic errors (2). Stochastic errors displace the individual measurements about a mean value in a more or less random manner, and they are estimated by the statistical variation among replicate measurements. They affect the sensitivity of the experiment, but this can be adjusted to any desired level by repeating the measurements a sufficient number of times. Systematic errors, on the other hand, displace the mean of the measurements and so affect the accuracy of the experiment. Their presence is much more difficult to detect, and their effect cannot be reduced by any expedient method as simple as increasing the number of measurements. Thus these two types of error differ fundamentally, both in their behavior and in their consequences. In the following discussion, each type of error will be considered separately.

III. Systematic Errors

In other fields, systematic errors are estimated directly from measurements made on or against a series of calibrated standards. Unfortunately, the standards that are available in cytophotometry either are oversimplified objects which incompletely test the system, or are biological specimens in which the reliability of the standard itself is deeply embedded in the very measurements that are to be standardized.

Some systematic errors operate as scaling factors; that is, they alter all results by a fixed percentage. When making absolute measurements, such errors contribute to the overall bias; but these scaling errors cancel out when (as is often the case in the cytophotometry of DNA) an experiment involves only the relative magnitude of measurements made under the same conditions.

In absorption cytophotometry these errors may occur in (1) the specimen, (2) the method, and (3) the instrument.

A. Specimen Errors

Specimen errors are associated with the preparation and presentation of the specimen on the slide. As a result of specimen errors the amount of chromophore misrepresents the amount of substrate originally present or else the chromophore is obscured by the optical behavior of the specimen. These errors have been discussed by a number of authors (6, 11, 40, 47, 49, 51) and they include:

(1) Loss of substrate in specimen preparation
(2) Nonspecific staining
(3) Changes in the stoichiometry of the staining reaction, including those that may occur with masking, pH, and ionic effects
(4) Changes in the specific absorptivity of the chromophore, such as may result from the orientation, interaction, and steric interference of chromophore molecules
(5) Anomalous effects so that the specimen does not act as a pure amplitude object because of phase changes and scattering due to refractive index mismatch, and because of optical polarizing activity
(6) Presence of other absorbing material in specimen or background
(7) Choice of inconsistent or inappropriate clear areas

The last three factors may act in an apparently random manner, and so also may contribute to the variation among replicate measurements.

As a group, specimen errors may be the largest source of uncertainty associated with cytophotometric measurements. Factors relating to the staining reaction are extremely difficult to control and are very unpredictable in their total effect. Scattering and phase effects can be minimized by carefully preparing and fixing the specimen so that refractive index differences within the specimen are diminished, and by matching the refractive index of the mounting oil to that of the specimen. The effect of scattering is also reduced by using an objective lens of high numerical aperture (6). Polarizing effects are generally small in biological material, but occasionally they can be significant (22). All inclined reflecting surfaces in the optical system act as partial polarizers; in addition, the sensitivity of an inclined photodetector surface is affected by the polarization of the incident light. Careful attention to instrument design, specimen preparation, staining, mounting, and the choice of object and clear area when measuring may control many specimen errors, but as a group they are extremely difficult to eliminate entirely and their presence should always be suspected when interpreting cytophotometric results.

B. ABSORPTION ERRORS

Absorption errors arise whenever there is heterogeneity in the three terms on the right of Eq. (1). Heterogeneity in the concentration of chromophore per unit projected area causes distributional error, the best known of these errors. There is a heterogeneity error that is part of condenser aperture error and is caused by variations in pathlength associated with conical illumination. And finally an error is caused by variation in the specific absorption of the chromophore across the spectral bandwidth used for the measurement. Each of these heterogeneity errors introduces a negative bias whose magnitude is a function of optical density and degree of heterogeneity, and each will be considered separately in this section.

1. Condenser Aperture Error

Conical illumination is the rule for microscope optics; and the sine of the half-angle of the cone is defined by the ratio of the condenser numerical aperture to the refractive index of the preparation. Conical illumination has two effects on the application of the Beer-Lambert law: (1) it biases either the mean pathlength through the specimen or the effective area of the measuring field, depending on the photometric method being used; and (2) it produces a potential heterogeneity of pathlength at every point in the specimen and hence within all measured transmissions. With most photo-metric methods, the former effect causes a large but exceedingly well behaved error that is independent of the specimen, whereas the latter effect causes a smaller error that is a function of optical density.

Consider first the effect of condenser aperture on mean pathlength or mean photometric area. The effect on pathlength is easily visualized in terms of a parallel-sided uniform disk that is perpendicular to the optic axis and that completely fills the photometric field (3). Clearly, the larger the angle a ray has to the optic axis, the longer will be its pathlength through the disk. Under such conditions, and allowing for the cosine fall-off in intensity with inclined illumination, the mean pathlength, $\bar{l}$, is given by integrating the weighted pathlength for each annulus of the cone and so is

$$\bar{l} = \left(\frac{\int_0^{\theta = \alpha} \sin\theta \cos\theta \sec\theta \, d\theta}{\int_0^{\theta = \alpha} \sin\theta \cos\theta \, d\theta} \right) l_0 = \left(\frac{2}{1 + \cos\alpha} \right) l_0$$

where α is the half-angle of the cone and l_0 is the axial pathlength.

The relationship between mean pathlength and angle is unpredictable for an irregularly shaped object completely within a photometric field, as in the two-wavelength case considered by Patau (39). For example, a tall thin cylinder rising perpendicularly from the microscope slide would present its longest dimension to the axial rays, and, in direct contrast to the case of the disk, the larger the inclination of a ray to the optic axis the shorter would be its pathlength through the cylinder. Patau argued that in the absence of distributional error the orientation of the object should not affect the measure-ments but that there would be an error associated with conical illumination because the effective field size decreases as a function of angle. The net effect is to produce a bias that is the same as the mean pathlength bias and to give measurements that are $[2/(1 + \cos\alpha)]$ times the correct value.

Recently, we have shown that the same condition exists for scanning cyto-photometry (28). In this case the essential parameter is the relationship between scanning density and the angle of illumination, the measurements are again increased by the factor $[2/(1 + \cos\alpha)]$, and the effect is independent of the size and shape of the specimen.

Thus with the two-wavelength method and with scanning cytophotometry (and with the plug methods using disklike objects) the apparent chromophore content is biased as a function of condenser numerical aperture. This relation-ship is shown in Fig. 1a. The effect will be significant only if absolute measure-ments are required or if the condenser aperture setting is changed during the course of an experiment.

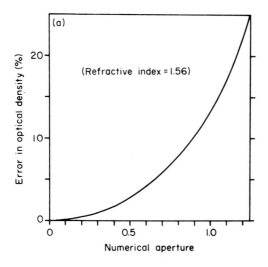

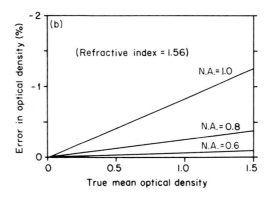

FIG. 1. Condenser aperture error. (a) Condenser aperture bias as a function of condenser numerical aperture. The percentage error is given by $100[2/(1 + \cos \alpha) - 1]$, where α is arcsin [Numerical aperture (n.a.)/Refractive index]. (b) Condenser aperture heterogeneity error as a function of true mean optical density for different condenser aperture settings. The change in transmission caused by off-axis illumination, defined as P, can be expressed by modifying the formula of Blout *et al.* (*3*) as follows:

$$ P = \frac{\int_0^{\theta = \alpha} \sin\theta \cos\theta \, (1 - 10^{kl_0(1 - \sec\theta)}) \, d\theta}{\int_0^{\theta = \alpha} \sin\theta \cos\theta \, d\theta} $$

E_T, the total percentage error in optical density, is given by

$$ E_T = - \frac{100 \log (1 - P)}{kl_0} $$

This error is a composite of aperture bias and aperture heterogeneity error, and the latter is calculated by using numerical integration to evaluate P and subtracting the aperture bias from the total aperture error. When the aperture heterogeneity error is isolated in this way, it is seen to be a roughly linear function of optical density and to be very small at low numerical apertures.

When the plug method or its variants are used on objects that are not disk-like, the aperture effect depends on the geometry of the object. For instance, a central plug measurement through a spherical object is virtually free of aperture errors, and so may differ by the full value of the aperture bias from measurements made on flattened objects.

Condenser aperture heterogeneity error occurs because the measured transmission at any point in the image is the average transmission of rays having different pathlengths through the object. It can be isolated by subtracting the aperture bias from the total aperture error (3) for any given value of numerical aperture. The condenser aperture heterogeneity error is a function of optical density and numerical aperture, and is shown in Fig. 1b. As can be seen, the error is small over the range of conditions used in cytophotometry and is unlikely to be a significant problem.

2. Chromatic and Glare Error

Chromatic error occurs whenever the specific absorptivity of the chromophore changes appreciably across the spectral bandwidth of the measuring light. In designing an instrument it is technically easy to reduce the spectral bandwidth to the point where the broad peaks of organic absorption curves give an insignificant error; but other pressures, such as the need to increase speed or reduce noise levels, may force the designer to compromise on spectral purity. The magnitude of chromatic error is given by the correct sum of optical densities for each spectral element (weighted by the energy flux of that element) minus the logarithm of the weighted sum of the transmissions for each spectral element (the quantity actually measured). If the spectral characteristics of the instrument and the spectral absorption curve of the chromophore are known, then the chromatic error is readily calculated as a function of optical density (29). In Fig. 2 an example of such a calculation is given for gallocyanin–chrome alum preparations as measured on CYDAC, a flying-spot, cathode-ray-tube scanner.

Chromatic error can be reduced by reducing the spectral bandwidth, or, when this is not possible, by using stains with absorption curves that are relatively flat in the relevant part of the spectrum. Chromatic error will have a similar effect on measurements made with either the scanning or the plug methods. However, in the two-wavelength method, chromatic error is partially compensated for because of its parallel effects on the definitive measurements and on the initial choice of wavelengths (39).

Glare includes both stray light and lens flare, and it results in a loss of contrast due to a disorganizing redistribution of light in the image. Howling and Fitzgerald (21) have discussed in detail the complex relationships which describe the nonuniformity of the glare flux as a function of the size and

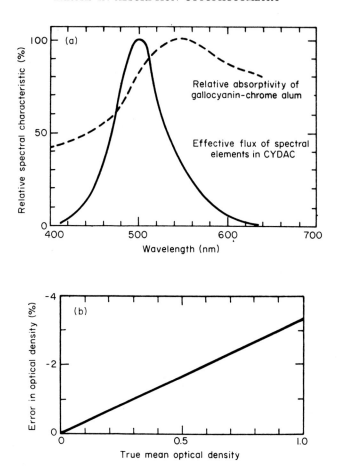

FIG. 2. The chromatic error for gallocyanin–chrome alum preparations as measured on the flying-spot scanner of CYDAC. These figures are taken from Mendelsohn *et al.* (*34*). (a) The spectral absorption curve for gallocyanin–chrome alum, and the spectral response of CYDAC. The first curve is used to compute the transmission for each element of the spectrum for a number of mean optical density values, and the second curve is used to weight these computed transmissions. The optical density, OD_{meas}, derived from the mean of the weighted transmission is then compared with the weighted mean of the optical density values for each spectral element, OD_{true}, to give the chromatic error. (b) The chromatic error, expressed as $100 [(OD_{meas} - OD_{true})/OD_{true}]$ and as a function of OD_{true}.

shape of the absorbing body and the size of the clear surround. However, if one assumes as a first approximation that the glare flux is uniform across the field, then the resulting error can be calculated by treating glare as a special case of chromatic error in which a fraction of the total light flux has a specific absorption of zero. As shown in Fig. 3, the error is a function of optical

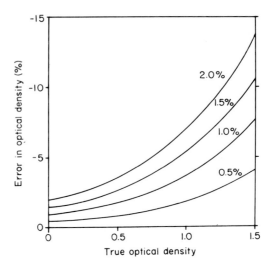

FIG. 3. Glare error, as a function of the optical density in the absence of glare, for different glare fractions. The error is given by $100[(OD_{meas} - OD_{true})/OD_{true}]$, where OD_{true} is the optical density in the absence of glare, $OD_{meas} = -\log[(T+g)/(1+g)]$, T is the transmission in the absence of glare, and g is the glare, as the fraction of the nonglare flux.

density and of the relative glare flux. The effects of glare are similar in scanning and plug cytophotometers, but the two-wavelength method partially compensates for glare in the same way that it compensates for chromatic error (10, 39, 40).

The glare flux in a system can be estimated by making plug measurements of intensity over an opaque object of about the same size as a typical specimen and mounted in a medium whose refractive index matches that of the object (40). Droplets of high concentrations of sudan black or crystal violet suspended in polyacrilamide gel satisfy these requirements (37a). Glare can be reduced by using coated optics of proper design, by using a darkened room to avoid extraneous light, by reducing the area of illumination in the plane of the specimen, and by reducing the condenser aperture.

3. Distributional Error

The third of the heterogeneity errors is distributional error. It involves two aspects of chromophore distribution: heterogeneity of concentration and heterogeneity of object thickness. In essence it deals with the residual heterogeneity when the contents of the photometric field are expressed as mass of chromophore per unit area.

The distributional error present in measurements made with the plug method and its variants depends on the amount of heterogeneity in the area sampled by the measuring spot, and this in turn will be a function of the size of the measuring spot and the texturing of the specimen. In general, the plug method is applied only to relatively homogeneous objects, and the distributional error associated with measurements will be small relative to the statistical uncertainty associated with the choice of area sampled and with the extrapolation of the estimate of area.

In the two-wavelength method, the error due to uncorrected distributional error is a function of the variation of the optical density of the specimen about its mean optical density. Patau (39) considered a number of model distributions, and showed that the distributional error was reduced by an order of magnitude compared to what it would have been without the two-wavelength approach.

Scanning cytophotometers combine a very large number of plug measurements into an integrated measurement of the entire object. Since each sample is converted to its logarithm independently, the distributional error in the integrated measurement is the sum of whatever distributional error is present within the individual samples. With modern equipment it is possible to scan the object with a wide range of spot sizes. Up to a point, it is clear that distributional error decreases as spot size decreases, but this relationship becomes obscure as one approaches the ultimate resolution of the light microscope and diffraction considerations become increasingly important.

By way of introduction to these diffraction effects, consider first the situation in a macroscopic scanning system in which the scanning spot is large compared to the resolution of the optics. Diffraction effects in such a system will be negligible and distributional heterogeneity in the object can be fully expressed in the image. If we think in terms of homogeneously absorbing, disklike objects with sharp edges, then the change in optical density at the boundary will be the only source of distributional error. As a scanning spot traverses such a boundary, the relative distributional error will describe a unimodal function whose maximum occurs when the boundary is approximately in the middle of the spot. The average distributional error associated with a single traverse is about two-thirds of the error obtained when the spot is centered on the boundary. This average value is shown in Fig. 4 as a function of the optical density of the homogeneous portion of the disk. The total relative distributional error for the entire object will depend on the number of boundary crossings. This can be summarized in the equation

$$DE_{total} = DE_{av\ trav}\ \frac{N_b}{N} \qquad (2)$$

where DE_{total} is the total relative distributional error, $DE_{av\ trav}$ is the average relative distributional error per traversal of a boundary, N is the number of scan points containing any part of the object, and N_b is the number of scan

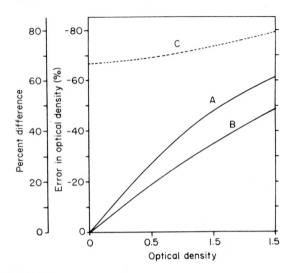

FIG. 4. Average distributional error associated with crossing a boundary. Curve A: Distributional error for a spot centered over the boundary. This is the same as the curve given by Mendelsohn (*33*). Curve B: Average distributional error as a square aperture moves orthogonally across the boundary. This curve was calculated by averaging the distributional error for all values of the ratio of clear field to chromophore from 0 to 1. Curve C: Curve B plotted as a percentage of curve A.

points containing any part of the boundary. The ratio of N_b to N is determined by the size and shape of the object and the size of the scanning spot.

In this diffraction-free system, optical aberrations and out-of-focus effects will blur the image of the boundary. Now when the spot crosses the boundary it will be sampling a gradual change in grayness instead of the sharp division between two distinct halves. Although this appearance of the image might suggest an absence of distributional error, the average distributional error associated with a boundary crossing remains unchanged. The light being sampled by the aperture still is composed of proportions of light that has passed through the clear area and light that has passed through the object. *To the extent that the blurring represents an averaging after these transmissions are defined by the object,* the average distributional error associated with crossing a boundary will be the same as before. Furthermore, as the image loses defintion a higher proportion of the total samples will involve the edge, thus increasing N_b/N and raising the total error accordingly. The net result of image degeneration is as if the scanning aperture had been altered so that its effective size now is greater than its geometric size.

Diffraction also causes a loss of definition in the image and so superficially resembles aberrations and out-of-focus effects in the imaging system. However, diffraction is a direct consequence of the inherent wave nature of light; and it is a moot point to what extent diffraction is analogous to these other factors in its effects on distributional error. There are three positions one could take on this question. First one could argue that the analogy is complete in the sense that heterogeneity of chromophore in the object leads to distributional error even when the heterogeneity is below the resolution limit of the optics and so is not apparent in the image. This position implies that exponential local absorption continues to hold at subresolvable dimensions, that distributional error is meaningful perhaps down to molecular levels, and that nothing can be done about such errors when using the light microscope. To the extent that this argument is correct the distributional error in scanning cytophotometry decreases continuously as the scanning spot size decreases, but approaches asymptotically a limiting residual error that can be neither sensed nor corrected because the effective spot size itself is diffraction limited.

The second and third positions are quite the opposite. In these cases one argues that there is no analogy to diffraction-free optics and that heterogeneity below the level of optical resolution does not cause distributional error. Now one is implying that spatial and absorptive resolution have similar restraints, that the exponential law applies not locally but across a resolution element, that light is unable to distinguish local effects within a resolution element, and that the wave front at such extremely small dimensions is continually averaging and hence does not cumulate distributional error. The distinction between the second and third positions involves the precise nature of absorptive resolution.

In the second model one argues that absorptive resolution is completely defined by the wavelength of the light (at the refractive index of the photometric field) and hence that unlike spatial resolution it is uncoupled from the numerical aperture of the imaging system. In this case, distributional error is eliminated only when absorptive and spatial resolution are identical; but this would require equality of the numerical aperture and the refractive index, a condition that is approached only by high quality immersion objectives. When the objective numerical aperture is less than the refractive index, spatial resolution will not be as sharp as absorptive resolution, transmissions will be averaged in the image, and some distributional error will occur regardless of how small the scanning aperture is made.

In the third position one argues that spatial and absorptive resolution go hand in hand and that regardless of the numerical aperture of the system the only heterogeneity of significance is the heterogeneity in the image. If this last interpretation is correct, then distributional error can be controlled in a scanning system by either decreasing the size of the scanning aperture or decreasing the numerical aperture of the objective.

The effect of these arguments on the magnitude of distributional error can be estimated for homogeneous sharply bounded objects using Eq. (2) and the relationships shown in Figs. 4 and 5. In the first model a single boundary crossing generates an average distributional error that is independent

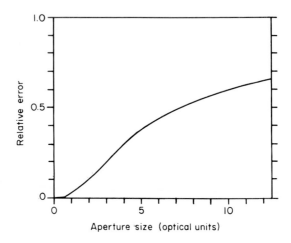

Aperture size (optical units)

FIG. 5. Distributional error of an edge, assuming a diffraction-limited image and the third absorption model, as described in the text. The spot is centered on the geometrical edge between two semi-infinite fields. One field is clear and the other has an optical density of 0.3. The classic edge function for incoherent illumination (25) is used to estimate transmission as a function of distance from the edge. Distributional error is calculated for scanning apertures of differing size by integrating these transmissions over various limits centered on the edge. The error is defined as the difference between 0.15 and the optical density derived from the mean of the integrated transmissions; it is expressed on the ordinate as a fraction of the error there would be in the absence of diffraction effects. On the abscissa, the size of the aperture is expressed in optical units; for an aperture whose width in the object plane is W microns, the optical width is

$$W\left(\frac{2\pi\text{n.a.}}{\lambda}\right) \text{optical units}$$

where n.a. is the numerical aperture of the objective, and λ is the wavelength of the illumination in microns. On this scale, the radius of the Airy diffraction pattern to the first minimum is 3.83 optical units.

The third absorption model assumes that the only source of distributional error is heterogeneity resolved in the image; therefore, the curve shown in this figure is derived from the gradient of the edge image. This is one of a family of curves corresponding to different optical densities, and in the range of 0.1 to 1.0 optical density these follow the given curve closely. With an increase in coherence of the illumination, the edge function shifts position and steepens slightly. The increased gradients will slightly raise the left-hand portion of the error curve under these circumstances.

The average distributional error associated with an edge crossing of a diffraction-limited image is obtained by using this curve to read off a scaling factor for curve B in Fig. 4.

of spot size. However, the effective spot size acts on the ratio N_b/N, and the smaller the spot the smaller will be the total relative distributional error. Eventually the effective spot size is defined by the Airy resolution disk for the system and both N_b/N and distributional error reach a limiting value.

In the third model the sharp edge of the object appears in the image as an edge function. A spot centered on the boundary will contain a gradient of intensities and the resulting distributional error will depend on the width of the spot in optical units (as defined in the legend of Fig. 5) and on the optical density of the object. The curve in Fig. 5 shows the fractional reduction in distributional error caused by this effect. It is calculated for a spot centered on the boundary and for an object of 0.3 optical density. In effect it gives the average distributional error of a boundary traverse for the third model divided by the corresponding error for either the first model or a diffraction-free system. To calculate the total relative distributional error using Eq. (2) one must keep in mind that N_b/N will be a function of the effective spot size and hence will increase as the objective numerical aperture decreases.

When the numerical aperture of the objective equals the refractive index of the photometric field, the second and third models are equivalent. At intermediate numerical apertures the second model will give errors that are intermediate between the third and the first model, and at zero numerical aperture the second model is equivalent to the first model (28).

Of these three models of absorptive resolution, the third is the most optimistic, while the second has been tacitly assumed in most approaches to cytophotometry. At present we know of no treatment which has rigorously examined these possibilities either experimentally or theoretically. Nor is it clear how these considerations relate to the relative merits of the available methods of photometry. However, although current theory does not allow us to choose rationally among these very different models, experimental measurements may allow the elimination of one or two of them; this is clearly an important area for further investigation.

4. *Absorption Law Failure*

In this discussion of absorption errors, it has been assumed that microscopic objects behave as pure amplitude objects, and that departures from the absorption laws can be explained by failure of the measuring system to meet the strict conditions necessary for application of the algebraic relationships embodied in Eq. (1). However, there are a number of considerations, some of which have been raised already, that make one question the accuracy of the absorption laws themselves when applied to objects whose size is comparable to the wavelength of light.

The absorption laws assume a pure amplitude absorber, but in practice this condition cannot be met in cytophotometry as there are inevitable changes in the electromagnetic field that accompany the absorption of light (4). These have the effect of small local changes in refractive index and lead to the scattering and refraction of light and to other nonabsorptive phenomena. When Caspersson (5) formulated the theoretical foundations for absorption cytophotometry he was concerned with these nonabsorptive effects particularly as they apply to the measurement of living cells in the ultraviolet. In his analysis, Caspersson applied the general scattering theory of Mie (36) and showed that nonabsorptive effects are relatively unimportant for large objects, but become increasingly severe as the size of the object approaches the wavelength of light. He also noted that the magnitude of these effects decreases when the refractive index of the surroundings closely matches that of the object and that the effect of nonabsorptive phenomena can be further decreased by using an objective lens with a wide collecting angle.

A number of authors have been concerned with the appearance of small objects under varying conditions of illumination and have used the scalar theory of image formation to calculate theoretical energy distributions for images of small apertures, small opaque disks, and small absorbing disks (38, 43–45). These calculations vividly demonstrate the effect of object size and degree of coherence of the illumination on the predicted appearance of the image. However, as is discussed by Welford (52), there are some appreciable discrepancies among the predictions. Charman (8, 9) analyzed the images of circular apertures and straight edges, and compared experimental measurements with theoretical predictions. For a low aperture objective, he found close agreement between his measurements and the predictions of Slansky (43), but for high aperture objectives there was a marked discrepancy, confirming the inapplicability of the scalar theory to wide angle systems. These measurements have not been extended to absorbing objects. However, Slansky (43) has used scalar theory to calculate the image profiles for one set of pure amplitude absorbers whose transmissivity was 0.5 (optical density, 0.301). His objects were homogeneous disks of different radii, and he assumed different degrees of coherence for the illumination. Distances were expressed in optical units (see legend to Fig. 5) and the degree of coherence was defined by

$$S = \frac{\text{Condenser numerical aperture}}{\text{Objective numerical aperture}}$$

Slansky's data can be used to calculate the integrated optical density by converting the intensity curves to optical density and then integrating over a circle. These values turn out to be significantly less than they should be, given the size and transmissivity of the objects (*25a*). For a disk whose radius is two optical units, the negative error is over 25% when $S = \infty$ (incoherent illumination), and it is about 5% when $S = 0.5$ (partially coherent illumination). As the size of the disk increases, the magnitude of the error lessens. These discrepancies may be associated with limitations in the scalar theory, with approximations necessary for Slansky's solution of the wave equations, or with the fact that integration could not be extended out to distances at which there were no further changes in total optical density. In any case, it is not clear why the discrepancies are a function of the degree of coherence, nor how they relate (if at all) to the problem of distributional error at the limit of optical resolution.

In summary, it seems fair to say that the absorptive process is reasonably well understood as it applies to homogeneous objects whose size is several times the optical resolution of the measuring system. But as the size of the object and its homogeneity decreases, *the application of current theory involves ever increasing uncertainties.* When crude cytophotometric methods are used, these uncertainties may be relatively unimportant; but as the methods of absorption cytophotometry are applied to ever more demanding problems, the general approximations that have sufficed in the past leave much to be desired.

C. INSTRUMENT ERRORS

In addition to specimen and absorption errors, cytophotometric measurements may contain systematic errors caused by nonlinearities in the photodetectors, in the associated electronics, and in the mechanical design of the instrument. With the application of modern technology to instrument design, errors from these sources should be minimal. A detailed discussion of them is outside the scope of this chapter, but it should be realized both that they may be a source of significant error and that it is relatively simple to test for their presence. Nonlinearities in the photodetector and associated circuits can be tested by altering the operating conditions, either by decreasing the light flux or by changing the photomultiplier dynode voltage. Comparable tests can be devised to evaluate such things as linearity of area measurement, uniformity of the intensity and spectral composition of light across the photometric field, and linearity of scanning rasters. These errors and their evaluation are different for different systems, and there is no substitute for understanding the details of one's own instrument.

IV. Model Calculations of Systematic Errors

As has already been mentioned, significant differences have been found in the DNA stain content of human leukocytes, and it is possible that these differences relate to absorption errors. The mechanical scanner of CYDAC was used to measure leukocytes that had been stained with gallocyanin–chrome alum after digestion with ribonuclease and it was found that, on the average, monocytes measure 16% higher than small lymphocytes and 13% higher than neutrophilic granulocytes (26). In the following treatment, simplified models of cells are used to simulate the potential absorption errors of this method of photometry and to relate these errors to the differences actually found among cells.

Let us represent the nuclei of monocytes, small lymphocytes, and neutrophils by three idealized flat objects, each containing the same amount of homogeneously distributed chromophore. The area of the monocyte nucleus is twice that of the other cells, the monocyte and the lymphocyte nuclei are circular, and the shape of the neutrophil nucleus is such that its perimeter ratio, N_b/N, is twice that of the lymphocyte.

First, let us consider the relative distributional error associated with measurements of these objects in a simple cytophotometer. This is simulated by assuming that a nucleus is fully contained within photometric fields of various sizes. The average transmission through each field is calculated,

TABLE I

ABSORPTION ERRORS CALCULATED FOR IDEALIZED CELLS

	Monocyte	Small lymphocyte	Neutrophil
A. Simple Cytophotometer			
Nuclear optical density	0.2	0.4	0.4
Nuclear area (square microns)	100	50	50
Distributional error (%)			
Field size = 1.5 × Nuclear area	−8.0	−16.5	−16.5
Field size = 2.0 × Nuclear area	−11.4	−22.3	−22.3
Field size = 3.0 × Nuclear area	−14.5	−27.1	−27.1
B. Scanning Cytophotometer			
Nuclear perimeter ratio (N_b/N)	0.14	0.19	0.38
Absorption errors (%)			
Condenser aperture (n.a. = 0.3)	0.96	0.96	0.96
Chromatic	−0.02	−0.03	−0.03
Glare (1.0%)	−1.26	−1.61	−1.61
Distributional			
(a) Diffraction-free	−1.07	−2.90	−5.79
(b) With diffraction (third model)	−0.38	−1.04	−2.09
Total			
(a) Diffraction-free	−1.39	−3.58	−6.47
(b) With diffraction (third model)	−0.70	−1.72	−2.77

converted to optical density, and multiplied by the area of the field. There is no correction for distributional error under such conditions, and the distributional errors shown in part A of Table I are large, related to optical density, related to the size of the field, and unrelated to the shape of the object.

Table I, part B, shows the results of a simulation of a scanning photometer having the specifications of the mechanical scanner of CYDAC. The scanning aperture is a circle of diameter 0.423 μ and the perimeter ratios given in the table were calculated for the circular objects by the following equation:

$$\frac{N_b}{N} = \frac{4Rr}{(R + r)^2}$$

where R is the radius of the object and r is the radius of the scanning aperture. The corresponding perimeter ratio for the neutrophil is by definition twice that of the lymphocyte. Distributional error has been calculated both in the absence and in the presence of diffraction, using the first and third models as described in the previous section. The diffraction-free distributional error is the product of the perimeter ratio and the mean boundary crossing error given in Fig. 4. The error when diffraction is present is the product of the diffraction-free error and the diffraction factor obtained from Fig. 5. At an objective numerical aperture of 1.0 and a wavelength of 566 nm, the optical size of a scanning aperture of 0.423 μ is 4.70 optical units and the corresponding diffraction factor is 0.36.

It can be seen that scanning has reduced the distributional error by an order of magnitude, and that diffraction effects could reduce it by another factor of 3 if the third model of absorptivity is correct. The effect of the other absorption errors is also given in Table I, part B for the following conditions: condenser numerical aperture 0.3, glare fraction 1.0%, and chromatic error as calculated for gallocyanin–chrome alum and the mechanical scanner of CYDAC (28). The total absorption errors are shown for both the first and third models of absorptivity. Note that distributional error is the major source of differences among cell types, that the errors are now a function of shape as well as optical density, and that the total errors with either of the absorptivity models are considerably smaller than the differences found experimentally.

When modeling, it is extremely important to match the starting assumptions to the problem at hand. For example, both the plug and the two-wavelength methods would give no distributional error with models based on homogeneous objects. If these methods are to be simulated, one must include internal heterogeneity of the chromophore, error functions for sampling and area estimates, and a realistic test situation for the choice of wavelengths. In the same vein, one can question the suitability of a homogeneous model as a test object for scanning cytophotometry. If internal heterogeneity is introduced without changing the conditions at the boundary of the object, as in the case of an internally textured disk, the distributional errors may increase. On the other hand, if internal heterogeneity reduces the gradients at the boundary, the total distributional error may fall below that of a homogeneous disk. A sphere is an example of this latter phenomenon, and a conical object is an even more extreme case. Obviously the modeling we have done is not definitive because it begins with simplified assumptions about the objects themselves. Nevertheless such modeling can give valuable insights into the way in which different errors interact, it gives order-of-magnitude estimates of prevailing errors, and under appropriate circumstances it defines the maximal errors that can be expected.

V. Experimental Evaluation of Systematic Errors

Systematic errors are best evaluated by calibrating an instrument against a series of standards whose properties cover the full operating range of the instrument. The paucity of rigorous standards has plagued cytophotometry since its onset; nevertheless, a number of indirect approaches are possible and have been used to test the magnitude of systematic errors. These approaches include using (1) model objects, (2) cells with biologically predictable behavior, (3) different instruments, and (4) changes in optical conditions.

(1) Many investigators have evaluated their systems by using model objects whose properties can be predicted by an independent method. Droplets of a dye solution suspended in a nonmiscible medium were first proposed by Caspersson (5) and are the most popular example of this type of model. The amount of dye in a spherical droplet is a function of its volume, and hence the total optical density of such droplets should bear a linear relation to the cube of the diameter. A limitation of this and all other model systems is that they differ from biological objects in ways that may prove to be significant.

(2) The expectation of DNA constancy is the basis of a frequently used biological test of cytophotometry, although constancy itself owes much of its credibility to cytophotometric measurements. Amphibian and avian erythrocytes have highly predictable behavior in this regard, but they have the limitation that they are uniform in appearance and hence may not test all aspects of a system. Measurements on liver parenchymal cells provide a more rigorous test. In this case the cells vary in size, and the larger cells are expected to have twice or four times the DNA content of the smaller cells. But in spite of their differences in size, liver cells are quite uniform in shape and texture and hence they too may not test a cytophotometer completely. As has already been mentioned, the leukocytes of peripheral blood are a good

example of cells that show wide differences in their morphology, texture, and darkness. Thus they would be expected to respond differently to systematic errors, particularly those that are a function of shape and optical density. However, leukocyte measurements depart significantly from constancy. When measurements on a biological test system agree with prediction, this is a good indication that the measurements have not been distorted by systematic errors; but when divergences from prediction are found, there may be no way of knowing whether they are due to systematic errors or to biological differences of one sort or another.

(3) When cells are measured on two instruments, it is unlikely that all systematic errors present in the measurements would be identical, particularly if the instruments are based on different cytophotometric principles. Thus, if comparative measurements show close agreement, confidence in each instrument is increased. But obviously such comparisons cannot indicate errors that are common to both instruments. Possibly the most effective use of comparisons between instruments is to calibrate an instrument of unknown performance against one whose behavior is well understood. Ideally, the cells used for interinstrument comparisons should cover the maximum possible range of measuring conditions. For instance, Mendelsohn and Richards (35) used a population of aneuploid, proliferating ascites tumor cells to compare two-wavelength and scanning measurements, and Mayall et al. (27) used both liver cells and leukocytes in their comparison of two-wavelength and scanning measurements. The first two populations show wide variation in area and in total optical density, and the last population emphasizes variation in darkness. In each case, there was close agreement between the two methods.

(4) When testing divergent cells such as leukocytes, the departures from constancy can be evaluated further by deliberate changes of the optical conditions. To the extent that errors are a function of optical density, they can be modified by changing wavelength and hence absorptivity. Similarly, the cells can be flattened or crushed (12), the numerical apertures of the condenser and objective can be changed, or gross changes can be made in the focus of the objective (30).

Recently we have been modifying the optical conditions as part of the evaluation of systematic errors associated with the mechanical scanner of CYDAC (28). In one aspect of these experiments, the effect of aperture error was demonstrated by comparing measurements made with the condenser aperture set to n.a. = 0.6 with those made under the standard condition of n.a. = 0.3. The mean of leukocyte measurements ,made with the larger aperture setting was found to be 3.5% greater than the mean of the standard measurements. This agrees well with the difference of 3.1% predicted on the basis of the change in aperture bias. When the measurements of any one of the 15 gallocyanin–chrome alum-stained leukocytes (5 monocytes, 5 small lymphocytes, and 5 neutrophils) were normalized by the respective population means, they showed close agreement and the differences among types and among cells within types remained unchanged. Thus, changing condenser aperture size had no effect on the relative values for these cells.

A similar approach was used to investigate the heterogeneity errors. All of these errors are a function of optical density, and so should change when the optical density of the specimen changes. The specific absorptivity of gallocyanin–chrome alum when measured in light from an interference filter centered at 433 nm is about half that measured with light from the 566-nm filter. When measurements made with the two filters are normalized and compared, no significant effect due to wavelength is found, and the differences among types and among cells within types remain unchanged.

If the differences found among cells were due to errors that were a function of optical density, then halving the optical density should have caused some comparable change in the relative differences among cells. Since no such change occurred, one must conclude either that there is an unknown compensating error that is a positive function of optical density, or, as seems much more likely, that the differences found among cells are due to factors that are not a function of optical density. This argument is further supported by the model calculations of the preceding section, since we were unable to generate sufficient differences among cell types by any of the methods of calculation for a scanning cytophotometer. Hence it seems probable that the differences found among these cells cannot be attributed to condenser aperture error, chromatic error, glare, or distributional error.

VI. Stochastic Errors

Reproducibility is measured by the standard deviation (or coefficient of variation) of replicate measurements, and as such it is an estimate of the stochastic errors (the errors that cause repeated measurements of the same object to fluctuate about a mean value). The statistical treatment of these errors is well understood, and their effect can be evaluated by using an analysis of variance in conjunction with an appropriate model (2, 15). The sensitivity of any set of measurements is determined by both the reproducibility and the redundancy of the set, and accordingly the following derivation can be used to determine the number of replicate measurements necessary for a give experiment.

The experimental sensitivity for a set of measurements may be defined as the minimum coefficient of variation that will be significant (i.e., will be resolved) at an arbitrarily chosen confidence level. It is related to R, the coefficient of variation among replicate measurements on the same cell; n.

the effective average number of replicate measurements; and the F value given in standard statistical tables for the degrees of freedom and the desired confidence level. When MS_{cells} is the mean square deviation (variance) due to cells and MS_{reps} is the mean square deviation (variance) due to replicates, the F ratio is given by

$$F = \frac{MS_{cells}}{MS_{reps}}$$

If a linear model is assumed, and $\bar{x}$ is the population mean, then

$$MS_{reps} = R^2\bar{x}^2$$

and

$$MS_{cells} = (nC^2 + R^2)\bar{x}^2$$

where C is the coefficient of variation due to cells. Thus

$$C^2 = \frac{(MS_{cells}/\bar{x}^2) - R^2}{n} = \frac{FR^2 - R^2}{n} = \frac{(F-1)R^2}{n}$$

and

$$C = R\left(\frac{F-1}{n}\right)^{\frac{1}{2}}$$

Hence, given the replication error, the number of cells, the number of replicates, the confidence level desired, and the F value from an appropriate table, the sensitivity is fully defined. The sensitivity can be improved by reducing the replication error or by increasing the number of replicates. In the latter case there is no limit to the degree of improvement, but as the number of replicates increases the effect on the sensitivity becomes progressively less.

Replication error is a composite of all the component replication errors for the total measuring system. Thus any of the following may contribute to the replication error: (1) dirt and background staining on the slide; (2) positioning of the cell and choice of clear field; (3) focus of the microscope, fluctuations in light intensity, and uneven illumination; (4) changes in wavelength of the illumination; (5) variations in sampling rate and sampling density with scanning cytophotometers and in measuring stop area for simple cytophotometers with variable stop size; (6) drift and noise in the photomultiplier tubes and in the electronic circuits. Obviously, the errors that are significant in a particular instrument will be determined by the design of the instrument and by its operating conditions.

The effect of component errors varies somewhat depending on what part of the system they act on. Thus any uncertainty affecting the area measurement will have a directly proportional effect on the overall error, but errors in the measurement of transmission must be transformed as a function of optical density before they can be assessed.

VII. Conclusion

In its three decades of growth, absorption cytophotometry has acquired a high degree of theoretical and practical competence. This is all to the good, but the neophyte and to some extent the veteran is apt to be misled by two aspects of the present-day situation. Armed with elegant equipment and commonly used techniques, they are likely to assume too much about the credibility of their results; and armed with the extensive literature on cytophotometry, they may assume that further theoretical development is both unnecessary and unlikely.

In writing this review we have hoped to discourage complacency and to encourage research into the frontier problems remaining in cytophotometry. With the advent of digital computers and the rapid growth of powerful electronic logic, one can expect to see cytophotometers embedded in a wide variety of complex machines [see, for instance, Wied *et al.* (*53*)]. New questions will be asked, old comfortable ways of eliminating errors may have to be dropped, and in return there will be the exciting new option to correct data in ways that were traditionally impossible. Let us hope we are ready for these challenges.

References

1. Atkin, N. B., and Richards, B. M., Deoxyribonucleic acid in human tumours as measured by microspectrophotometry of Feulgen stain: A comparison of tumours arising at different sites. *Brit. J. Cancer* **10**, 769–786 (1956).
2. Bartels, P. H., Sensitivity and evaluation of microspectrophotometric and microinterferometric measurements. *In* "Introduction to Quantitative Cytochemistry" (G. L. Wied, ed.), pp. 93–105. Academic Press, New York, 1966.
3. Blout, E. R., Bird, G. R., and Grey, D. S., Infra-red microspectroscopy. *J. Opt. Soc. Am.* **40**, 304–313 (1950).
4. Born, M., and Wolf, E., "Principles of Optics." Macmillan, New York, 1964.
5. Caspersson, T., Über den chemischen Aufbau der Strukturen des Zellkernes. *Skand. Arch. Physiol.* **73**, Suppl 8, 1–151 (1936).
6. Caspersson, T., "Cell Growth and Cell Function." Norton, New York, 1950.
7. Caspersson, T., and Lomakka, G. M., Scanning microscopy techniques for high resolution quantitative cytochemistry. *Ann. N.Y. Acad. Sci.* **97**, 449–463 (1962).
8. Charman, W. N., Some experimental measurements of diffraction images in low-resolution microscopy. *J. Opt. Soc. Am.* **53**, 410–414 (1963).
9. Charman, W. N., Diffraction images of circular objects in high-resolution microscopy. *J. Opt. Soc. Am.* **53**, 415–419 (1963).

10. Cleland, K. W., Sources of error in two-wavelength microspectrophotometry. *Australian J. Biol. Sci.* **18**, 1057–1080 (1965).
11. Davies, H. G., and Walker, P. M. B., Microspectrometry of living and fixed cells. *Progr. Biophys. Biophys. Chem.* **3**, 195–236 (1953).
12. Davies, H. G., Wilkins, M. H. F., and Boddy, R. G. H. B., Cell crushing: A technique for greatly reducing errors in microspectrometry. *Exptl. Cell Res.* **6**, 550–553 (1954).
13. Deeley, E. M., An integrating microdensitometer for biological cells. *J. Sci. Instr.* **32**, 263–267 (1955).
14. Freed, J. J., and Engle, J. L., Development of the vibrating-mirror flying spot microscope for ultraviolet spectrophotometry. *Ann. N.Y. Acad. Sci.* **97**, 412–430 (1962).
15. Garcia, A. M., Studies on DNA in leukocytes and related cells of mammals. I. On microspectrophotometric errors and statistical models. *Histochemie* **3**, 170–177 (1962).
16. Garcia, A. M., A one-wavelength, two-area method in microspectrophotometry for pure amplitude objects. *J. Histochem. Cytochem.* **13**, 161–167 (1965).
17. Garcia, A. M., Cytophotometric studies on haploid and diploid cells with different degrees of chromatin coiling. *Ann. N.Y. Acad. Sci.* **157**, 237–249 (1969).
18. Gurbanov, V. P., A method of photometry of cytological objects in two light beams of different areas. *Federation Proc.* **25**, T883–T888 (1966).
19. Hale, A. J., The leucocyte as a possible exception to the theory of deoxyribonucleic acid constancy. *J. Pathol. Bacteriol.* **85**, 311–326 (1963).
20. Hale, A. J., and Cooper, E. H., Studies on DNA Replication in leukaemic and non-leukaemic leukocytes. *In* "Current Research in Leukaemia" (F. G. J. Hayhoe, ed.), pp. 95–107. Cambridge Univ. Press, London and New York, 1965.
21. Howling, D. H., and Fitzgerald, P. J., The nature, significance, and evaluation of Schwarzschild-Villiger (SV) effect in photometric procedures. *J. Biophys. Biochem. Cytol.* **6**, 313–337 (1959).
22. Inoué, S., and Sato, H., Deoxyribonucleic acid arrangement in living sperm. *In* "Molecular Architecture in Cell Physiology" (T. Hayashi and A. G. Szent-Györgyi, eds.), pp. 209–248. Prentice-Hall, Englewood Cliffs, New Jersey, 1966.
23. Jansen, M. T., A simple scanning cytophotometer. *Histochemie* **2**, 342–347 (1961).
24. Kelly, J. W., Color film cytophotometry. *In* "Introduction to Quantitative Cytochemistry" (G. L. Wied, ed.), pp. 247–280. Academic Press, New York, 1966.
25. Martin, L. C., "The Theory of the Microscope." American Elsevier, New York, 1966.
25a. Mayall, B. H., Unpublished observations (1968).
26. Mayall, B. H., Deoxyribonucleic acid cytophotometry of stained human leukoctyes. I. Differences among cell types. *J. Histochem. Cytochem.* **17**, 249–257 (1969).
27. Mayall, B. H., Edwards, R. Q., Bateson, R. C., Connolly, J. R., and Mendelsohn, M. L., DATEM, a dual-beam automatic two-wavelength microspectrophotometer. *Ann. N.Y. Acad. Sci.* **157**, 225–236 (1969).
28. Mayall, B. H., and Mendelsohn, M. L., Deoxyribonucleic acid cytophotometry of stained human leukocytes. II. The mechanical scanner of CYDAC, the theory of scanning photometry, and the magnitude of residual errors. *J. Histochem. Cytochem.* (1970). In preparation.
29. Meehan, E. J., "Optical Methods of Analysis." Wiley (Interscience), New York, 1964.
30. Mendelsohn, M. L., The two-wavelength method of microspectrophotometry. I. A microspectrophotometer and tests on model systems. *J. Biophys. Biochem. Cytol.* **4**, 407–414 (1958).

31. Mendelsohn, M. L., The two-wavelength method of microspectrophotometry. III. An extension based on photographic color transparencies. *J. Biophys. Biochem. Cytol.* **4**, 425–431 (1958).
32. Mendelsohn, M. L., The two-wavelength method of microspectrophotometry. IV. A new solution. *J. Biophys. Biochem. Cytol.* **11**, 509–513 (1961).
33. Mendelsohn, M. L., Absorption cytophotometry: Comparative methodology for heterogeneous objects, and the two-wavelength method. *In* "Introduction to Quantitative Cytochemistry" (G. L. Wied, ed.), pp. 201–214. Academic Press, New York, 1966.
34. Mendelsohn, M. L., Mayall, B. H., Prewitt, J. M. S., Bostrom, R. C., and Holcomb, W. G., Digital transformation and computer analysis of microscopic images. *In* "Advances in Optical and Electron Microscopy" (R. Barer and V. E. Cosslett, eds.), Vol. 2, pp. 77–150. Academic Press, New York, 1968.
35. Mendelsohn, M. L., and Richards, B. M., A comparison of scanning and two-wavelength microspectrophotometry. *J. Biophys. Biochem. Cytol.* **5**, 707–709 (1958).
36. Mie, G., Beiträge zur Optik trüber medien, speziell kolloidaler Metallösungen. *Ann. Physik* [4] **25**, 377–455 (1908).
37. Ornstein, L., The distributional error in microspectrophotometry. *Lab. Invest.* **1**, 250–262 (1952).
37a. Ornstein, L., Personal communication (1968).
38. Osterberg, H., and Smith, L. W., Diffraction theory for images of disk-shaped particles with Köhler illumination. *J. Opt. Soc. Am.* **50**, 362–369 (1960).
39. Patau, K., Absorption microphotometry of irregular-shaped objects. *Chromosoma* **5**, 341–362 (1952).
40. Pollister, A. W., and Ornstein, L., The photometric chemical analysis of cells. *In* "Analytical Cytology" (R. C. Mellors, ed.), pp. 431–518. McGraw-Hill, New York, 1959.
41. Rudkin, G. T., Microspectrophotometry of chromosomes. *In* "Introduction to Quantitative Cytochemistry" (G. L. Wied, ed.), pp. 387–407. Academic Press, New York, 1966.
42. Sandritter, W., Methods and results in quantitative cytochemistry. *In* "Introduction to Quantitative Cytochemistry" (G. L. Wied, ed.), pp. 159–182. Academic Press, New York, 1966.
43. Slansky, S., Images d'un disque clair ou sombre en éclairage partiellement cohérent. *Rev. Opt.* **39**, 555–577 (1960).
44. Smith, L. W., Diffraction images of disk-shaped particles computed for full Köhler illumination. *J. Opt. Soc. Am.* **50**, 369–374 (1960).
45. Som, S. C., Diffraction images of annular and disk-like objects under partially coherent illumination. *J. Opt. Soc. Am.* **57**, 1499–1509 (1967).
46. Swift, H., The desoxyribose nucleic acid content of animal nuclei. *Physiol. Zool.* **23**, 169–198 (1950).
47. Swift, H., and Rasch, E., Microphotometry with visible light. *Phys. Tech. Biol. Res.* **3**, 353–400 (1956).
48. Tonkelaar, E. M. den, and van Duijn, P., Photographic colorimetry as a quantitative cytochemical method. I. Principles and practice of the method. *Histochemie* **4**, 1–9 (1964).
49. Walker, P. M. B., Ultraviolet absorption techniques. *Phys. Tech. Biol. Res.* **3**, 401–487 (1956).
50. Walker, P. M. B., Leonard, J., Gibb, D., and Chamberlain, P. J., A scanning and integrating spectrophotometer for very small objects. *J. Sci. Instr.* **40**, 166–172 (1963).

51. Walker, P. M. B., and Richards, B. M., Quantitative microscopical techniques for single cells. *In* "The Cell" (J. Brachet and A. E. Mirsky, eds.), Vol. 1, pp. 91–138. Academic Press, New York, 1959.
52. Welford, W. T., The Mach effect and the microscope. *In* "Advances in Optical and Electron Microscopy" (R. Barer and V. E. Cosslett, eds.), Vol. 2, pp. 41–76. Academic Press, New York, 1968.
53. Wied, G. L., Bahr, G. F., and Bartels, P. H., Automatic analysis of cell images by TICAS. *In* "Automated Cell Identification and Cell Sorting" (G. L. Wied and G. F. Bahr, eds.), pp.195–360. Academic Press, New York, 1970.

RECENT DEVELOPMENTS IN GALLOCYANINE–CHROME ALUM STAINING

Günter Kiefer

DEPARTMENT OF PATHOLOGY, UNIVERSITY OF FREIBURG, GERMANY

I. Accompanying Protein

The suitability of the gallocyanine–chrome alum staining for the quantitation of nucleic acids has thoroughly been tested. A few years ago Sandritter *et al.* (*17*) described the conditions under which quantitative staining is achieved and reported good correlation between the DNA content of a cell nucleus and the amount of bound dye after the cell had been treated with ribonuclease. Experiments with a large number of cell types obtained both from mammalian and nonmammalian vertebrates showed that the quantity of dye measured parallels the quantity of DNA and not the protein content of the nucleus. This is only true for somatic cells, while the staining intensity of fully differentiated sperm was less than one should expect from their DNA content. In this respect, gallocyanine–chrome alum behaves in the same way as other basic dyes (*1, 12*). There seem to be two reasons why sperm stain less intensely than they should: The first is that they contain a protein which is much more basic than the histones of somatic nuclei. The second reason could be the dense packing and ordered arrangement of the nucleoproteins in the sperm head (*19*). There exist only quantitative differences between sperm and somatic cells in respect to their dye capacity, for the same factors, i.e., composition and structure of the accompanying proteins, are influencing the staining intensity of the somatic cell nuclei as well. Therefore, it was thought that sperm would provide a good model system for studying the role of protein in nucleic acid staining.

One method to abolish the competing influence on nucleic acid staining of protein amino groups is to block these groups by acetylation. This method has already been used by Deitch (*3*) and Rigler (*14*) in order to increase staining intensity and to make the results more reproducible. In our experiments ethanol–acetone (1:1, v/v) fixed cells were treated for 1 hr at room temperature with a mixture of acetic anhydride and pyridine (4:6, v/v). (A more intensive acetylation procedure using pure acetic anhydride at 60°C for 18 hr was found unsuitable since nucleic acids were partially extracted under these conditions.)

If we block protein-bound amino groups by acetylation the basophilia of the cell nuclei increases as one should expect from the previously mentioned reasons and from experimental results (Table I). The amount of methylene

TABLE I

AMOUNT OF DYE (IN PERCENT OF CONTROL ± STANDARD DEVIATION) BOUND TO RAT THYMUS LYMPHOCYTE NUCLEI ($N = 50$) AFTER ACETYLATION

Staining	Control	Acetylation
Methylene blue	100 ± 8.7	132.1 ± 13.7
Gallocyanine–chrome alum	100 ± 15.1	97.3 ± 10.4

blue bound to thymus lymphocytes after acetylation is more than 30% higher than without the blockage of the competing protein groups (3). If we do the same experiment substituting gallocyanine–chrome alum for methylene blue we observe no increased binding of the dye. This failure to increase the uptake of gallocyanine–chrome alum could be demonstrated with a number of cell types, including sperm (Table II). We can interpret the data in the following way: Basic groups of protein do not act as competitive groups if one uses gallocyanine–chrome alum for the demonstration of nucleic acids in fixed tissues contrary to their role in the staining with ordinary basic dyes. This is possibly due to lower pH and higher ionic strength of the dye solution used for staining, as well as longer staining time and much stronger binding of gallocyanine–chrome alum to the nucleic acid. It is not possible to increase the gallocyanine binding of sperm by acetylation to the expected value of half that of diploid nuclei. This means that the high amino group content of sperm protein is not the cause for the weak staining, for these groups should be blocked after acetylation. Therefore, the extremely dense packing of the nucleoproteins in sperm, i.e., steric factors are to be considered as the main reason for weak staining of sperm. If the proteins are partially removed an increase in stainability can be observed (13, 16). Steric factors within the denatured nucleoproteins seem to play a role in the staining of somatic cells, too. It has been calculated that not more than eight out of ten nucleic acid phosphate groups are occupied by gallocyanine–chrome alum

TABLE II

GALLOCYANINE–CHROME ALUM CONTENT OF RNASE-TREATED
NUCLEI FROM DIFFERENT TISSUES AFTER ACETYLATION
(IN PERCENT OF NONACETYLATED CONTROL)[a]

Tissue	Control	Acetylated
Rat		
Thymus	100.0 ± 14.1	97.0 ± 10.5
Rabbit		
Sperm	100.0 ± 18.2	104.6 ± 21.8
Rooster		
Sperm	100.0 ± 16.1	99.3 ± 14.1
Erythrocytes	100.0 ± 11.1	85.0 ± 10.1
Newt		
Erythrocytes	100.0 ± 14.1	98.4 ± 11.2
Leukocytes	100.0 ± 10.2	98.4 ± 6.9
Frog		
Erythrocytes	100.0 ± 15.5	97.8 ± 12.1
Spleen	100.0 ± 13.2	85.3 ± 13.2

[a] Number of nuclei measured: 50 each.

(17). The acetylation experiments show that this proportion cannot be raised by blocking the protein amino groups (10).

II. Stoichiometry with RNA

After we had seen that gallocyanine–chrome alum can be used for the quantitative cytochemical determination of DNA we wondered whether the dye could be used for the quantitative staining of RNA as well. This appeared important since we already have the Feulgen reaction for DNA but no equally satisfying reaction for RNA. To test the stoichiometry of gallocyanine for RNA is more difficult than for DNA, since the RNA content varies considerably between cells. It is, therefore, necessary to know the RNA content of an individual cell before measuring the amount of gallocyanine–chrome alum bound. The only way to measure the RNA content of a single cell is by UV microphotometry, which can be supposed to be sufficiently independent from cytophotometry of stained preparations.

So, we measured the nucleic acid content of single cells in a smear preparation using microphotometry at the wavelengths 265, 280, and 313 nm. The nucleic acid content was calculated according to a formula proposed by Sandritter in 1958 (15). Most of the experiments were done with isolated

rat liver cells fixed by freeze substitution. Part of the smears were treated
with RNase, DNase, or trichloroacetic acid, respectively, to remove DNA
or RNA or both. After the cells had been measured in UV light they were
stained with gallocyanine–chrome alum and remeasured in visible light (*9*).
The results indicate that dye-binding capacity parallels nucleic acid content
(Fig. 1). About 70–80% of total nucleic acid in liver cells is RNA. So gallo-

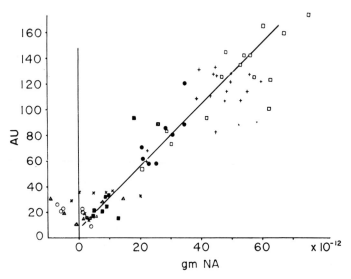

FIG. 1. Proportionality between nucleic acid content and gallocyanine–chrome alum
content from isolated rat liver cells after different treatments: ▲, RNase; □, DNase;
■, RNase + DNase; ×, PCA 10%, 4°C, 20 hr; ○, TCA 5%, 90°C, 15 min; ●, TCA 1%,
90°C, 15 min; + untreated cells [after Kiefer *et al.* (*9*)].

cyanine–chrome alum probably binds in a stoichiometric relationship to
this nucleic acid. This can be directly demonstrated by removing DNA
with DNase. In this case cells from another animal were used as those which
remained untreated. The nucleic acid content of these cells was obviously
higher, but the proportion of bound dye to RNA content is the same as in
untreated cells and in cells from which nucleic acids were partially removed
by acid extraction. We have calculated the regression coefficient for each
treatment group of cells and could not find a significant difference between
these coefficients. These investigations showed that the amount of gallocyanine–chrome alum bound by a cell is proportionate to its nucleic acid
content, be it DNA or RNA. This means that gallocyanine–chrome alum
can be used for the quantitative cytochemical determination of RNA.

If we extract all or nearly all nucleic acid from the cell, some UV absorption
still persists. If we now use the measuring values in Sandritter's formula,

a "negative" nucleic acid content results in some cases. This means that the accuracy of the method is limited, which is especially the case when we deal with low nucleic acid concentration. The second point is that the regression line does not go through the origin of the coordinate system. This means that there is a light loss in the visible region of the spectrum which cannot be attributed to nucleic acids. There are three reasons for this: the first factor is scattering within the stained cell and the second one is unspecific staining by gallocyanine–chrome alum and a third might be an error in nucleic acid calculation from UV data. It does not seem possible to differentiate between these three possibilities, but we can say that this maybe-error does not exceed 10% of the integrated extinction of a gallo-cyanine–chrome alum-stained liver cell.

III. Separate DNA and RNA Determination

Until now three different methods have been published for the separate evaluation of DNA and RNA within the same cell.

(1) In Caspersson's Institute total nucleic acid content is measured by UV absorption. Thereafter, DNA alone is determined in the same cell after Feulgen reaction. Knowing the relationship of intensity of Feulgen stain to UV absorption of DNA it is possible to calculate the fraction of UV absorption caused by RNA (2).

(2) The second method has been published by Arline Deitch (3). Total nucleic acid is measured after methylene blue staining, then the dye and RNA are removed by hydrochloric acid. After a second staining with methylene blue the amount of dye bound to DNA can be determined. Both methods (1) and (2) need two measurements with a staining step in between. Therefore, exact relocation of the cell is necessary.

(3) The third method does not involve this complication. It is the fluoro-metric determination of DNA and RNA after acridine orange staining. In this case two successive measurements are taken in the red and the green region of the emission spectrum (14).

The point at which we started our considerations was the fact that gallo-cyanine–chrome alum staining withstands elution with water and alcohol and is only very slowly removed from the stained tissue by treatment with acids. So we wondered whether it would be possible to combine a Feulgen reaction with a preceding gallocyanine–chrome alum stain. This turned out to be possible (11). In a cell stained in this way, nucleus and cytoplasm are stained with gallocyanine–chrome alum, while the nucleus shows the additional Feulgen stain. It is then possible to determine the amounts of both dyes by measuring at two different wavelengths where the absorbances of both dyes are sufficiently different from each other. In practice the absorption maxima of the two dyes should lie at sufficiently different wavelengths to allow for accuracy. The absorption maximum of gallocyanine–chrome alum is around 570 nm, which is close to the maximum of the often used para-rosaniline in the Schiff reagent. Therefore coriphosphine O was substituted for pararosaniline. This acridine dye can be used as a "Schiff-type" reagent (7). Its absorption maximum is around 450 nm as in many other acridine dyes, while it seems to be more resistant to photodecomposition than other dyes of this group (8). In a double-stained nucleus both peaks can be demonstrated. We used the wavelengths 480 and 550 nm for our measurements. At each of these wavelengths both dyes absorb, so we cannot take the absorption at 480 nm as a measure for the coriphosphine content, nor that at 550 nm as a measure for the gallocyanine–chrome alum content. One has to make some short calculations for correction.

The amount of coriphosphine expressed in arbitrary units $AU_{Cori\ 480}$ is therefore

$$AU_{Cori\ 480} = \frac{f_2 AU_{tot\ 550} - f_1 f_2 AU_{tot\ 480}}{1 - f_1 f_2}$$

$$f_1 = \frac{AU_{GC\ 550}}{AU_{GC\ 480}}$$

$$f_2 = \frac{AU_{Cori\ 480}}{AU_{Cori\ 550}}$$

while the gallocyanine content ($AU_{GC\ 550}$) is

$$AU_{GC\ 550} = AU_{tot\ 550} - AU_{Cori\ 480}/f_2$$

$AU_{tot\ 550}$ and $AU_{tot\ 480}$ mean the total integrated absorbances at 550 and 480 nm, respectively. Coriphosphine corresponds to the DNA content, while gallocyanine–chrome alum corresponds to total nucleic acid content. For calculating the RNA content it is necessary to know how much gallocyanine–chrome alum is bound to DNA. For this we prepared a second slide of cells treated with RNase and then stained together with the first slide. From this second preparation we obtained the amount of gallocyanine–chrome alum bound to a certain amount of DNA, which can be expressed as the cori-phosphine content:

$$AU_{GC,\ DNA,\ 550} = f_3\ AU_{Cori\ 480}$$

We therefore can calculate the RNA content expressed as $AU_{GC,\ RNA,\ 550}$

$$AU_{GC,\ RNA,\ 550} = AU_{GC\ 550} - f_3\ AU_{Cori\ 480}$$

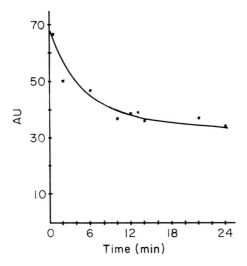

FIG. 2. Decrease of dye amount of gallocyanine–chrome alum-stained thymus nuclei during hydrolysis in 1 N NCl at 60°C.

By this method we determined the amount of DNA and RNA within the same cell by measuring a double-stained cell at two different wavelengths.

This is the theoretic basis of the method. It remains to be proved that both staining reactions behave stoichiometrically when used in the combined stain. Gallocyanine is partially lost from the tissue during the Feulgen hydrolysis (Fig. 2). It could be demonstrated that the decrease in gallocyanine–chrome alum is very similar in a number of cell types (Table III). This means that a proportionality between RNA content and gallocyanine–chrome alum after hydrolysis still exists. The question whether the dye is

TABLE III

GALLOCYANINE–CHROME ALUM CONTENT (AU_{550}) OF DNASE-EXTRACTED CELLS BEFORE AND AFTER FEULGEN HYDROLYSIS

	N	Before hydrolysis	After hydrolysis	% Decrease
Liver	50	26.3	11.7	55.52 ± 4.30
Ehrlich ascites	50	129.7	55.1	57.44 ± 3.57
L cells	50	366.5	167.1	54.41 ± 2.35
Leukocytes	50	23.6	11.6	50.85 ± 3.21

split off from the nucleic acid or whether the intact nucleic acid–dye complex is removed from the cell, is unsettled, but this is without any importance for the problem.

Further, experiments showed that gallocyanine–chrome alum-stained nuclei bind less Feulgen–coriphosphine O than unstained nuclei (Table IV). How-

TABLE IV

CORIPHOSPHINE CONTENT (AU_{480}) OF CELLS WITH AND WITHOUT PRECEDING GALLOCYANINE–
CHROME ALUM STAINING

	N	Without GC	N	With GC	With GC / Without GC
Liver					
Diploid	50	53.35 ± 1.92	50	20.69 ± 2.01	0.36
Tetraploid	50	107.14 ± 4.74	50	44.29 ± 2.68	0.41
Thymus					
Diploid	50	50.31 ± 2.36	50	21.90 ± 1.96	0.43
Tetraploid	50	101.31 ± 1.00	50	43.08 ± 2.83	0.42
Frog					
Erythrocytes	50	47.82 ± 3.37	50	21.58 ± 2.47	0.42
Chick					
Erythrocytes	50	21.23 ± 1.09	50	10.11 ± 2.06	0.47

ever, proportionality between DNA content and coriphosphine O content is maintained. These experiments demonstrate that it is possible to determine both DNA and RNA content of cells by use of the combined gallocyanine–chrome alum and Feulgen staining.

IV. Chemistry of the Gallocyanine–Chrome Alum Complex

Most recently, considerable progress has been made in the chemistry of the gallocyanine–chrome alum dye solution (6). Thin-layer chromatography and electrophoresis revealed a number of different compounds within the dye lake, depending on the duration of boiling and the proportion of gallo-cyanine to chrome alum in the preparation. It appears that the Einarson formula (4) and a boiling time of 10–20 min give optimal results. Under these conditions only one main component was present besides uncomplexed dye and chrome alum, while prolonged heating led to the appearance of a greater number of compounds. From preparative chemical experiments it was deduced that a single chromium atom is linked to two gallocyanine molecules. This is in contradiction to the suggestions of Einarson (*4*) and Harms (*5*) in referring to the composition of the gallocyanine–chrome alum complex. Unfortunately, we still do not know how gallocyanine–chrome alum is bound to nucleic acids. In this respect it is important to note that in stained tissue, chromium has been shown to have the same localization as nucleic acids and gallocyanine (*18*).

REFERENCES

1. Alfert, M., Cytochemische Untersuchungen an basischen Kernproteinen während der Gametenbildung, Befruchtung und Entwicklung. *In* "Chemie der Genetik," 9th Mosbacher Colloq., pp. 73–84, Springer, Berlin, 1959.
2. Caspersson, T., Forber, S., Foley, G. E., and Killander, D., Cytochemical observations on the nucleolus-ribosome system. *Exptl. Cell Res.* **32**, 529–552 (1963).
3. Deitch, A. D., A method for the cytophotometric estimation of nucleic acids using methylene blue. *J. Histochem.* **12**, 451–461 (1964).
4. Einarsson, L., On the theory of gallocyanine-chrome alum staining and its application for quantitative estimation of basophilia. A selective stain of exquisite progressivity. *Acta Pathol. Microbiol. Scand.* **28**, 82–102 (1951).
5. Harms, H., "Handbuch der Farbstoffe für die Mikroskopie." Staufenverlag, Kamp-Lintfort, 1957.
6. Horobin, R. W., and Murgatroyd, L. B., The composition and properties of gallocyanin-chrome alum stains. *Histochem J.* **1**, 36–54 (1968).
7. Kasten, K. F., The chemistry of Schiff's reagent. *Intern. Rev. Cytol.* **10**, 1–100 (1960).
8. Keeble, S. A., and Jay, R. F., Fluorescent staining for the differentiation of intracellular ribonucleic acid and deoxyribonucleic acid. *Nature* **193**, 695–696 (1962).
9. Kiefer, G., Kiefer, R., and Sandritter, W., Cytophotometric determination of nucleic acids in UV-light and after gallocyanine chromalum staining. *Exptl. Cell Res.* **45**, 247–249 (1967).
10. Kiefer, G., Kiefer, R., and Sandritter, W., Cytophotometric determination of the binding of basic dyes by DNA following acetylation. *Histochemie* **14**, 65–71 (1968).
11. Kiefer, G., Zeller, W., and Sandritter, W., Eine Methode zur histochemischen Bestimmung von RNS und DNS an der gleichen Zelle. *Histochemie*, **20**, 1–10 (1969).
12. Lison, L., Variation de la basophilie pendant la maturation de spermatozoide chez le rat et sa signification histochimique. *Acta Histochem.* **2**, 47–67 (1955).
13. Posalaky, Z., Kiefer, G., and Sandritter, W., Quantitative histochemische Untersuchungen über die säurefeste Anfärbung des Spermienkopfes. *Histochemie* **4**, 312–321 (1964).
14. Rigler, R., Jr., Microfluorometric characterization of intracellular nucleic acids and nucleoproteins by acridine orange. *Acta Physiol. Scand.* **67**, Suppl. 267, (1966).
15. Sandritter, W., Ultraviolettmikrospektrophotometrie. *In* "Handbuch der Histochemie" (W. Graumann and K. Neumann, eds.), Vol. 1, Part I, pp. 220–338. Fischer, Stuttgart, 1958.
16. Sandritter, W., Jobst, L., Rakow, L., and Bosselmann, K., Zur Kinetik der Feulgenreaktion bei verlängerter Hydrolysezeit. *Histochemie* **4**, 420–437 (1965).

17. Sandritter, W., Kiefer, G., and Rick, W., Gallocyanin chrome alum. *In* "Introduction to Quantitative Cytochemistry" (G. L. Wied, ed.), pp. 295–326. Academic Press, New York, 1966.
18. Sims, R. T., and Marshall, D. J., Location of nucleic acids by electron probe X-ray microanalysis. *Nature* **212**, 1359 (1966).
19. Wilkins, M. H. F., Molecular configuration of nucleic acids. *Science* **140**, 941–950 (1963).

ENVIRONMENTAL CONDITIONS FOR OPTIMAL FEULGEN HYDROLYSIS

Sally B. Fand

DEPARTMENT OF PATHOLOGY, WAYNE STATE UNIVERSITY SCHOOL OF MEDICINE AND VETERANS
ADMINISTRATION HOSPITAL, ALLEN PARK, MICHIGAN

We have been measuring DNA in whole nuclei of human pituitary imprints by light absorption in the Feulgen stain at 550–555 mμ on the commercial version of the Deeley integrating microdensitometer (3). Variable intensity levels of DNA coloration have been worrisome especially when one attempted to go from case to case. Of the possible number of factors which can be implicated in inconstancy of staining level, our first thoughts have always been concerned with machine variation and our suspicions have often been justified with the integrating microdensitometer. But it appeared to us that there was considerable virtue in excluding or minimizing errors produced by variables in the Feulgen procedure. A 1966 report of DeCosse and Aiello (2) seemed to suggest that 5 N HCl hydrolysis at room temperature offered broad and stable peak values of a slightly stronger coloration than the standard short 1 N acid hydrolysis at 60°C.

This report summarizes a series of experiments extending over the past 4 years in which we consecutively evaluated the intensity and characteristics of coloration for (a) the standard 60°C Feulgen hydrolysis vs. hydrolysis in 5 N HCl at room temperature; (b) 5 N hydrolysis at varying times and temperatures; (c) varying acid strength, and for selected temperatures and times. We followed this particular sequence of experiments because we noted inconstant results in different laboratories, made obvious with our move from New York to Michigan in 1967. More rigorous definition and control of room temperature hydrolysis was needed. In the course of this reappraisal it seemed clear that maximal and minimal acid strengths, as well as incubation times and incubation temperatures should be studied in detail. We then, somewhat arbitrarily, selected upper and lower bounds based on our earlier experience, choosing 5 N, 2 hr, and 60°C on the one hand and 0.5 N, 1 min, and 25°C on the other.

TABLE I

Analysis of Effects of Hydrolysis Time and Temperature on 5 N Feulgen Staining: Allocation of 98 New York Experiments (1966–1967)

Temperature of incubation (°C)			
4	25	37	60
Number of experiments: 18	26	34	20

Time of incubation (hr)							
0.1	0.1–0.2	0.25–0.5	0.6–1	1.5–3	4–6	16–24	26–42
Number of experiments: 23	10	15	15	13	8	8	6

TABLE II

Analysis of Effects of HCl Molarity and Hydrolysis Time and Temperature on Feulgen Staining: Allocation of 108 Michigan Experiments (1967–1968)

Molarity of HCl						
0.5	1.0	2.0	3.0	3.5	4.0	5.0
Number of experiments: 9	15	18	17	15	12	22

Temperature of incubation (°C)				
25	37	45	50	60
Number of experiments: 6	46	8	28	20

Time of incubation (min)					
10	20	30	45	60	120
Number of experiments: 2	59	36	4	6	1

Tables I and II depict the experimental points explored in the 206 runs which form the basis of Sections II, B and C. Results of our initial studies were presented in April, 1967, at the 18th Annual Meeting of the Histochemical Society in Chicago (Abstract 31), and published in the proceedings of that meeting (6). Some of the experiments described in Section II, B were included in a preprint circulated in July, 1967. The extension of these studies to select an optimal acid strength for a better defined temperature and a shorter time is presented here for the first time.

I. Methods

Imprints from the cut surface of freshly obtained human anterior pituitary glands were made directly on receiving the tissue from postmortem examination. The preparations, on long coverslips, were quick-frozen in isopentane over dry ice and then freeze-substituted with ethanol (16). When the substitution was complete they were transferred to alcoholic formalin for fixation prior to Feulgen staining. The time of fixation in the alcoholic formalin was kept at 16–24 hr.

Imprints from other tissue were occasionally used to compare our results with pituitary cells with results from tissues used by other workers; we have noticed neither a systematic variation for parenchymal cells of man nor any different response to the various hydrolysis conditions examined in this report.

For determination of the amount of Feulgen stain in pituitary nuclei we used the Barr and Stroud Integrating Microdensitometer with a tungsten light source; net absorption readings were made at 550 mμ; three nuclear readings and three background readings were made for each of the 25 to 100 nuclei comprising an experiment. The microdensitometer was used at an absorption range of 20%; extinction, 0.75; wedge setting 5 ± 1; lamp voltage 9.0.

II. Results

A. Standard Feulgen Hydrolysis vs. Hydrolysis in 5 *N* HCl at Room Temperature

Figure 1 shows mean DNA values for several runs of 1 *N* HCl hydrolysis at 60°C. A relatively sharp peak of values at 10–20 min is noted. In Fig. 2 data are presented on the same cases subjected to 5 *N* acid hydrolysis at room temperature. There is a broad time span at which maximum or near-maximum values are achieved. The 5 *N* room temperature hydrolysis

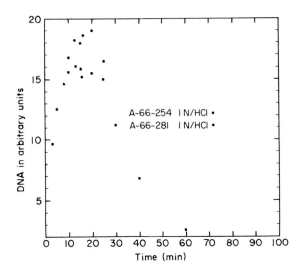

FIG. 1. 1 *N* HCl hydrolysis at 60°C, 1–60 min.

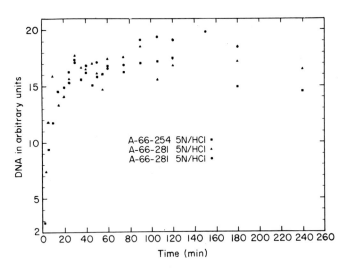

FIG. 2. 5 *N* HCl hydrolysis at 25°C, 1–240 min.

produces maximally stained nuclei over a much longer time space. Figures 3, 4, and 5 illustrate mean values, together with variances, for 1 *N* HCl hydrolysis at 60°C, and 5 *N* HCl hydrolysis at 25°C. The difference is in the order of $+13.5\%$ in favor of the 5 *N* with a range from $+5.2\%$ to $+17.5\%$. Calculations on the data given by DeCosse and Aiello (*2*) yield a difference of 11.9%.

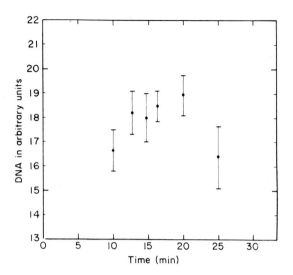

FIG. 3. 1 N HCl hydrolysis at 60°C, 10–25 min. Variance is delimited by vertical lines above and below each point.

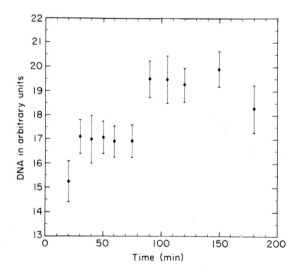

FIG. 4. 5 N HCl hydrolysis at 25°C, 20–180 min. Variance charted as in Fig. 3.

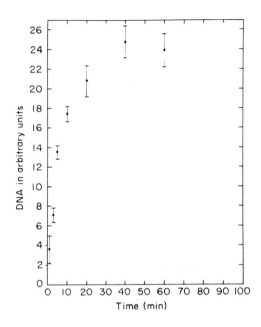

FIG. 5. 5 *N* HCl hydrolysis at 25°C, 1–60 min. Variance charted as in Fig. 3.

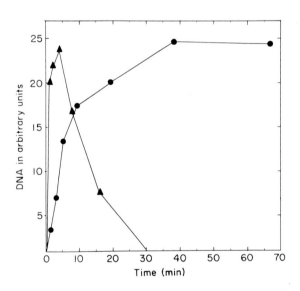

FIG. 6. Comparison of 5 *N* HCl hydrolysis over time at two different temperatures: ●, at 25°C; ▲, at 60°C.

B. Hydrolysis with 5 *N* HCl at Varying Time and Temperature

An interesting comparison of the effect of temperature on the 5 *N* acid hydrolysis curve is noticeable in Fig. 6. The sharp peak of 60°C 5 *N* hydrolysis compares unfavorably with the broad maximum in room temperature acid of the same normality, but the maximum value achieved is almost as high. Use of 60°C 5 *N* acid does not seem advantageous since there is no

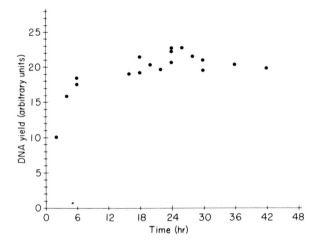

Fig. 7. 5 *N* HCl hydrolysis at 4°C, 2–42 hr.

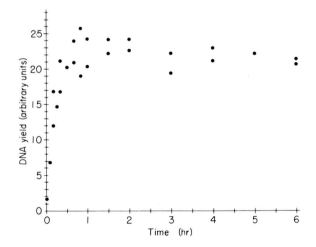

Fig. 8. 5 *N* HCl hydrolysis at 25°C, 0 6 hr.

gain in maximum absorption while there is, instead, a substantial shortening of time at which the maximum is maintained. Determinations of absorption curves for representative nuclei in these experiments have not revealed any wavelength shifts, although the 5 *N* HCl hydrolysis at room temperature tends to yield a wider maximum in a few instances from 535–575 mμ.

Figures 7 through 10 summarize the results obtained in 98 cases of 5 *N* HCl hydrolysis studied at four different temperatures, with times varied

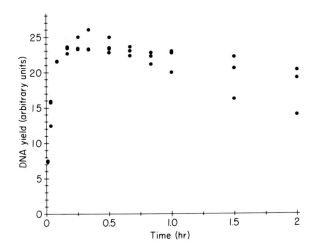

FIG. 9. 5 *N* HCl hydrolysis at 37°C, 0–2 hr.

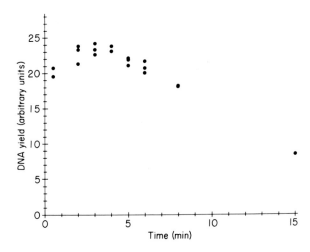

FIG. 10. 5 *N* HCl hydrolysis at 60°C, 1–15 min.

accordingly. There is a maximum absorption value for every temperature. In fact, there is little reason other than convenience for choosing among the various temperature and time conditions (with 5 *N* HCl) which combine to produce a maximum. Convenience, however, would clearly lead us to omit 4° and 60°C, the first because of the considerably prolonged time of hydrolysis required, the second because of the danger of working with 5 *N* HCl at 60°C and because short shifts in time would seriously attenuate the recorded DNA level.

A closer examination of the 25° and 37°C figures, however, suggests that we might gain somewhat more by introducing a third variable, namely the acid strength.

C. Optimal HCl Normality for Selected Temperatures and Times

Experiments described in this section represent about 100 additional studies covering a range of HCl normalities, with temperatures largely limited to three levels and hydrolysis time restricted to 15 or 20 min. Figure 11 summarizes results for 15-min hydrolysis; Fig. 12, for 20-min incubation. These results demonstrate clearly a maximum stain yield at 37°C and also suggest that an acid strength between 2 and 5 *N* might be superior. Figure 13 recapitulates the comparison between room temperature hydrolysis at two

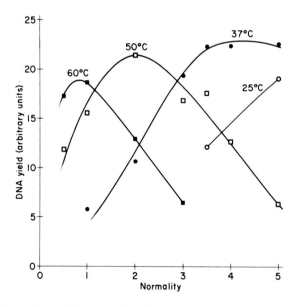

Fig. 11. Comparison of 15 min hydrolysis curves at 25°, 37°, 50°, and 60°C, 0.5 to 5 *N* HCl.

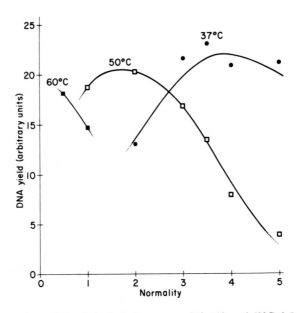

Fig. 12. Comparison of 20 min hydrolysis curves at 37°, 50°, and 60°C, 0.5 to 5 N HCl.

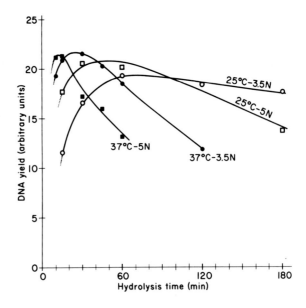

Fig. 13. Comparison of 25° vs. 37°C hydrolysis curves at 3.5 and 5 N HCl, 15–180 min.

acid levels (3.5 and 5 N) and 37°C runs at the same acid strengths. Here, too, the results seem to favor the 37°C runs.

Figure 14 demonstrates the widened range of temperatures with higher acid strengths where absorption values equal or exceed those obtained in the same case run under standard hot Feulgen hydrolysis conditions. It also demonstrates that room temperature hydrolysis renders generally low values.

III. Discussion

Cytochemists have been using the Feulgen reaction to demonstrate DNA for 44 years (7), and many variables, such as fixative used, type of Schiff

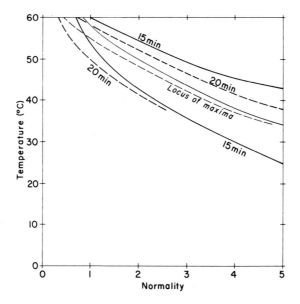

FIG. 14. Charts illustrating broadened range of permissible temperatures for maximum DNA yield with higher acid strengths at 15 and 20 min.

reagent, and organ and species effects have been investigated in considerable detail (5, 8, 10, 14). With but few dissonant voices (9), there has been surprising agreement about the use of hot 1 N HCl hydrolysis, and this is the method cited in almost all standard texts (1, 11, 13).

Increasingly, in the past several years, a number of laboratories have addressed themselves to the question of optimal conditions for the production of aldehydes, focusing on the hydrolytic segment of the Feulgen reaction

(2, 4, 6, 12, 15). These studies have been aided by the availability of accurate measuring devices and methods *(14)*.

Our underlying hypothesis in this study has been that the major direction of error in Feulgen hydrolysis is to decrease the intensity of staining. With this simple assumption we set out to evaluate and to begin to quantitate the effect of several environmental variables. The constraints in the selection of optimal hydrolysis conditions were not only that a maximum be obtained but that slippage off the point would have relatively small effects. Put another way, we wish to get the highest values under conditions that are least sensitive to small changes.

IV. Summary

Reproducible, maximum DNA coloration in the Feulgen reaction is provided by 3.5 N HCl hydrolysis for 15–20 min at 37°C. This procedure is recommended for general use.

REFERENCES

1. Barka, T., and Anderson, P. J., "Histochemistry: Theory, Practice, and Bibiliography," pp. 97–99. Harper, New York, 1963.
2. DeCosse, J. J., and Aiello, N., Feulgen hydrolysis: Effect of acid and temperature. *J. Histochem. Cytochem.* **14**, 601–604 (1966).
3. Deeley, E. M., An integrating microdensitometer for biological cells. *J. Sci. Instr.* **32**, 263–267 (1955).
4. Deitch, A. D., Wagner, D., and Richart, R. M., The effect of hydrolysis conditions and fixation on the intensity of the Feulgen reaction. *J. Histochem. Cytochem.* **14**, 779 (1967) (abstr.).
5. Deitch, A. D., Wagner, D., and Richart, R. M., Conditions influencing the intensity of the Feulgen reaction. *J. Histochem. Cytochem.* **16**, 371–379 (1968).
6. Fand, S. B., Genco, K. K., and Holland, R. H., Qualitative and quantitative utility of room temperature 5N HCl Feulgen hydrolysis. *J. Histochem. Cytochem.* **14**, 778 (1967) (abstr.).
7. Feulgen, R., and Rossenbeck, H., Mikroskopisch-chemischer Nachweis einer Nucleinsäure vom Typus der Thymonucleinsäure und die darauf beruhende Elektiv Färbung von Zellkernen in mikroskopischen Präparaten. *Z. Physiol. Chem.* **135**, 203–248 (1924).
8. Hale, A. J., The leucocyte as a possible exception to the theory of deoxyribonucleic acid constancy. *J. Pathol. Bacteriol.* **85**, 311 (1968).
9. Itikawa, O., and Ogura, Y., The Feulgen reaction after hydrolysis at room temperature. *Stain Technol.* **29**, 13–15 (1954).
10. Kasten, F. H., The chemistry of Schiff's reagent. *Intern. Rev. Cytol.* **10**, 1 (1960).
11. Lillie, R. D., "Histopathologic Technic and Practical Histochemistry," 3rd ed., pp. 148–150, Mc-Graw Hill, New York, 1965.
12. Mayall, B. H., Variability in the stoichiometry of deoxyribonucleic acid stains. *J. Histochem. Cytochem.* **14**, 762 (1967) (abstr.).

13. McManus, J. F. A., and Mowry, R. W., "Staining Methods: Histologic and Histochemical," pp. 75–76. Harper, New York, 1963 (reprint).

14. Mendelsohn, M. L., Absorption cytophotometry: Comparative methodology for heterogeneous objects, and the two-wavelength method. *In* "Introduction to Quantitative Cytochemistry" (G. L. Weid, ed.), pp. 202–213, Academic Press, New York, 1966.

15. Murgatroyd, L. B., A quantitative investigation into the effect of fixation, temperature and acid strength upon the Feulgen reaction. *J. Roy. Microscop. Soc.* [3] **88**, 133–139 (1967).

16. Simpson, W. L., An experimental analysis of the Altmann technic of freezing-drying. *Anat. Record* **80**, 173 (1941).

POTENTIALITIES OF CELLULOSE AND POLYACRYLAMIDE FILMS AS VEHICLES IN QUANTITATIVE CYTOCHEMICAL INVESTIGATIONS ON MODEL SUBSTANCES

P. van Duijn and M. van der Ploeg

HISTOCHEMICAL LABORATORY OF THE DEPARTMENT OF PATHOLOGY, UNIVERSITY OF LEIDEN, THE NETHERLANDS

I. Introduction

Quantitative cytochemical determination of cellular compounds rests on two prerequisites. Specific staining procedures with reproducible and, if possible, known stoichiometry are necessary, and in addition, microspectrophotometers should be available for the accurate measurement of the quantity of chromophores formed in the biologic structure as a result of the staining procedure.

Considerable progress has already been made in the field of cytophotometry, but the chemistry of most staining procedures is still only superficially known. In the early days of quantitative cytochemistry, disappointing test tube results with the Feulgen reaction induced Caspersson (9) to concentrate on the natural absorbance of tissues. This approach, though it avoids the need of staining methods, has its limitations. It necessitates operation in the ultraviolet region where stray light errors are difficult to overcome and where specificity is limited to nucleic acids and proteins in general. Enzymic extraction procedures performed to achieve greater specificity, may interfere in an unknown way with the stray light conditions. Therefore, to exploit the full scope of cytochemistry, reliable staining reactions are of paramount importance. For the development of specific cytochemical staining procedures, sufficient knowledge of the underlying chemistry is essential. Since biologic material is of a complex and partly unknown composition, it is difficult to elucidate the chemical nature of the staining

223

reactions by investigating exclusively biologic objects. Blocking reactions performed on these objects often lead to conflicting results. This is to be expected since in most cases the specificity of blocking reagents is inferred from organic chemistry without considering the special cytochemical conditions under which blocking occurs. Often, independent proof of the proposed stoichiometry is also lacking. The prospects for elucidating the chemistry of a staining method are improved when the reaction is studied on pure compounds. Test tube experiments, however, are not always relevant since there are several factors in a cytochemical system differing from those in a homogeneous biochemical one. During cytochemical staining, for instance, the moving of slides from one bath to another is a matter of seconds. In a homogeneous system, the equivalent procedures are precipitation or dialysis, which take much more time. Furthermore, the compound to be stained in the biologic object is part of a cellular structure and thus can be present in a specialized steric configuration.

From these considerations, it follows that for the development of reliable quantitative cytochemical staining reactions, it is indispensable to use models which not only contain the investigated substance in a purified form but also permit its investigation under conditions which approximate those present when sections or smears are stained.

In a series of investigations initiated some years ago, we have attempted to find a versatile procedure which satisfied these conditions and at the same time allowed independent enzymic or chemical analysis of the compound to be stained by macrochemical methods. A system of general applicability was developed using macrofilms of cellulose or polyacrylamide as vehicle for the substance to be investigated, in combination with a specially developed film colorimeter.

Other approaches also have been advocated. Both microdroplets obtained by homogenizing or spraying procedures (22, 41, 64, 65) as well as microfilms smeared onto slides or cut from fixed blocks of material (16, 63, 66, 74) have been used as models. For quantitative studies these types of models have some drawbacks. There is always the danger that part of the compound leaches away from the model during preparation, fixation, or staining. Furthermore, mass determinations of microdroplets by interference microscopy, and thickness measurements of microfilms, do not easily reach the accuracy of macromethods. For the macrofilms to be described in this chapter, possible leaching or chemical change of the compound is traced relatively easily, since aliquots of the films can be analyzed with chemical procedures during any stage of assay. The problem of accurate determination of absorbance in relation to mass was solved by weighing the part of film measured in the colorimeter.

Several modifications of the approach provided a means of studying the stoichiometry and specificity for a great range of staining reactions. Macromolecules can be trapped into polyacrylamide with retention of their original structure. Low molecular weight compounds were covalently linked to cellulose. In this way, problems of uncontrolled loss, encountered in models where these highly diffusable compounds are absorbed into filter paper, could be avoided. The films possess a good mechanical strength and can be handled easily without the danger of damaging them. The success of the carrier film model, in our experience, is due to the fact that this type of film combines two indispensable properties. The incorporated compounds can be investigated cytochemically and, on the other hand, the films can be assayed independently following analytic procedures.

II. Preparation of Films

A. CELLULOSE FILMS

Although other materials can be used, in principle, up till now only cellulose and polyacrylamide have been applied as materials for cytochemical model films. Cellulose films can be prepared from an alkaline solution of cellulose xanthogenate (viscose) with the aid of acid-containing "developing" media. Before development, the viscose is spread with a chromium-plated steel cylinder into a thin layer (0.1–0.5 mm) over a glass plate (57); the distance between cylinder and glass plate and the velocity of the cylinder determine the thickness of the films (30–150 μ). Variations in the ultra-structure of the films (e.g., internal pore width and transparency), can be produced by changing the viscose preparation procedure, by diluting the viscose, by drying the viscose spread before development, or by changing the composition of the developing medium. Films of good transparency and favorable mechanical properties can be obtained by developing the viscose in a mixture of 1 volume of formic acid (99%) and 3 volumes of methanol at $-20°C$ (57).

Two types of cellulose films have been used for cytochemical model studies of cellular components. When a macromolecular compound is to be tested, it is mixed with the viscose before it is spread and developed. Due to the alkalinity of the viscose only compounds which are stable at high pH can be incorporated in this way. So far, this approach has been used only for DNA. Feulgen staining of the films revealed the presence of small DNA-containing vacuoles lying within the cellulose matrix. Apparently, the DNA solution and the viscose, despite vigorous agitation, did not mix homogeneously but formed a coacervate of DNA-containing spheres in the viscose [see Albertson (2)]. For this and other reasons, in later studies on nucleic acids the cellulose was replaced by polyacrylamide.*

Cellulose films on the other hand are very well suited for the cytochemical study of low molecular compounds. For instance, staining reactions of amino acid side groups of proteins and the bases of nucleic acids can be studied in this way. The substance to be stained is chemically linked to the cellulose molecules in the film (after development). This method has been used to bind the $-O-CH_2CH_2-NH_2$ group as well as amino acids (1) and nucleotides (30) to hydroxyl groups in the cellulose molecules.†

These cellulose films are preferable to models consisting of filter paper impregnated with the dissolved test material. The filter paper models, which in some cases have been reported to be used in semiquantitative studies, for quantitative work have the drawback that unknown amounts of compound may be lost during fixation or staining. The cytochemical results obtained with cellulose model films containing covalently bound groups can be correlated with the results of titrimetric or spectroscopic procedures in aliquots of the films. The possibility of relating the results of cytochemical staining to other analytic procedures makes these cellulose films excellent models for quantitative investigations.

In principle, the chemical coupling method could also be applied to macromolecules, as several procedures permit the linking of proteins to cellulose (52) or to other carbohydrates such as Agarose (61) or Sephadex (40).

B. POLYACRYLAMIDE FILMS

The second type of film which is of general applicability in model studies consists of polyacrylamide (18, 58, 59). The films are prepared from mixtures of a monomer solution containing acrylamide and N,N'-methylenebis(acrylamide) mixed at a neutral pH with solutions of the macromolecular compounds to be investigated. Polymerization, initiated by persulfate, is allowed to proceed between glass plates, resulting in the formation of films of poly-

---

* In a series of biochemical investigations, Goldman et al. (26) used cellulose nitrate (collodion) films as a support for the enzyme papain. A cellulose nitrate film was impregnated with a solution of the enzyme which was immobilized in the outer parts of the films by extensive cross-linking with bis(diazo)benzidinedisulfonic acid. The membrane-coated microcapsules developed by Chang (10) have not yet been utilized for cytochemical model studies either.

† The synthesis of SH-containing cellulose films as described by Hardonk and van Duijn (30) in later investigations could not be reproduced. The reported positivity of the acrolein–Schiff reaction, which was considered to indicate a successful synthesis, probably was due to insufficient washing out of the acrolein. SH-Containing films can be prepared by coupling cysteine to cellulose as reported by Ahsmann and van Duijn (1), and reducing the S—S bridges.

acrylamide gels. The pore size of the films can be varied (*15, 57*) and is chosen in such a way that macromolecules such as nucleic acids, proteins, and carbohydrates become immobilized in the gel matrix, while low molecular compounds can diffuse in and out more or less freely. When polymerization is performed near 0°C, it only induces a slight degree of inactivation in the enzymes so far studied. This procedure results in excellent cytochemical models which offer a range of applications. In studies of enzymes, for instance, not only can purified enzymes or isoenzymes be incorporated, but also intact cellular organelles or sonicated cells. The influence of the simultaneous presence of added nonenzymic compounds can also be studied. Biochemical applications of enzymes trapped in polyacrylamide gels were described independently by Bernfeld and Wan (*6*) and Hicks and Updike (*34*).

III. Instrument Requirements for Quantitative Measurements of Films

The absorbance of stained films can be measured quantitatively on a film colorimeter which has been developed by modification of a Zeiss spectro-photometer PMQ II (*19, 59*). The main part of the modification is a film holder consisting of two metal or Perspex plates (Fig. 1), which are screwed together and can be moved vertically in the photometer cell in such a way that alternatively one or another of two apertures can be brought to exactly the same position in the light beam. The film is placed between the plates, and covers the lower aperture; the upper aperture is used as a reference. A similar system can be mounted on the cover plate of the photometer cell compartment of other spectrophotometers as described for the Unicam SP 500 (*58*).

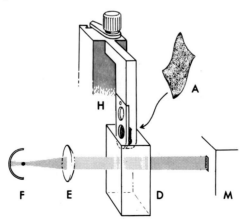

FIG. 1. Holder for films: A, piece of film; H, film holder; M, monochromator; D, photometer cell; E, achromatic focusing lens; F, photomultiplier.

The absorbance at a certain wavelength will depend not only on the concentration of the chromophore but also on film thickness, and must therefore be referred to the weight of film present in the light beam during measurements. For this purpose, the piece of film which filled the measuring aperture is punched out (Fig. 15) after the absorbance measurement, dried and weighed (see Appendix). When the concentration of the investigated compound is known—or can be analyzed (e.g., phosphorus in DNA-containing films)—it is possible to relate absorbance to the amount of compound that was stained.

When measuring films with high absorbances, the accuracy of the measurement can be improved by using a reference aperture with a smaller diameter (Fig. 14, r_2). In this way the absorbance during measurements is reduced with the artificially created negative absorbance for the measuring aperture. The negative absorbance depends on the size difference between r_2 and r_1, and can be measured accurately for all wavelengths used. Provided that the spectral purity of the spectrophotometer allows it, absorbances as high as 2.5 can be determined accurately with this device.

To minimize light scattering, cellulose films are transferred to xylene, and polyacrylamide films are dehydrated through graded mixtures of glycerol and ethanol and also transferred to xylene or to benzyl alcohol. Benzyl alcohol ($n_D^{20} = 1.540$, b.p., 206°C) must be removed by more volatile media such as ethanol or water before the films are dried. Transfers must, of course, always involve miscible fluids (see Appendix).

Polyacrylamide films transferred to ethanol shrink to about half of their size in water; there is less shrinkage, however, in glycerol. When polyacrylamide films containing water are placed in aqueous mixtures containing more than 80% (v/v) ethanol, they become opaque as water is trapped in the polyacrylamide. For that reason glycerol–alcohol mixtures have to be used to eliminate the last traces of water when these films are to be transferred to benzyl alcohol or xylene. This transfer is sometimes necessary, even when the films show only small scattering in water, because theoretically the values for the molar absorptivity [in older nomenclature: molar extinction coefficient (38)] for a dye complex in water can be different from those of nonpolar media used for mounting sections or smears and it may be desirable to study the stained films in both polar and nonpolar solvents. Polyacrylamide films when measured in water in general show little scattering, but when whole cell sonicates are incorporated, scattering does occur. This phenomenon has also been observed by biochemists studying the spectra of pigments present in cells [e.g., in erythrocytes (72) or in chloroplasts (3)] in aqueous suspensions. Scattering has been given considerable attention in densitometry of photographic plates where silver grains are embedded in a gelatin matrix (79). True absorbances, to a great extent free of the inherent scatter of the film, can be measured by inserting a piece of opal glass (Fig. 14, O) directly behind the film in the photometer cell. For an explanation of this apparently paradoxic effect see Tupper and Weaver (*79*) and Shibata (*72*) and for the ultraviolet part of the spectrum Amesz *et al.* (*3*). In densitometric practice, the true densitometric value corrected for scatter is called "totally diffuse density." Large variation in thickness over the measured area of film theoretically can cause somewhat lower absorbance readings due to distributional error. Such errors can become significant only when holes are present in the measured area of the film. In that case large errors can arise if the measuring aperture is not homogeneously illuminated or if the photomultiplier does not show a uniform response to light passing through the apertures. Procedures for testing illumination and photomultiplier response by moving a small aperture through the measuring field have already been described (*19*, *59*). When, however, extreme variation in thickness and holes in the parts of film to be measured are avoided, strict adherence to these conditions is not necessary. In that case, even when only part of the area is actually measured, its absorbance is representative for the whole film present in the measuring aperture.

IV. Experimental Results with Model Films

Following the approach outlined above, only a limited number of staining reactions have been studied in some detail so far. The first object of study has been the Feulgen procedure for DNA. Experience has also been gained with staining reactions for amino groups, protein side-chain groups and with the PAS procedure. Recently the approach was extended to enzymes by studying the kinetics of alkaline phosphatase constrained in a polyacrylamide film. Encouraging results, which are also of importance for electron·microscopic cytochemistry, have been obtained in the study of the Gomori technique for acid phosphatase.

A. THE FEULGEN PROCEDURE

The initial investigations were carried out with DNA incorporated in cellulose films. In aliquots of the films, DNA-phosphorus was analyzed before and after staining (*57*). A linear relation was found between absorbance and the amount of DNA present *after* the staining procedure (Fig. 2), both with the classic pararosaniline-SO_2 Schiff reagent (*28*) and with the thionine-SO_2 reagent (*17*).* The correlation between absorbance and *initial* DNA content of the films is less perfect, due to the fact that a somewhat

---

* The Schiff reagent must be filtered with care to avoid that even small particles of activated carbon contaminate the nuclei and interfere with the precise measurement of spectra.

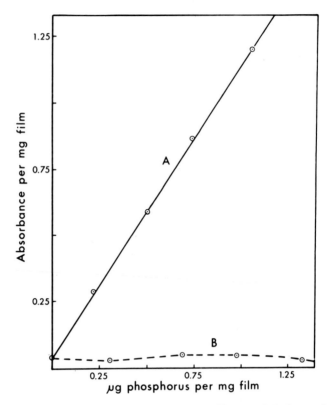

FIG. 2. The relationship between absorbance per milligram of Feulgen-stained DNA-containing cellulose films and the phosphorus content per milligram of stained film. Redrawn from Persijn and van Duijn (57). A, measurements. B, extracted film (blank).

variable percentage of DNA (20–30%) is lost during hydrolysis in 1 *N* hydrochloric acid at 60°C. When the Feulgen procedure is performed on fixed nuclei in smears or tissue sections, a certain amount of DNA is probably also lost, although reliable data concerning this effect do not exist. Comparison of the absorbance spectra of DNA-containing cellulose films and cell nuclei subjected to the same staining procedure, revealed differences in the shape of the curve and in the wavelength of maximum absorbance (Fig. 3). The spectrum of the stained DNA-cellulose films on the other hand was identical to that of a Feulgen-stained solution of DNA. In search of a better model for this reaction, the cellulose film was abandoned. It was found that the incorporation of deoxyribonucleoprotein (DNP) together with 5% bovine albumin in polyacrylamide resulted in a film which, when subjected to the Feulgen procedure, showed a spectrum nearly identical to that of nuclei, with an absorbance maximum at 565 nm. Polyacrylamide can be prepared

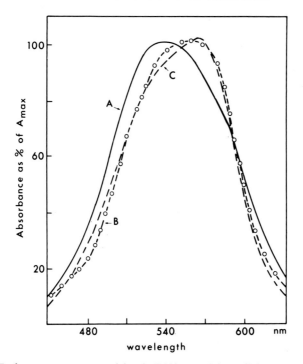

FIG. 3. Feulgen spectra measured in: A, DNA-containing cellulose films; B, (DNP + bovine albumin)-containing polyacrylamide films; C, chicken erythrocyte nuclei. Redrawn from Hardonk and van Duijn (33).*

at a neutral pH; in this way the alkalinity of viscose and its possible denaturating effect on DNP can be avoided. Another advantage of this model is the homogeneous mixing of DNP throughout the supporting film. The incorporation of extra protein proved necessary in order to obtain a "cell-like" model, since the Feulgen absorbance spectra of polyacrylamide films containing only DNA or DNP, showed maxima at 545–550 nm (Table I).

Attempts were made to explain the differences in the spectra obtained with the different models (31). Since it was realized that the steric position of the aldehyde groups (which originate in DNA during hydrolysis) might

* In Persijn and van Duijn (57) and Hardonk and van Duijn (33) the spectra of DNA-cellulose have been compared with spectra measurements in large areas of Feulgen-stained sections mounted in the film holder. Although it was realized that the spectra showed depression of the higher absorbances (19), the position of the absorbance maximum was considered to be reliable. This conclusion was disproved by later cytophotometric measurements on individual chromatin particles in nuclei. Unequal effects of the distributional error in different parts of the stained thymus sections probably are responsible for the shift of the position of the absorbance maximum.

TABLE I

ABSORBANCE MAXIMA (nm) OF FEULGEN-STAINED MODEL FILMS AND NUCLEI[a]

	A	B
DNA solution	545	—
DNA–cellulose	545	545
DNA–polyacrylamide	545	—
DNP–polyacrylamide	545–550	—
(DNA + bovine albumin)–polyacrylamide	560	—
(DNP + bovine albumin)–polyacrylamide	565	—
Deoxyadenylic acid–cellulose	562	581
Deoxyguanylic acid–cellulose	562	581
Liver nuclei of rat	565	—
Kidney nuclei of rat	565	—
Sperm nuclei of bull	565	—

[a] The maxima in column B were measured after posttreatment with a formaldehyde–sulfite mixture. Data from Hardonk and van Duijn (33).

be important, films were prepared by coupling phosphoric acid groups of purine-nucleotides to the hydroxyl groups of cellulose. These films, as expected, could be stained with the Feulgen procedure. The spectra of the resulting pararosaniline–nucleotide–cellulose complex were quite similar to those of stained cell nuclei and DNP–albumin films. As an explanation of differences between Feulgen spectra of DNA–cellulose and (DNP + albumin)–polyacrylamide, it was postulated that because of the greater mobility of the apurinic acid coil, all three amino groups of the pararosaniline molecule did react in the first model (as in DNA solutions). In the case of (DNP + albumin)–polyacrylamide film, cell nuclei, and deoxyribonucleotide–cellulose, however, the more rigid position of the aldehyde groups apparently prevents this. This supposition is strengthened by the observation that treatment with formaldehyde–sulfite after staining does not change the spectra of DNA–cellulose, but causes a shift from 562 to 581 nm for other models (Table I).

In further studies the actual number of pararosaniline molecules bound per nucleotide unit was determined for several models. For this purpose the dye was extracted with hot perchloric acid according to the method of Hiraoka (35). Using a calibration curve obtained with weighed amounts of pararosaniline, the actual number of bound dye molecules was calculated, the number of nucleotides being derived from phosphorus analysis on aliquot parts of the films. The resulting pararosaniline: nucleotide ratio for deoxyadenylic acid–cellulose was 1:15, for deoxyguanylic acid–cellulose 1:19, for DNA–cellulose 1:11, and for (DNP + albumin)–polyacrylamide 1:10 (32). If all aldehyde groups had reacted, this ratio theoretically should have been 1:1 for the first two models, and 1:2 for the others, since only the purine nucleotides are hydrolized to form free deoxyribose aldehyde groups. The reason that only a small part of the theoretically possible number of aldehyde groups combine with pararosaniline during the Schiff reaction, is at present still obscure. Perhaps more knowledge of the complicated reaction of pararosaniline and aldehyde can be obtained by using dyes with fewer free amino groups or by further study of the reactions of solitary nucleotides bound to cellulose compared with those of apurinic acid in which neighboring aldehyde groups are present.

A much debated and practically important question is to what extent protein influences the stoichiometry of the Feulgen reactions.

Caspersson (9) and recently Ishida (39) questioned the validity of the Feulgen procedure for quantitative cytochemical determinations, since model studies with mixtures of DNA and protein in solution showed a significant influence of the protein on the staining intensity. It is doubtful, however, whether the results of these test tube experiments have relevance to cytochemical situations.

In the test tube the colored reaction product of pararosaniline is measured in the presence of a great excess of the colorless components of the Schiff reagent. Under cytochemical conditions the reaction product successively comes into contact with a sulfite rinse and distilled water. This may cause the recoloration of bound molecules which do not recolor in the Schiff reagent itself, meaning that the effects of proteins demonstrated in test tube experiments do not necessarily invalidate the use of the Feulgen procedure for cytochemical purposes. The results obtained with DNA- and (DNP + albumin)-containing films, point to the fact that apurinic acid trapped in protein structures has an absorbance spectrum different from that of apurinic acid freely movable. However, a strong influence of the presence of protein on the stoichiometry of the Feulgen reactions has not been found (Table II). These results with models, however, do not exclude the possibility that the degree of coiling of the chromatin in the cell has some influence on the

TABLE II

INFLUENCE OF PROTEIN ON STOICHIOMETRY OF FEULGEN REACTION[a]

Type of film	$A_{max}/\mu g$ P	Molar ratio of P/pararosaniline
DNA–cellulose	0.95 ± 0.01	9.6 ± 0.2
(DNP + albumin)–polyacrylamide	1.23 ± 0.02	9.6 ± 0.3

[a] Tabulated are the averages of six determinations $\pm$ S.E. Data from Hardonk and van Duijn (32).

stoichiometric relations. Small differences could be caused by varying steric positions of the deoxyribose aldehyde groups after hydrolysis of DNA, in compact or less compact chromatin, to different permeability gradient factors (*24, 47, 51*), or to differences in optimal hydrolysis conditions for different objects.

B. The Periodic Acid–Schiff Procedure

Investigation of the periodic acid–Schiff (PAS) procedure has been carried out using cellulose or glycogen-containing polyacrylamide films. A linear relation has been found between the amount of glycogen incorporated in the film and the absorbance per milligram of film (Fig. 4) (*60*). A significant influence of different reaction conditions on the resulting absorbance curve was found. This aspect will have to be investigated further.

The relation between periodic acid consumption and amount of para-rosaniline (molecules) bound during the staining reaction was also studied. Assuming the formation of two aldehyde groups by one molecule of HIO_4,

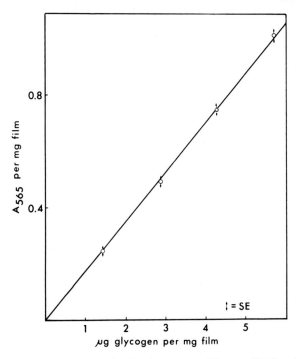

Fig. 4. The relationship between the absorbance per milligram of PAS-stained glycogen-containing polyacrylamide films, and the amount of glycogen incorporated per milligram of film. Plotted are the averages of six values ± SE.

we find that stained glycogen in polyacrylamide exhibits a pararosaniline–aldehyde ratio of 1:10. This deviation from what is theoretically expected is similar to that observed for the Feulgen reaction.

A similar experiment on cellulose films revealed a pararosaniline–aldehyde ratio of 1:200 (20). There is no obvious reason to account for this difference in behavior of aldehyde groups formed in the glucose units of glycogen to similar units in the cellulose molecules. In the latter, additional structural factors may be involved. Cellulose molecules in films show birefringence and are partly present as crystallites (46, 50). It seems possible that HIO_4 can penetrate and react with cellulose molecules in the crystalline part of the films, while the larger pararosaniline molecules can not. This example illustrates that conclusions from model studies should be drawn with caution and that the combination of an inert film into which a reactive substance is incorporated, is preferable to a system in which the model compound is at the same time carrier substance.

C. Studies on Models Consisting of Simple Organic Molecules Linked to Cellulose

In our opinion this type of model, although not yet studied in great detail, promises to resolve a number of controversies about the specificity of staining and blocking reactions of, e.g., side groups in proteins and nucleic acids. Attempts to develop reliable staining procedures of known specificity and stoichiometry in this field are still in their infancy.

The proposed mechanisms for many staining reactions have been " demonstrated" with blocking reactions, the specificity of which is cited from organic chemistry. However, the results in homogeneous organic chemical systems, although in some cases providing valuable indications, supply insufficient evidence for staining mechanisms in a heterogeneous cytochemical system. Equally, blocking reactions carried out on biologic objects are of little help, since the biologic object is chemically very complex and involves many poorly understood variables. Studies with these reactions on cells and tissues seldom lead to unequivocal results.

Quantitative investigations on models consisting of simple organic groups bound to a cellulose support, in principle, can yield the necessary information.

1. Aminocellulose as a Model

As an illustration, some results with a model system of aminocellulose are reported here. Amino groups can be bound to cellulose by treating cellulose films with 2-aminoethylsulfuric acid at 100°C in strong alkali (30). The reaction proceeds as follows:

$$\text{Cellulose-OH} + \text{HO}_3\text{SO-CH}_2\text{CH}_2\text{-NH}_2 \rightarrow \text{Cellulose-O-CH}_2\text{CH}_2\text{-NH}_2$$

The amount of amino groups bound per weight of cellulose can be varied by changing the reaction time or the amount of aminoethylsulfuric acid. Films treated with hot alkali in the absence of aminoethylsulfuric acid serve as controls. The aminocellulose films can be studied by several analytic methods. Theoretically, titration of the number of amino groups present per milligram of dry film is possible, since the stoichiometric relation for acid–base titrations is 1:1. Titration of structurally bound amino groups, however, can only be performed in the presence of a neutral salt in order to suppress the Donnan effect (67). In addition, the amino groups are present together with carboxyl groups formed by autoxidation of cellulose molecules. Thus some of the amino groups produce zwitterions such as are present in proteins. Because of these complications, the direct titration has not been found to be completely reliable. It is possible, however, to determine the amount of bound amino groups by their capacity to react with picric acid. This procedure, used in cellulose chemistry as a part of a method to determine aldehyde groups (29), in principle also gives a stoichiometric relation of 1:1. The amount of picric acid bound to the film is determined colorimetrically after extraction (1). As in the case of pararosaniline extracted from Schiff-stained models, picric acid binding therefore can be expressed in molar amounts per milligram of film. Aliquots of these aminocellulose films can be stained with one of the cytochemical staining procedures for amino groups. Experiments were carried out with the dinitrofluorobenzene

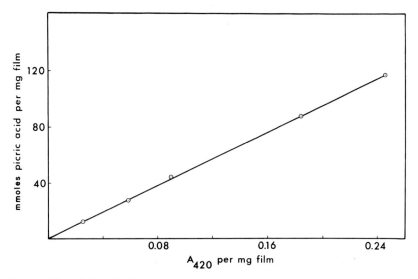

FIG. 5. The relationship between the amount of picric acid reacting per milligram of aminocellulose film and the absorbance per milligram of film after reaction with the DNFB reagent. Data from van Dalen et al. (13).

(DNFB) reagent (*44, 70*) on films containing different numbers of amino groups. A good correlation was obtained when the absorbances per milligram of treated film were plotted against the molar quantities of amino groups (Fig. 5). The slopes of curves as presented in Fig. 5 can be used to calculate the molar absorptivity of the cytochemically bound dinitrophenylamino groups, provided the stoichiometric relation is known. To calculate the stoichiometry of the reaction, however, the molar absorptivity must be known. This fundamental difficulty is inherent in the quantitation in absolute units of any cytochemical staining reaction for which the reaction product cannot be extracted.

From the organic chemistry literature it is known that the molar absorptivity of the dye derivative in most cases will differ significantly from the value calculated for the original reactant. It is, however, generally not possible to isolate and purify cytochemical end products for the determination of their molecular weight and molar absorptivity, without changing that absorptivity.

2. *Determination of the Molar Absorptivity of Cytochemical Reaction Products*

A procedure has been developed to solve this problem, which can be illustrated for the reaction of DNFB with amino groups. Films of different

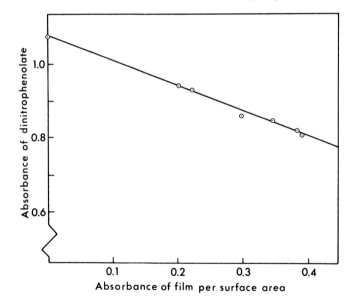

FIG. 6. The relationship between the absorbance at 360 nm of dinitrophenolate obtained by hydrolysis of the DNFB-containing media, and the absorbances per square millimeter at 355 nm of the total pieces of aminocellulose film after reaction with the DNFB reagent. Data from van Dalen *et al.* (13).

amino content were reacted with a known quantity of DNFB reagent. In order to determine the amount of molecules that have reacted with the pieces of film, the decrease in the concentration of DNFB reagent is measured colorimetrically* (Fig. 6). The absorbance of the reacted film is measured with the film colorimeter, after which the absorbance of the film per surface area can be calculated for the total piece of film. By dividing this value by the number of molecules that have reacted with that piece of film, the molar absorptivity per surface area for the cytochemically bound dinitrophenyl-amino group can be calculated without having isolated it. At the wavelength of 420 nm a value of 5900 was obtained (1). Knowing this molar absorptivity, the measured absorbances of the films can be expressed in molecules of DNFB bound by a given film. The latter proved to be nearly identical to the number of amino groups that reacted with picric acid in aliquots of the same film. From these observations it was concluded in both cases that the stoichiometry of the reaction was about 1:1 (13).

When aminocellulose films are subjected to the ninhydrin–Schiff procedure (62), about 70% of the amino groups are converted to aldehyde groups. It is of interest to note that here again only 1 in 10 to 15 aldehyde groups reacts with pararosaniline.

Recently, a new model system was developed by linking the amino group of amino acids covalently to carboxyl groups of oxidized cellulose films (1). Detailed quantitative investigation of staining and blocking reactions can be performed on this type of model. The results, supplemented with studies on pure protein incorporated into polyacrylamide films, could in our opinion provide a much stronger basis for quantitative protein cytochemistry.

D. ENZYME-CONTAINING POLYACRYLAMIDE FILMS

Few attempts to quantitate enzyme activity by cytophotometry of stained individual cells have been published [for reviews, see Glenner (25) and Shugar (73)]. Felgenhauer and Glenner (21) studied the potentialities for quantitation of an azo dye method for aminopeptidase activity in kidney sections. Compared with enzyme kinetics in a homogeneous biochemical system, complications could arise when diffusion gradients of reagents influence the reaction velocity in cells. The theoretic requirements for such a quantitation have been discussed by Holt and O'Sullivan (36, 56).

---

* It should be noted that DNFB will slowly be hydrolyzed when kept in alkaline solution. Care was taken that all remaining DNFB was converted totally into the well-defined dinitrophenolate before the colorimetric measurements. It was unnecessary to know the molar absorptivity of dinitrophenol if the percent decrease was determined and the molar concentration of the original DNFB reagent was known. For this reason chemically pure DNFB was used.

The problem of slow diffusion can be studied quantitatively on model films. The initial attempt to utilize polyacrylamide films as models for cytochemical enzyme reactions was made in a study of the properties of 5,6-dihydroxyindole as a new substrate for peroxidases (58). Lactoperoxidase and horseradish peroxidase can be incorporated into polyacrylamide films with retention of their activity and could be studied cytochemically. The results indicated that 5,6-dihydroxyindole is an excellent peroxidase substrate for light and electron microscopy, because of sharp localization and high electron density of the amorphous end product of the enzyme reaction (12). Some model film investigations were made to find the conditions for an optimal reaction procedure (Table III).

TABLE III

CYTOCHEMICAL ALKALINE PHOSPHATASE ACTIVITY IN POLYACRYLAMIDE FILMS EXPRESSED AS A_{575} PER MILLIGRAM OF FILM OF 0.11-MM THICKNESS[a]

Films containing purified alkaline phosphatase per ml monomer mixture (mg)	Incubation media containing (mg per ml)			
	Naphthol AS-MX phosphate	Variamine Blue RT (mg)		
		2	3	4
1.1	1	0.46	0.46	0.48
1.1	2	0.54	0.60	0.61
1.1	3	0.62	0.62	0.64
2.2	1	0.80	0.84	0.96
2.2	2	1.00	1.10	1.18
2.2	3	1.21	1.25	1.30
3.3	1	1.22	1.31	1.33
3.3	2	1.56	1.59	1.64
3.3	3	1.64	1.85	1.86

[a] Data from van Duijn et al. (18).

A theory concerning the influence of slow penetration of reagents on cytochemical enzyme kinetics and quantitation was developed, relying on the theoretical framework developed in the field of technological catalysis. The experimental conditions for the quantitation of alkaline phosphatase were studied using a model consisting of polyacrylamide films into which a commercially available preparation of purified calf intestinal phosphatase was incorporated (18).

The theory of reaction kinetics for enzymes constrained in a matrix can be expressed more clearly using the following definitions. For a catalytic process operating within a structure, an *effectiveness factor* (η) can be defined as the ratio between the reaction rate when the catalyst is present in a matrix and the potential rate for a catalyst in solution (71). It is clear that when the diffusion gradient is small η approaches 1. The quantitative effects of factors governing the diffusion of reactants into structure-bound catalysts are formalized in the Thiele modulus (71):

$$\phi = \tfrac{1}{2}d(k_r C^{m-1}/D_{eff})^{\tfrac{1}{2}} \tag{1}$$

in which d is the thickness of the film, k_r is the rate constant of the reaction, C is the concentration of the reactant in the bulk medium, m is the order of the reaction with which the reactant is converted, and D_{eff} is the effective diffusion coefficient for the reaction in the film. D_{eff} characterizes the ability for diffusive flow of reactant through a unit of volume of film in which the catalyst is constrained. The factor $\tfrac{1}{2}$ is added because diffusion into the films is possible from both surfaces. The relationship between η and ϕ for a flat film has been theoretically derived by Wheeler (82). From Fig. 7 it can be concluded that η approaches 1 for ϕ smaller than a certain value which depends on the order of the reaction.

The steady state velocity of an enzymic reaction can be formulated in the Michaelis-Menten equation as

$$v = k_{+2} E_0 \frac{C}{K_m + C} \tag{2}$$

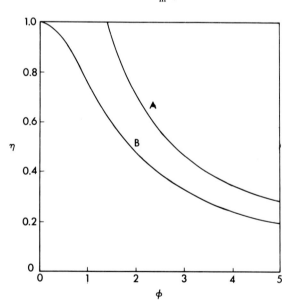

FIG. 7. Effectiveness factor as a function of the Thiele diffusion modulus. Curves for diffusion in a film. A, Zero-order reaction; B, first-order reaction. Redrawn from Wheeler (82).

where k_{+2} is the turnover number, E_0 is the enzyme activity, C is the concentration of the substrate, and K_m is the Michaelis-Menten constant. At saturating substrate concentrations, $C \gg K_m$ and v becomes

$$V = k_{+2}E_0 \tag{3}$$

Under these conditions the reaction is of zero-order in relation to substrate, and by substitution of $k_r = k_{+2}E_0$, Eq. (1) becomes

$$\phi = \tfrac{1}{2}d\,(k_{+2}E_0/CD_{eff})^{\frac{1}{2}} \tag{4}$$

At substrate concentrations where the enzymic reaction is of first-order $(C \ll K_m)$ Eq. (2) becomes

$$v = \frac{k_{+2}E_0C}{K_m} \tag{5}$$

Substitution of $m = 1$ and $k_r = k_{+2}E_0/K_m$ in Eq. (1) gives

$$\phi = \tfrac{1}{2}d\,(k_{+2}E_0/K_mD_{eff})^{\frac{1}{2}} \tag{6}$$

Equations (4) and (6) are identical, except that in (6) C is replaced by K_m (substrate concentration which gives half-maximum velocity).

Consequently we find that for both types of reaction order, the smallest value of ϕ can best be achieved by choosing thin films, with low enzyme concentrations and a substrate with a low turnover number for the enzyme under investigation. The substrate should also have as high an effective diffusion coefficient as possible. The latter not only depends on the nature of the substrate, but also on the cross section of the pores in the polyacrylamide films, which can be influenced, e.g., by the concentration of monomers in the polymerizing mixture. In biologic objects pore width may depend on the nature of the fixation. From these and other considerations it can be concluded that an effectiveness factor approaching the value of 1 can be obtained if the system in which the enzyme is constrained has appropriate parameters with respect to enzyme concentration, film thickness, diffusion coefficient, Michaelis-Menten constant, and turnover number of the substrate.

In order to test these conclusions and to determine the conditions for which η approaches the value of one for alkaline phosphatase, model film experiments were performed with a cytochemical azo dye coupling and a biochemical method (Fig. 8). First, several combinations of substrates and diazonium salts were tested with respect to nonspecific background staining and enzyme inhibition. The diazonium sulfate of 4-aminodiphenylamine (Variamine Blue RT) produced a very low background staining and did not irreversibly inhibit alkaline phosphatase when used in combination with

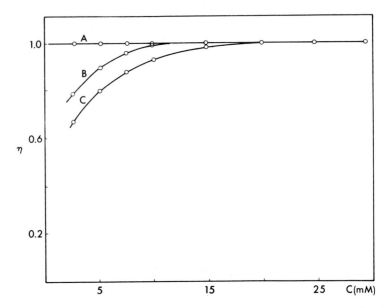

FIG. 8. The relationship between effectiveness factor and substrate concentration. Each value presents the average of several biochemically determined alkaline phosphatase activities in polyacrylamide film (containing purified alkaline phosphatase) related to the activities of the same amount of enzyme in solution. A, for film thickness 0.07 mm; B, for film thickness 0.14 mm; C, for film thickness 0.18 mm. Redrawn from van Duijn *et al.* (*18*).

naphthol AS-MX phosphate. Other azo dyes are known to inhibit enzyme activity. Fast Garnet GBC causes a decrease in aminopeptidase activity in the first 10 min of the reaction (*21, 37*).

These model studies revealed among other things a hitherto unknown effect of the *in situ* environment of the enzyme on the absorbance spectrum of some azo dye reaction products. When leukocytes or films containing complete sonicate were incubated in media containing naphthol AS-MX phosphate and Fast Red TR a precipitate resulted with an absorbance peak at 565 nm. When films containing purified alkaline phosphatase were incubated, a reaction product with a main peak at 525 nm and a lower peak at 560 nm resulted (*43*). This difference probably is due to the presence of cellular lipoproteins in the cells, whose removal by acetone fixation results in an azo dye with a spectrum identical to that of the reaction product obtained after incubation of films containing purified enzyme. The combination of naphthol AS-MX phosphate and Variamine Blue RT is one of the few media resulting in azo dye with a spectrum independent of the presence of lipoproteins.

E. INVESTIGATION OF METAL SALT METHODS FOR ACID PHOSPHATASE

Films can also be applied as a model system for the quantitative investigation of cytochemical reactions used in electron microscopy, provided the amount of electron-dense reaction product can be measured. The electron density in most cases is caused by heavy metal atoms, which can be determined quantitatively with analytic procedures or with light microscopic cytochemical methods if available. Preliminary investigations were performed on the electron microscopic method for acid phosphatase (5, 27, 42). Contrary to what is generally presumed, the amount of brown reaction product, resulting from the conversion of lead phosphate into lead sulfide, can be measured accurately with the film colorimeter when it is formed under well-defined conditions (7).

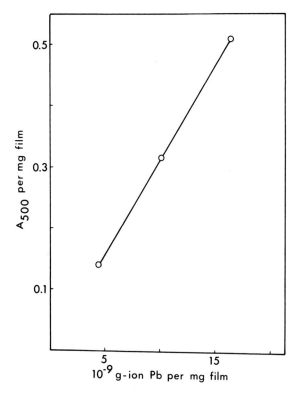

FIG. 9. The relationship between the absorbance per milligram of film, and the amount of lead determined per milligram of film. Polyacrylamide films were incubated in a Gomori medium for acid phosphatase after being impregnated with solutions of inorganic phosphate. Parts of the films were stained in a sulfide-containing medium, and aliquots were assayed for lead. Data from Cornelisse *et al.* (*11*).

The totally diffuse density (79) of the precipitate of lead sulfide is linearly related to the independently determined amount of lead present in the films (Fig. 9). Thus a number of factors important in the heavy metal acid phosphatase methods for light and electron microscopy could be studied and evaluated quantitatively by use of the film colorimeter.

It is of interest to note that apart from the enzymic reaction itself, the ability of several media to capture liberated phosphate can also be investigated using model films. For this purpose, pure polyacrylamide films were impregnated with dilute inorganic phosphate solutions of different concentrations. After 1 hr, parts of these films were lightly blotted between filter paper to remove adhering liquid, and immersed in media containing ionic lead and agitated for 1 min. At the moment of immersion of the films, a dynamic situation begins which is highly analogous to what occurs *in situ* during the enzyme reaction. The phosphate ions begin to diffuse out of the films and the lead ions from the medium begin to penetrate the films. When the ions meet, a lead phosphate precipitate is formed. The amount of lead phosphate that precipitates in the films can be determined by measuring the absorbance of the brown reaction product after its conversion with

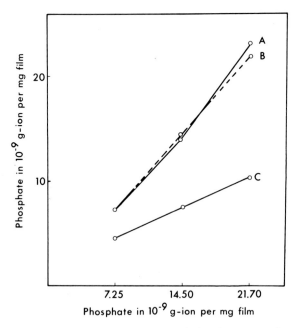

FIG. 10. The relationship between the amount of phosphate *assayed* per milligram of film and the amount of phosphate *impregnated* per milligram of film. A, Determined after impregnation; B, theoretical curve; C, determined after impregnation and immersion in a Gomori medium. Film thickness 0.2 mm. Data from Cornelisse *et al.* (*11*).

sulfide ions. It is also possible to dissolve the lead phosphate in nitric acid and measure the amount of lead colorimetrically (68). In this way, the percentage of the phosphate originally present in the film which has been captured by the medium, can be calculated (Fig. 10). Media of different composition with respect to lead concentration, type of buffer, pH, etc., can be examined with respect to their potentialities in capturing phosphate. This is, of course, not the only parameter that determines the suitability of a medium; an increase in lead ion concentration, which is favorable for high

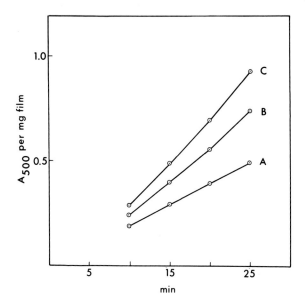

FIG. 11. The relationship between the absorbance per milligram of acid phosphatase containing film and reaction time of polyacrylamide films containing guinea pig leukocyte sonicate, as revealed after incubation at pH 5.0 in media prepared according to: A, Gomori (27); B, Lake (42); C, Barka and Anderson (5). Every value is the average of five measurements. Redrawn from Brederoo et al. (8).

trapping efficiency, will progressively inhibit the enzymes. It is therefore also necessary to study the enzyme reaction itself, and the influence of, for instance, lead-ion concentration in different media, on the turnover number of the enzyme (Fig. 11). The influence on the enzyme activity of prefixation procedures used for electron microscopy, can also be investigated with these model films (8). By continued study along these lines, methods can be developed that, in combination with available instruments for quantitative electron microscopy (4), could make enzyme quantitation at the ultrastructural level a reality.

V. Chemical Calibration of Cytochemical Staining by Use of Model Films

When a model system has been studied and modified so that both film and the biologic object react identically to variations in a particular staining method the model can be accepted as quantitatively reliable for that method. After the chemical reaction mechanism of the staining procedure has been investigated sufficiently, such a "cell-like" model can be used for calibration purposes. In this respect, a film, on which standard analytic methods can be applied relatively easily, may be considered to form a bridge between biochemical assay and cytochemical staining. So far, the calibration procedure has been investigated for the Feulgen staining and for the azo dye coupling method for alkaline phosphatase.

A. CALIBRATION OF THE FEULGEN PROCEDURE

1. *Photographic Colorimetry*

The amount of chromophore present in objects of microscopic dimensions can be assayed rather accurately with the method of photographic colorimetry (*49*, *54*, *76*). With this method *relative* amounts of Feulgen-stained DNA in cell nuclei have been determined (*77*). When the cells are stained together with films containing DNA the amount of DNA in individual nuclei

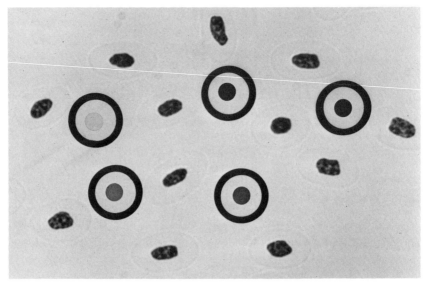

FIG. 12. Chicken erythrocyte nuclei and DNA-containing cellulose disks (magnification of nuclei in relation to disks is 793×) pictured on one photographic plate.

can be expressed in *absolute* units (*23, 53, 78*). For this purpose, smears or imprints of the cells are stained by the Feulgen procedure together with DNA-containing films. Using light at a wavelength of 550 nm, photographs were made of cells, and of punched-out disks of stained DNA-containing films (embedded and mounted between thin glass slides) placed directly in front of the photographic plate. In this way, the nearly unmagnified disks are imaged on the same photographic plate in areas where no cell images are present (Fig. 12).

The magnification ratio between microscopic object and disks can be determined. Under certain conditions, e.g., working in the straight part of the Hurter-Driffield curve of the photographic plate, the local and integrated densities of such negatives are true reflections of the absorbances in the objects, except for some stray light errors. Accepting the model as sufficiently "cell-like" with respect to the Feulgen reaction, the amount of DNA in cell nuclei can be expressed in amounts of DNA-phosphorus, using the density values of disks and nuclei in the negative, after the phosphorus in stained films is determined by analysis.

Integrated densitometry of the photographic plate can be replaced by photographic colorimetry in this procedure. An enlargement of the negative is projected onto normal black and white photographic paper. The exposed paper is converted to a blue colored print by use of a special developer containing 2,4-dichloro-1-naphthol and *N,N*-diethyl-*p*-phenylenediamine sulfate (*49*). The dye can be eluted from selected parts of this blue print, after which the absorbance of the resulting blue solution can be measured macrocolorimetrically. Under specified conditions (*23, 53, 76*) the absorbance per milligram of paper is proportional to the integrated densities in the original object. In this way, the amount of dye in the blue pictures of the highly magnified nuclei can be expressed in DNA, using the amount of dye present in the blue pictures of the disks in which the DNA content is known as a calibration. A close agreement was found between the average values obtained with this method for 6–20 nuclei and values obtained by biochemical investigation of cell suspensions (about 10^7 cells) (Table IV). It is likely that there were errors due to stray light in the microscope, spectral differences between the stained objects, and the DNA–cellulose films used in these experiments, and possible differences in the stoichiometry of the reactions in biologic objects and models. Some of these errors, assumed to be in the 10–20% range, may be cancelled out, suggesting a better accordance than actually obtained. This needs further investigation using (DNP + albumin)–polyacrylamide films as calibrating models.

2. Two-Wavelength Cytophotometry

The calibration procedure is not limited to the photographic colorimetry method. A similar investigation on DNA of cell nuclei was carried out using

TABLE IV

AMOUNT OF DNA PER NUCLEUS IN 10^{-12} GM[a]

Object	Method		
	Biochemistry	Photographic colorimetry	Two-wavelength cytophotometry
Chicken erythrocyte	2.2–2.6 (*14, 45, 80*)	2.6	2.9
		2.6*	—
Rat spleen	6.6 (*69*)	—	—
Liver diploid	—	—	6.9
Liver tetraploid	—	—	13.4
Kidney	5.5 (*69*)	5.7	6.6
Leukocyte	6.6 (*69*)	6.1	6.5

[a] The data tabulated under photographic colorimetry and two-wavelength cytophotometry are the mean values of 6–20 nuclei measured after Feulgen–pararosaniline staining. Preparations stained following the Feulgen–thionine–SO₂ procedure resulted in the value marked *. Data in the last two columns are from den Tonkelaar and van Duijn (*78*) and Struyck and van Duijn (*75*).

two-wavelength cytophotometry; the absorbance of the models being measured with the film colorimeter. To calculate the amount of DNA in cell nuclei, the law of Lambert and Beer is used in its cytochemical form (*81*)

$$M = \frac{AO}{K} \tag{7}$$

where M is the amount of DNA, A is absorbance, O is the area measured, and K is a constant depending on the nature of the staining reaction. When slides and films are stained together, and K is accepted to be the same for micro-objects and films (macro-system). We now can write

$$K = \frac{A_{mi}O_{mi}}{M_{mi}} = \frac{A_{ma}O_{ma}}{M_{ma}} \tag{8}$$

in which the subscripts mi and ma indicate micro and macro system, respectively. Rearrangement of Eq. (8) gives

$$M_{mi} = \frac{A_{mi}O_{mi}}{A_{ma}O_{ma}} \, M_{ma} = A_{mi}O_{mi} \, \frac{M_{ma}}{A_{ma}O_{ma}} \tag{9}$$

M_{mi} now can be calculated since all other data in Eq. (9) are known from measurements with the cytophotometer and film colorimeter. Data obtained by application of this method are in good accord with results from biochemical assay (Table IV).

B. CALIBRATION OF AN AZO DYE COUPLING METHOD FOR ALKALINE
   PHOSPHATASE ACTIVITY

A model system consisting of polyacrylamide films into which cell sonicate is incorporated was employed to investigate the quantitative calibration of a cytochemical azo dye method for alkaline phosphatase activity in neutrophilic leukocytes (59). In Fig. 13 the role of the model film is schematically

	SONICATE	FILM	CELL PREPARATION
BIOCHEMICAL ASSAY	B_s	B_f	
CYTOCHEMICAL ASSAY		C_f	C_c

FIG. 13. Activity relations in the calibration procedure.

presented. Leukocytes, which were obtained from exudates in the abdominal cavity of guinea pigs, were applied as a test object because they stain rather strongly within a reasonable period of time, and can easily be harvested in quantity after intraperitoneal injection of hypertonic NaCl. The sonicate containing films, together with air-dried unfixed smears of the original leukocyte suspension, were assayed in a medium containing naphthol AS-MX phosphate and 4-aminodiphenylamine diazonium sulfate. The enzyme activity of the films was also studied biochemically, using disodium phenyl phosphate as a substrate. Media were selected from model film studies (see Section IV, D) to give optimal quantitative results.

For the cytochemically stained film, the absorbance per surface area, per minute incubation, per milligram can be determined from measurements with the film colorimeter. Taking into account the number of cells counted per gram of original suspension and the weight of sonicate incorporated per milligram film, absorbance per cell, per surface area, per minute incubation (C_f) can be calculated. The stained cell preparations can be measured with the two-wavelength spectrophotometer. Cytochemical activity in the cell preparation (C_c) can then be expressed as absorbance per cell, per surface area, per minute incubation. Following similar procedures, the biochemical activities (expressed, for example, as moles of phenol released per cell, per minute incubation time), can be achieved for the sonicate (B_s) and for films (B_f).

TABLE V

RELATIONSHIP BETWEEN BIOCHEMICAL AND CYTOCHEMICAL ALKALINE PHOSPHATASE ACTIVITY[a]

Expt.	$B_s \times 10^8$	$B_f \times 10^8$	$\dfrac{B_f}{B_s} \times 100\%$	$\dfrac{B_f}{C_f} \times 10^3$	$C_f \times 10^6$	$C_c \times 10^6$	$\dfrac{C_c}{C_f} \times 100\%$
1	7.44	6.40	86	5.8	10.86	6.86	63
2	7.19	6.54	91	4.6	14.05	8.65	62
3	7.05	6.90	98	3.2	21.65	12.59	58
4	8.21	7.47	91	5.6	13.22	7.82	59
5	(9.24)	(9.52	(103)	4.8	(19.68)	9.42	(48)

[a] B_s, Average absorbance at 575 nm per leukocyte per minute incubation time in sonicate.

B_f, Average absorbance at 575 nm per leukocyte per minute incubation time in sonicate-containing films.

B_f/B_s, B_f as percentage of B_s.

B_f/C_f, Relationship between biochemical and cytochemical activity in sonicate-containing films.

C_f, Total absorbance of azo dye in sonicate-containing films, calculated per leukocyte per minute incubating time per mm².

C_c, Total absorbance of azo dye in cell preparations, calculated per leukocyte per minute incubation time per mm².

C_c/C_f, C_c as percentage of C_f.

Data from van der Ploeg and van Duijn (59).

In the ideal situation, B_s and C_f will be identical with B_f and C_c, respectively.* Different values of B_s and B_f will be obtained when the incorporation of sonicates into the films results in a decreased, biochemically assayed alkaline phosphatase activity. It may be pointed out that even a considerable decrease of enzyme activity during polymerization of the film need not invalidate the films for calibration provided the relation between the biochemical and cytochemical activities remains unchanged. A difference in cytochemical activities between the model (C_f) and the cell preparation (C_c) can result when part of the enzyme in the cell preparations is less accessible to the cytochemical reagents, or when part of the enzyme has leached out of the cells. The amount of dye formed in the cytoplasm of individual leukocytes (C_c) was measured with a two-wavelength cytospectrophotometer (Table V, C_c); the azo dye in the films was measured with the film

---

* Even when $B_f = B_s$ and $C_f = C_c$ it is theoretically still possible that only a part of the molecules which are active in the one procedure shows activity in the other. If isoenzymes are present, and one or more of them cannot react with one of the substrates, they will not be detected without further investigation. In principle, it is possible to detect such a difference in affinity of the enzyme to biochemical or cytochemical substrate by comparing the results obtainable with films into which isolated isoenzymes are incorporated.

colorimeter (C_f); and the biochemical activities (B_s and B_f) were determined colorimetrically. With reference to the relation between biochemical and cytochemical enzyme activity in the films (B_f/C_f), the enzymic activity in the cells (C_c) could be calibrated and expressed in moles of phenol liberated per cell per minute incubation. Independent biochemical assay was performed on a sonicate of the counted cell suspension to check the reliability of the method (Table VI). The mean value of cytochemically assayed alkaline

TABLE VI

ALKALINE PHOSPHATASE ACTIVITY IN NEUTROPHILIC LEUKOCYTES OF GUINEA PIG EXUDATE, EXPRESSED AS 10^{-14} MOLE OF PHENOL FORMED PER CELL PER MINUTE INCUBATION TIME $\pm$ S.E.[a]

Expt.	A	B
1	$5.8 \pm 2\%$	$3.1 \pm 8\%$
2	$5.6 \pm 3\%$	$3.1 \pm 7\%$
3	$5.5 \pm 3\%$	$3.1 \pm 6\%$
4	$6.4 \pm 2\%$	$3.5 \pm 7\%$
5	$(7.3 \pm 2\%)$	$3.6 \pm 5\%$
Mean	$5.8 \pm 2\%$	$3.3 \pm 8\%$

[a] A, Average activity of $\frac{1}{2}$–3×10^6 cells assayed biochemically in sonicate. B, Calibrated average activity of 60 leukocytes in cell preparations assayed cytochemically. Data from van der Ploeg and van Duijn (59).

phosphatase activity in the smears proved to be about 60% of the mean biochemical activity of the cell suspensions. This discrepancy—caused by a too low value of C_c—could be due to the fact that at least part of the enzyme is located in granules which are less accessible to the cytochemical reagents. Some enzyme also could have leached out of the cell smears during incubation. Cytochemical activity in films containing sonicate (where most of the intracellular membranes were supposedly disrupted) was higher than in whole cell preparations. Preliminary experiments using a fixation procedure described by Melnick (48), resulted in a higher mean cytochemical activity, suggesting that methods can be found which increase the cytochemically assayed activity in cell smears. The variation in reaction yield during different incubations, which is probably due to inhomogeneity of the diazonium salt (59), demonstrates the importance of calibration procedures for cytochemical methods when inhomogeneous reagents are used.

Although the method needs further investigation with regard to fixation procedures, the results demonstrate that cytochemical quantitation of enzyme activity is possible. Films into which intact cell organelles are incorporated can also be used to study the problems of latency of enzyme activities. Films containing purified enzymes or isoenzymes could be employed in order to define more exactly their respective turnover numbers for cytochemical and biochemical substrates.

VI. Concluding Remarks

The study of a cytochemical staining reaction on a biologic object gives incomplete and often contradictory information, since the chemical composition of biologic material is complicated and insufficiently known. It is usually not possible to obtain the necessary information from studies with solutions of model compounds in a test tube. For the development of reliable quantitative cytochemical staining procedures, a model system allowing the study of compounds or groups in a fixed structure is indispensable.

In this chapter, suitable procedures have been described for the preparation of model films carrying compounds which vary in size from amino acids to macromolecules such as deoxynucleoprotein. That enzymes can be incorporated, holds promise for quantitative studies in this field. The films are of good mechanical strength and of homogeneous composition. In combination with the film colorimeter, the model offers a reliable base for cytochemical studies.

The films proved to be useful for the investigation of reaction kinetics and diffusion gradients of especially enzyme substrates. The influence of fixation and the activation of latent enzymes inside cellular structures can be studied after incorporation of cell organelle suspensions. By combining biochemical and cytochemical assay, quantitation and calibration of cytochemical staining results in biochemical units is achieved. Though only a few staining reactions have been studied so far, the quantitation of many more enzymes seems to be possible.

A new method has been developed for the determination of the molar absorptivity of reaction products bound in a stained object without requiring the isolation and purification procedures necessitated in biochemistry. From a practical point of view the films also can be applied for the standardization of commercially available reagents. It was repeatedly observed that cytochemical reagents claimed to be identical varied considerably in quantitative behavior.

It is our conviction that many cytochemical procedures for light and electron microscopy can be developed into reliable analytic tools by investigation along lines described in this chapter. It must be realized, however, that even the best models only approximate the situation in the biologic object. The value and validity of every model always have to be tested in repeated confrontation with the biologic material. Investigations are necessary in which as many parameters of the staining procedure as possible are varied. Only when the effects of these variations are the same in model and in biologic object, can it be concluded that the compound to be stained is present under approximately the same conditions in both systems. If such a situation is reached, the resulting cytochemical staining method can be considered to have a similar reliability as well-established biochemical methods. It can then be applied with confidence and can serve for the ultimate aim of histochemistry and cytochemistry: the quantitation of compounds in tissues and cells *in situ*.

APPENDIX

A. The Preparation of Viscose

A solution of 64.8 gm of sodium hydroxide (analytical grade) in 295 ml of distilled water is cooled to 15°C, after which 10 gm of cotton-wool is added. The temperature, which due to the reaction tends to rise, is kept at 17°–18°C. After 1 hr the excess of sodium hydroxide solution is removed by decantation and the cotton-wool is packed in a sheet of polyvinyl chloride filter and pressed between two flat Perspex plates to a final weight of 32–33 gm. The resulting material, after being cut with scissors into pieces measuring a few millimeters, is then placed in a 300-ml bottle. The rubber stopper of the bottle is pierced with a glass capillary of about 1-mm diameter, which allows some air to enter but prevents excessive drying of the contents. The bottle is first held at 20°C for 55 hr and then 15 hr at 4°C. After this "ripening" period 3 ml of carbon disulfide (analytical grade), precooled at 4°C, is added. The bottle is rapidly closed with a ground-glass stopper and slowly revolved in a water bath at $20° \pm 0.1°C$ for 135 min. At the end of this period, the excess of CS_2 is expelled with an air stream. The product is dissolved in a solution of 4.18 gm of sodium hydroxide in 100 ml of distilled water, and kept at room temperature for 48 hr with occasional stirring. By centrifugation at 25,000 g for 45 min at 15°C, unchanged cellulose fibers are removed from the viscose. Stored at 4°C, the viscose can be kept for use during a period of about 6 weeks.

Caution: The alkali-containing cotton-wool and the viscose are aggressive to eyes and skin.

B. The Preparation of Cellulose Films

A quantity (7–10 gm) of viscose is placed on a glass plate and spread in a thin layer by sliding over the plate surface a chromium-plated steel cylinder of 14.1-mm diameter (except at the two ends where the diameter is 15.0 mm). To obtain a more compact film ultrastructure, the viscose layer can be dried at 50°–60°C for 40–60 min before being "developed." The common procedure, however, is to immerse the layer on the glass plate directly in a freshly prepared mixture of 1 volume of formic acid (90%) and 3 volumes of methanol, this mixture having been precooled to −20°C beforehand. The low temperature slows down the rate of development and prevents formation of irregularities caused by too rapid H_2S formation in or below the film. The resulting film is lifted from the plate as soon as possible by means of a knife, and left in the "developer" for 1 hr at room temperature. The film is then given three washes in methanol of 10 min duration each. If necessary, films can be fixed in a mixture of 1 volume of 40% aqueous formaldehyde and 9 volumes of methanol during 24 hr, followed by washing in three 10-min changes of methanol. Films prepared following this procedure are completely transparent in media of appropriate refractive index. To prevent growth of fungi, films are kept in an aqueous solution of methanol (20%, v/v) at 4°C when storage for longer periods is necessary. The thickness of these films is in the range of 100–150 μ. When the viscose is dried before developing, the thicknesses of the films are 50–80 μ.

Very thin films can be obtained by dipping a wire loop into the viscose and developing the membrane enclosed within the loop.

Caution: The mixture of formic acid and methanol is very aggressive to eyes and skin.

C. The Preparation of Cellulose Globules

A mixture of 3 volumes of 1-octanol and 2 volumes of chloroform is cooled to 4°C and saturated with 1 N sodium hydroxide in a separatory funnel. The organic layer is centrifuged at 200 g at 4°C during 5 min. In a 25-ml flask of a homogenizer (Edmund Bühler, Tübingen, Germany, type ho, 220V, 260W) 10 ml of this mixture is cooled by melting ice. By means of a syringe 0.5 ml of viscose is added, and immediately thereafter the contents of the flask is vigorously homogenized (at speed 10 for 1 min). The suspension is then poured rapidly into a large excess of formic acid (99%) which is being violently agitated by a magnetic stirrer. The mixture is stirred at room temperature for $\frac{1}{2}$–1 hr. The resulting globules are sedimented by centrifugation during 10 min at 1000 g. They are washed three times with acetone and three times with methanol, and can be stored in aqueous methanol (20%, v/v). Most of the globules appear to be spheres with diameters varying from 0.5 to 200 μ. Reactive groups can be bound covalently as described in the text for the cellulose films. After appropriate staining, the globules may be mounted in aqueous or nonaqueous mounting media on slides and used as a microscopic standard system. Differential sedimentation in sucrose solutions may be used to obtain populations with a more limited range of diameters. The globules have a tendency to clump. It is therefore advisable to sonicate between successive sedimentations.

D. The Preparation of Polyacrylamide Films

The following reagents are mixed in a test tube: 3000 mg of acrylamide (Eastman Organic Chemicals, Rochester, N.Y.); 120 mg of N,N'-methylenebis(acrylamide) (Eastman); 15 μl of N,N,N',N'-tetramethylethylenediamine (TEMED) (Eastman); 12 μl of 3-dimethylaminopropionitrile (DMAPN) (Koch and Light Lab., Ltd, Colnbrook, Bucks, Great Britain); 775 mg of 2-amino-2-(hydroxymethyl)propane-1,3-diol (Tris); and 3 ml of 2 N HCl. The pH is adjusted to 7.0 ± 0.1, after which the volume is brought to 6.7 ml with distilled water. One part of a solution containing proteins or other macromolecules is mixed thoroughly with two parts of the monomeric solution, avoiding the formation of air bubbles.

Then, polymerization is initiated by mixing carefully 1 part of a freshly prepared 2–5% ammonium persulfate solution with 10 parts of the above solution. The polymerization is allowed to proceed between glass plates for 1 hr at 4°C, because at room temperature there is more danger of enzyme inactivation by local heat production during the polymerization process.

To remove possible traces of inhibitory metal ions, the glass plates are placed in a solution of disodium ethylenediaminetetraacetate (0.1 M) and then rinsed in distilled water and dried before being used.

After the polymerization, the glass plates are pryed apart and the film is washed for 30 min in three changes of distilled water at 4°C.

When the above method is used for the preparation of films containing DNP and albumin, these macromolecules tend to precipitate in the salt-containing mixture. No difficulties, however, are encountered when these macromolecules are incorporated using the following procedure.

Dissolve 8–25 mg of DNP in 0.9 ml of twice-distilled water during 14 hr at 4°C. Adjust the pH of a 30% bovine albumin (demineralized, Povite, Amsterdam, Holland) solution in twice-distilled water with some drops of TEMED to 8.5, and mix 1 ml of this solution with 2.5 ml of a solution containing 750 mg of acrylamide and 40 mg of N,N'-methylenebis(acrylamide) in twice-distilled water, which was brought to pH 10 with TEMED.

Add this mixture slowly to the DNP solution, mix carefully, remove air by bubbling N_2 gas for 5 min through the mixture, centrifuge at 20,000 g for 20 min to remove gas bubbles and undissolved material, and polymerize directly afterwards by thorough but rapid mixing with 1.3 ml of an aqueous solution containing 1% ammonium persulfate and 0.2% DMAPN.

Polyacrylamide films in general can be measured while placed in water especially when the opal glass is used to correct for scatter. Films, which are measured in xylene for the sake of transparency or because the corresponding cell smears have been mounted in nonaqueous media, can be given the following treatments.

From distilled water they are transferred to the following mixtures in which they are kept at room temperature under gentle shaking for about 12 hr each.

3 times ethanol: glycerol (5 : 5, v/v)
1 time ethanol: glycerol (6 : 4, v/v)
1 time ethanol: glycerol (7 : 3, v/v)
3 times ethanol: xylene (5 : 5, v/v)
2 times xylene

All organic liquids are of analytic grade. The films can be stored in the dark at 4°C until they will be measured.

After measurement the punched-out disks are returned to water via a change of ethanol and an ethanol: water mixture for 24 hr each. They are dried overnight at 120°C, and then equilibrated to room temperature and humidity above a saturated solution of $CaCl_2$.

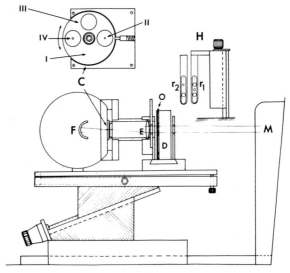

Fig. 14. Diagram of apparatus for absorbance measurements of model films: H, film holder; C, rotating disk; D, photometer cell; E, achromatic focusing lens; F, photomultiplier; M, monochromator; O, opal glass placed in photometer cell when films causing scatter are to be measured; r_1, reference aperture having the same diameter as the one in which piece of film is positioned. Each of the two apertures of the film holder can be brought to exactly the same position in the light beam; r_2, reference aperture used when high absorbances are to be measured. For further details see text. Redrawn from van der Ploeg and van Duijn (59).

E. SPECTROPHOTOMETRY AND PUNCHING OF FILMS

Spectrophotometry is carried out in an apparatus shown in Fig. 14. The light from the monochromator (M) is concentrated by a small focusing lens (E, $f = 4$ cm and $d = 1.8$ cm)—after having passed through the film in the photometer cell (D)—to the point where alternatively can be brought the apertures II, $\phi = 1$ mm; III, $\phi = 10$ mm; or IV, $\phi = 2$ mm. Apertures II and IV are filled with a piece of opal glass. During the changing of the film position I is used to shut off the light. All parts of the setup can be aligned so that the light beam passes through the centers of the lenses. The light coming from the monochromator must illuminate the aperture in front of the photometer cell as homogeneously as possible. The small aperture (II) serves to adjust the beam during the homogeneity tests which are carried out regularly after a new light source has been brought into the lamp housing. The opal glass in IV is used to obtain a diffuse distribution of the light on the photomultiplier. This procedure ensures that light passing

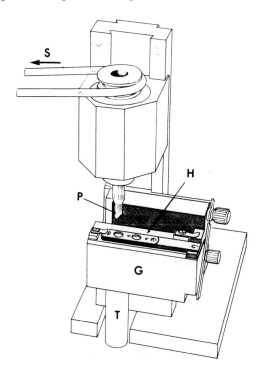

FIG. 15. Punching apparatus for films: G, guiding block; H, film holder in position for punching; P, internally sharpened punch of silver steel; S, motor; T, a partly visible rod of Teflon supporting the film during mounting or punching procedure. Drawn from van der Ploeg and van Duijn (59).

through different parts of the film contributes equally to the transmission measurement. It is possible to correct for light scatter in the films by use of a piece of opal glass (O) placed in the photometer cell directly behind the film (see text). In that case the open aperture III is used instead of IV because otherwise light intensity at the photomultiplier cathode will be too low. After the measurement the holder is placed in the punching apparatus (Fig. 15) and the measured part of film is punched out.

The silver steel punch is internally ground and has a scalloped edge which proved to give the best punching results with different types of film, provided the punching edge is very sharp.

The punch is connected to a motor (S) (400 rpm), and can be brought down by hand to the position where it hits the film. After some practice, undamaged disks also can be made by hand-punching. During mounting of the film in the holder and during the punching procedure the film is supported by a small Teflon rod (T) which is inserted from the bottom of the guiding block.

ACKNOWLEDGMENT

The authors gratefully acknowledge the critical reading of the manuscript by Dr. A. Deitch.

REFERENCES

1. Ahsmann, W. B. A. M., and van Duijn, P., Amino acid cellulose films as models for histochemical protein reactions. *Histochemie* **12**, 285–288 (1968).
2. Albertson, P. -A., "Partition of Cell Particles and Macromolecules," p. 13. Wiley, New York, 1960.
3. Amesz, J., Duysens, L. M. N., and Brandt, D. C., Methods for measuring and correcting the absorption spectrum of scattering suspensions. *J. Theoret. Biol.* **1**, 59–74 (1961).
4. Bahr, G. F., Quantitative electron microscopy: An introduction to general quantitative measurements with the electron microscope with special reference to dry mass determinations. *In* "Introduction to Quantitative Cytochemistry" (G. L. Wied, ed.), pp. 137–149. Academic Press, New York, 1966.
5. Barka, T., and Anderson, P. J., Histochemical methods for acid phosphatase using hexazonium pararosanilin as coupler. *J. Histochem. Cytochem.* **10**, 741–753 (1962).
6. Bernfeld, P., and Wan, J., Antigens and enzymes made insoluble by entrapping them into lattices of synthetic polymers. *Science* **142**, 678–679 (1964).
7. Brederoo, P., Daems, W. Th., van Duijn, P., and van der Ploeg, M., Cytochemical investigations of the lead method for acid phosphatase by use of a model system. *Proc. Roy. Microscop. Soc.* **3**, 153 (1968).
8. Brederoo, P., Daems, W. Th., van Duijn, P., and van der Ploeg, M., Quantitative investigations on the effect of aldehyde fixation on acid phosphatase activity. *Proc. 4th Reg. Conf. (Eur.) Electron Microscopy, Rome, 1968* Summary Repts., **2**, pp. 79–80, Tipografia Poliglotta Vaticana, Roma, 1968.

9. Caspersson, T., Die quantitative Bestimmung von Thymonucleinsäure mittels fuchsinschwefliger Säure. *Biochem. Z.* **253**, 97–110 (1932).

10. Chang, T. M. S., and Poznansky, M. J., Semipermeable microcapsules containing catalase for enzyme replacement in acatalasaemic mice. *Nature* **218**, 243–245 (1968).

11. Cornelisse, C. J., Brederoo, P., and van der Ploeg, M., Unpublished material.

12. Daems, W. T., van der Ploeg, M., Persijn, J. -P., and van Duijn, P., Demonstration with the electron microscope of injected peroxidase in rat liver cells. *Histochemie* **3**, 561–564 (1964).

13. Dalen, J. P. R. van, Ahsmann, W. B. A. M., and van Duijn, P., A method for the determination of the molar absorbance of structure-linked chromophores. *Histochem. J.* (1970) (in press).

14. Davidson, J. N., Les nucléoprotéines et la croissance des tissus. *Bull. Soc. Chim. Biol.* **35**, 49–66 (1953).

15. Davis, B. J., Disc electrophoresis. II. Method and application to human serum proteins. *Ann. N.Y. Acad. Sci.* **121**, 404–427 (1964).

16. Deitch, A., A method for the cytophotometric estimation of nucleic acids using methylene blue. *J. Histochem. Cytochem.* **12**, 451–461 (1964).

17. Duijn, P. van, A histochemical specific thionin-SO_2 reagent and its use in a bicolor method for deoxyribonucleic acid and periodic acid positive substances. *J. Histochem. Cytochem.* **4**, 55–63 (1956).

18. Duijn, P. van, Pascoe, E., and van der Ploeg, M., Theoretical and experimental aspects of enzyme determination in a cytochemical model system of polyacrylamide film containing alkaline phosphatase. *J. Histochem. Cytochem.* **15**, 631–645 (1967).

19. Duijn, P. van, den Tonkelaar, E. M., and Hardonk, M. J., An improved apparatus for quantitative cytochemical model studies and its use in an experimental test of the two-wavelength method. *J. Histochem. Cytochem.* **10**, 473–480 (1962).

20. Duijndam, W. A. L., and van Duijn, P., Unpublished material.

21. Felgenhauer, K., and Glenner, G. G., Quantitation of tissue-bound renal aminopeptidase by a microdensitometric technique. *J. Histochem. Cytochem.* **14**, 53–63 (1966).

22. Gahrton, G., Microspectrophotometric quantitation of the periodic acid Schiff (P.A.S.) reaction in human neutrophil leukocytes based on a model system of glycogen droplets. *Exptl. Cell Res.* **34**, 488–506 (1964).

23. Gaillard, J. L. J., van Duijn, P., and Schaberg, A., Photometric determination of DNA in human chromosomes. *Exptl. Cell Res.* **53**, 417–439 (1968).

24. Garcia, A. M., and Iorio, R., Studies on DNA in leukocytes and related cells of mammals. V. The fast green-histone and the Feulgen-DNA content of rat leukocytes. *Acta Cytol.* **12**, 46–51 (1968).

25. Glenner, G. G., Enzyme histochemistry. *In* "Neurohistochemistry" (C. W. M. Adams, ed.), pp. 109–160. Elsevier, Amsterdam, 1965.

26. Goldman, R., Silman, M. J., Caplan, S. R., Kedem, O., and Katchalski, E., Papain membrane on a collodion matrix, preparation and enzymic behaviour. *Science* **150**, 758–760 (1965).

27. Gomori, G., "Microscopic Histochemistry, Principles and Practice," p. 193. University of Chicago Press, Chicago, Illinois, 1952.

28. Graumann, W., Zur Standardisierung des Schiffschen Reagens. *Z. wiss. Mikroskop.* **61**, 225–226 (1953).

29. Green, J. W., Determination of carbonyl groups. *Methods Carbohydrate Chem.* **3**, 49–54 (1963).

30. Hardonk, M. J., and van Duijn, P., Synthesis and properties of model systems, with their use in studying the Schiff reaction in histochemistry. *J. Histochem. Cytochem.* **12**, 533–537 (1964).

31. Hardonk, M. J., and van Duijn, P., The mechanism of the Schiff reaction as studied with histochemical model systems. *J. Histochem. Cytochem.* **12**, 748–751 (1964).

32. Hardonk, M. J., and van Duijn, P., A quantitative study of the Feulgen reaction with the aid of histochemical model systems. *J. Histochem. Cytochem.* **12**, 752–757 (1964).

33. Hardonk, M. J., and van Duijn, P., Studies on the Feulgen reaction with histochemical model systems. *J. Histochem. Cytochem.* **12**, 758–767 (1964).

34. Hicks, G. P., and Updike, S. J., The preparation and characterization of lyophilized polacrylamide enzyme gels for chemical analysis. *Anal. Chem.* **38**, 726–730 (1966).

35. Hiraoka, T., Quantitative extraction of fuchsin from Feulgen-stained nucleoprotein. *J. Biophys. Biochem. Cytol.* **3**, 525–544 (1957).

36. Holt, S. J., and O'Sullivan, D. G., Studies in enzyme cytochemistry. I. Principles of cytochemical staining methods. *Proc. Roy. Soc.* **B148**, 465–480 (1958).

37. Hopsu, V. K., and McMillan, P. J., Quantitative characterization of a histochemical enzyme system. *J. Histochem. Cytochem.* **12**, 315–324 (1964).

38. Hughes, H. K., Suggested nomenclature in applied spectroscopy. *Anal. Chem.* **24**, 1349–1354 (1952).

39. Ishida, M. R., A cytochemical study of nucleic acids in plant cells. III. Effects of protein on Feulgen reaction *in vitro*. *Cytologia* **24**, 107–114 (1959).

40. Kay, G., and Crook, E. M., Coupling of enzymes to cellulose using chloro-5-triazines. *Nature* **216**, 514–515 (1967).

41. Kelly, J. W., and Carlsson, L., Protein droplets, especially gelatin, hemoglobin and histone, as microscopic standards for quantitation of cytochemical reactions. *Exptl. Cell Res.* **30**, 106–124 (1963).

42. Lake, B. D., The histochemistry of phosphatases: The use of lead acetate instead of lead nitrate. *J. Roy. Microscop. Soc.* [3] **85**, 73–75 (1966).

43. Lojda, Z., van der Ploeg, M., and van Duijn, P., Phosphates of the naphthol AS series in the quantitative determination of alkaline and acid phosphatase activities "in situ" studied in polyacrylamide membrane model systems and by cytospectrophotometry. *Histochemie* **11**, 13–32 (1967).

44. Maddy, A. H., 1-Fluoro-2:4-dinitrobenzene as a cytochemical reagent. *Exptl. Cell Res.* **22**, 169–180 (1961).

45. McIndoe, W. M., and Davidson, J. N., The phosphorus compounds of the cell nucleus. *Brit. J. Cancer* **6**, 200–214 (1952).

46. Mann, J., Crystallinity of cellulose. *Methods Carbohydrate Chem.* **3**, 114–119 (1963).

47. Mayall, B. H., Variability in the stoichiometry of deoxyribonucleic acid stains. *J. Histochem. Cytochem.* **15**, 762–763 (1967).

48. Melnick, P. J., Leukemic leukocytes examined with histochemical enzyme technics. *Proc. 3rd Intern. Congr. Histochem. Cytochem., New York*, 1968, Summary Repts., pp. 176–177. Springer, Berlin, 1968.

49. Mendelsohn, M. L., The two-wavelength method of microspectrophotometry. III. An extension based on photographic color transparencies. *J. Biophys. Biochem. Cytol.* **4**, 425–431 (1958).

50. Meyer, K. H., "Natural and Synthetic High Polymers," p. 298. Wiley (Interscience), New York, 1950.

51. Mirsky, A. E., and Ris, H., Variable and constant components of chromosomes. *Nature* **163**, 666–667 (1949).

52. Mitz, M. A., and Summaria, L. J., Synthesis of biologically active cellulose derivatives of enzymes. *Nature* **189**, 576–577 (1961).
53. Mulder, M. P., van Duijn, P., and Gloor, H. J., Replicative organization of DNA in polythene chromosomes of Drosophila Hydei. *Genetica* **39**, 385–428 (1968).
54. Ornstein, L., The distributional error in microspectrophotometry. *Lab. Invest.* **1**, 250–265 (1952).
55. Ornstein, L., Disc electrophoresis. I. Background and theory. *Ann. N.Y. Acad. Sci.* **121**, 321–349 (1964).
56. O'Sullivan, D. C., Quantitative potentialities in enzyme cytochemistry. Modified Michaelis-Menten rate law applicable when a substrate diffuses slowly into an enzyme site. *J. Theoret. Biol.* **2**, 117–128 (1962).
57. Persijn, J. -P., and van Duijn, P., Studies of the Feulgen reaction with the aid of DNA incorporated cellulose films. *Histochemie* **2**, 283–297 (1961).
58. Ploeg, M. van der, and van Duijn, P., 5,6-Dihydroxy indole as a substrate in a histochemical peroxidase reaction. *J. Roy. Microscop. Soc.* [3], **83**, 415–423 (1964).
59. Ploeg, M. van der, and van Duijn, P., Cytophotometric determination of alkaline phosphatase activity of individual neutrophilic leukocytes with a biochemically calibrated model system. *J. Histochem. Cytochem.* **16**, 693–706 (1968).
60. Ploeg, M. van der, and van Duijn, P., Unpublished material.
61. Porath, J., Axén, R., and Ernback, S., Chemical coupling of proteins to agarose. *Nature* **215**, 1491–1492 (1967).
62. Rappay, Gy., and van Duijn, P., Aminoethylcellulose membranes as model for a quantitative cytochemical study of the ninhydrin-Schiff reaction. *Abstr. 2nd Intern. Congr. Histochem. Cytochem., Frankfurt, Germany, 1964*, Summary Repts., p. 188. Springer, Berlin, 1964.
63. Rasch, E., and Swift, H., Microphotometric analysis of the cytochemical Millon reaction. *J. Histochem. Cytochem.* **8**, 4–17 (1960).
64. Rigler, R., Jr., Microfluorometric characterization of intracellular nucleic acids and nucleoproteins by acridine orange. *Acta Physiol. (Scand.)* **67**, Suppl. 267, p. 10 (1966).
65. Ritzén, M., Quantitative fluorescence microspectrophotometry of catecholamine in formaldehyde products. *Exptl. Cell Res.* **44**, 505–519 (1966).
66. Rosselet, A., and Ruch, F., Cytofluorometric determination of lysine with dansylchloride. *J. Histochem. Cytochem.* **16**, 459–466 (1968).
67. Samuelson, O., Determination of carboxylic groups. *Methods Carbohydrate Chem.* **3**, 31–38 (1963).
68. Sandell, E. B., "Colorimetric Determination of Traces of Metals," p. 282. Wiley (Interscience), New York, 1944.
69. Sandritter, W., Ultraviolettmikrospektrophotometrie. *In* "Handbuch der Histochemie" (W. Graumann and K. Neumann, eds.), Vol. 1, Part I, pp. 220–338. Fischer, Stuttgart, 1958.
70. Sanger, F., The free amino groups of insulin. *Biochem. J.* **39**, 507–515 (1945).
71. Satterfield, C. N., and Sherwood, T. K., "The Role of Diffusion in Catalysis," pp. 56–63, Addison-Wesley, Reading, Massachusetts, 1963.
72. Shibata, K., Spectrophotometry of translucent biological materials—opal glass transmission method. *Methods Biochem. Anal.* **7**, 77–109 (1959).
73. Shugar, D., Quantitative staining in histo- and cytochemistry. *Progr. Biophys. Biophys. Chem.* **12**, 153–210 (1962).
74. Singer, M., and Morrison, P. R., The influence of pH, dye and salt concentration on the dye binding of modified and unmodified fibrin. *J. Biol. Chem.* **175**, 133–145 (1948).
75. Struyck, C. G., and van Duijn, P., Unpublished material.

76. Tonkelaar, E. M. den, and van Duijn, P., Photographic colorimetry as a quantitative cytochemical method. I. Principles and practice of the method. *Histochemie* **4**, 1–9 (1964).
77. Tonkelaar, E. M. den, and van Duijn, P., Photographic colorimetry as a quantitative cytochemical method. II. Determination of relative amounts of DNA in cell nuclei. *Histochemie* **4**, 10–15 (1964).
78. Tonkelaar, E. M. den, and van Duijn, P., Photographic colorimetry as a quantitative cytochemical method. III. Determination of the absolute amount of DNA in cell nuclei. *Histochemie* **4**, 16–19 (1964).
79. Tupper, J. L., and Weaver, K. S., Quantitative evaluation of the developed image. *In* "The Theory of the Photographic Process" (C. E. Kenneth Mees, ed.), pp. 809–846. Macmillan, New York, 1963.
80. Vendrely, R., and Vendrely, C., Arginine and deoxyribonucleic acid content of erythrocyte nuclei and sperms of some species of fishes. *Nature* **172**, 30–31 (1953).
81. Walker, P. B. M., Ultraviolet microspectrophotometry. *Gen. Cytochem. Methods* **1**, 163–217 (1958).
82. Wheeler, A., Reaction rates and selectivity in catalyst pores. *Advanc. Catalysis* **3**, 249–327 (1951).

THE POTENTIAL OF QUANTITATIVE CYTOCHEMISTRY IN TUMOR AND VIRUS RESEARCH*

Frederick H. Kasten †

DEPARTMENT OF ULTRASTRUCTURAL CYTOCHEMISTRY, PASADENA FOUNDATION FOR MEDICAL
RESEARCH, PASADENA, CALIFORNIA

Twenty years have passed since Arthur W. Pollister opened up the field of quantitative cytochemistry, working in the visible light region in his laboratory at Columbia University. His work and that of his graduate students during the late 1940's and early 1950's is well known. Most of the analytic data of this period were obtained by using the Feulgen cytochemical reaction for DNA. Surprisingly enough, the same reaction still serves as our most useful technique for cytophotometric analyses in cellular pathology. This appears to be the case despite the fact that we are 45 years "post-Feulgen" and the chemistry of the reaction is not fully explained nor is the physical nature of the DNA–dye complex understood. This is the "enigma" I spoke of in 1962 at the First Feulgen Memorial Lecture in Giessen (9).

Passing to the main subject of this chapter, I would like to emphasize at the outset a point with which I think many of you would agree, namely that quantitative cytochemistry is only one of many tools available for the solution of cellular pathologic problems. Whenever possible, other experimental approaches such as fluorescence microscopy, radioautography, phase-contrast microscopy, and electron microscopy should be employed. I would like to discuss some personal results on the following three general problems in the area of tumor and virus research, which I feel may be instructive

* Supported by U.S. Public Health Service Research Grants NB-03113 from the National Institute of Neurological Diseases and Blindness and CA-07991 from the National Cancer Institute and by a grant from The Milheim Foundation for Cancer Research.
† Present address: Department of Anatomy, Louisiana State University Medical Center, New Orleans, Louisiana.

examples of the association of quantitative cytochemistry with other experimental approaches:

(1) Quantitative DNA alterations during the process of azo dye-induced liver carcinogenesis

(2) Experimental studies of the cell cycle of chemically synchronized tumor cells *in vitro*

(3) Virus–cell interaction involving DNA viruses

I. Experimental Liver Carcinogenesis

One of the fundamental problems in experimental carcinogenesis is to detect the earliest possible chemical alterations which would help pinpoint the site of action of the carcinogen. The well-known azo dye, butter yellow or dimethylaminoazobenzene (DAB), when fed to rats induces liver cancer

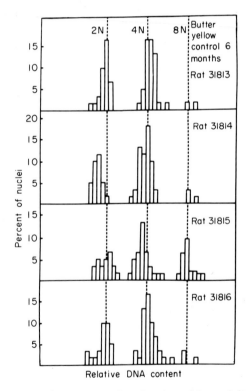

FIG. 1. Feulgen-DNA histograms from liver imprints of four adult rats. Typical polyploid distributions are seen. The 2N or diploid position is set in each case by measuring a group of kidney nuclei from the same animal.

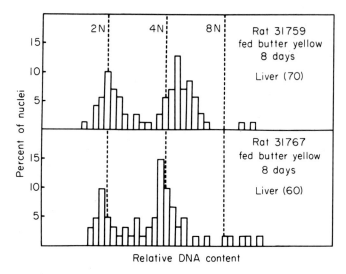

Fig. 2. Feulgen-DNA histograms from liver imprints of two rats which had been on the DAB carcinogenic diet for 8 days. At this early stage, atypical DNA values are detected in the triploid range. The 2N values are determined as in Fig. 1.

in a large proportion of animals in less than a year. This system appeared ideal to test with Feulgen cytophotometry the hypothesis that the carcinogen interacts with DNA in the precancerous stages. Previous investigations were not in agreement (3, 4, 23, 24). The work to be summarized was done in collaboration with C. Vendrely and M. Riviere and was presented at the First International Histochemistry Symposium in Warsaw in 1962 (21).

Treated and control rats were sacrificed at regular intervals during a 6-month period. Small random pieces of kidney and liver were excised and used for standard histopathologic sections and imprint preparations for Feulgen staining. Since the chemical carcinogen apparently acts specifically on the liver and has no influence on kidney cells, this made it practical to use the kidney cells from the same animal to establish the diploid DNA value. By placing them on the same slide, external variations between the two cellular populations were minimized with respect to fixation and Feulgen staining. Measurements of 20–30 kidney nuclei proved an adequate sample to establish the diploid or 2N value. At least 60 hepatic cell nuclei were measured on each slide.

The imprint technique employed here is simple and involves rubbing the cut surface of the fresh organ on the surface of a glass slide. If the wet surface becomes thick with clumps of tissue, it is useful to spread this by rubbing another slide over the area. However, too vigorous a rubbing will break the nuclei. Liver imprints are made in one area and kidney imprints in another area of the same slide. Slides are air-dried and then fixed by formaldehyde vapor in a closed staining dish. The imprint technique leads to very little distributional error, judging from the relative DNA results which were obtained using careful measurements at a single wavelength. The DNA distributions in livers of four rats are shown in Fig. 1. The 2N, 4N, and 8N values calculated from kidney determinations coincide very well with the positions of the three prominent liver ploidy classes. The classes are distinct and well separated from each other. Atypical polyploid classes are already evident in liver populations from animals on the diet only 8 days (Fig. 2). The most striking change is the presence of some cells with intermediate DNA values, primarily in the triploid range. This altered DNA pattern persists and is further exaggerated in most animals on the diet for many weeks.

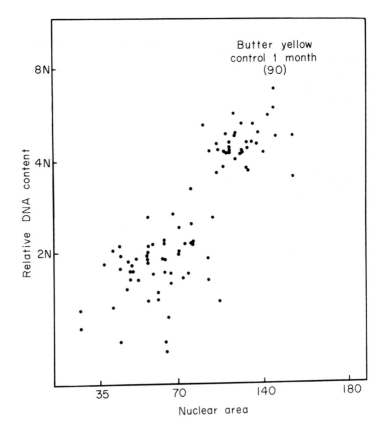

FIG. 3. Feulgen-DNA data from 90 liver nuclei of a control rat are plotted as relative DNA versus nuclear area. Two main groups appear segregated.

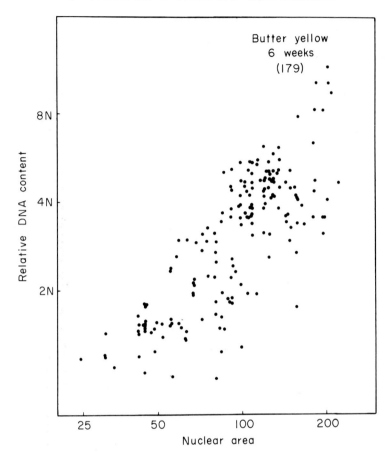

FIG. 4. Feulgen-DNA data from 179 liver nuclei of a DAB-fed rat are plotted as in Fig. 3. The lack of discrete groups reflects the presence of some nuclei in the process of DNA synthesis.

Not only do the diploid and tetraploid peaks tend to become obliterated, but the octoploid group likewise is dispersed. This difference in DNA population patterns is seen especially well when the data are plotted in terms of individual DNA values against nuclear area (Figs. 3 and 4). There are no longer typical, discrete, cell populations during the precancerous period, especially after 1 month or longer on the diet. There is a progressive increase in nuclear size and DNA content of cells with continued exposure to the carcinogen; mitoses and nuclei with bizarre star-shaped chromatin patterns are seen.

   A few exceptional animals on the carcinogenic diet displayed normal DNA histograms during the precancerous period (Fig. 5). As a tentative explanation, it is suggested that each one of these rare individuals was "resistant" to the action of the carcinogen and would not have later developed liver cancer. The suggestion seems reasonable in view of other reports that some treated animals always fail to develop liver cancer, but the hypothesis suggested here certainly needs to be tested. The possibility of predicting the eventual development or nondevelopment of malignancy in this system by DNA measurements is an exciting prospect.

Photomicrographs of Feulgen-stained nuclei from imprints are shown in Fig. 6. Magnifications are identical so that one can see size alterations readily. Control liver nuclei (Fig. 6a) show at least two size classes in the photographed field. An example of an extraordinarily large nucleus is shown from an animal on the diet for 3 weeks (Fig. 6b). The DNA content of this nucleus is 16N or greater. Preneoplastic liver tissues contain large nuclei as well as dividing cells and peculiar nuclei with a prophasic appearance. In conclusion, this investigation of precancerous stages of carcinogenesis reveals that there is increased DNA proliferation in liver, beginning approximately 1 to 2 weeks after animals are placed on the DAB diet. It is suggested that the carcinogen acts quickly on DNA control mechanisms, e.g., by releasing DNA synthesis inhibitors. Continued DNA synthesis and mitotic phenomena may be prerequisites for the later development of tumors.

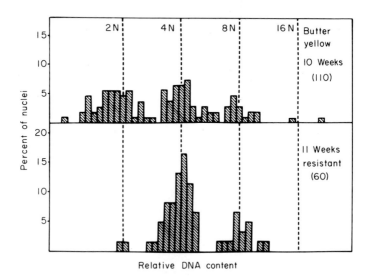

FIG. 5. The upper histogram shows the Feulgen-DNA values from the liver of an animal on the carcinogenic diet for 10 weeks. Nuclei with intermediate amounts of DNA are observed between the main polyploid classes. A few nuclei with 16N amounts of DNA are seen. The lower histogram from the liver of an animal on the diet 11 weeks shows a normal DNA pattern. This animal is thought to be "resistant."

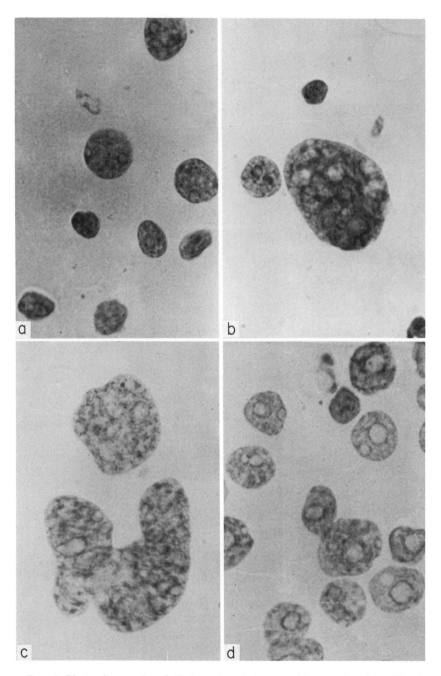

FIG. 6. Photomicrographs of Feulgen-stained liver nuclei from imprints. Magnifications are all 1260 ×. (a) Control; (b) 3 weeks on DAB diet showing giant nucleus; (c) hepatoma nuclei; (d) another hepatoma showing hypertrophied nucleoli.

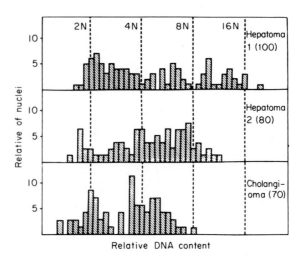

FIG. 7. Feulgen-DNA histograms from three liver tumors induced by DAB feeding. The 2N values are determined as in Fig. 1. The two hepatomas present a scattered DNA distribution without a dominant mode. The cholangioma has diploid and tetraploid modes.

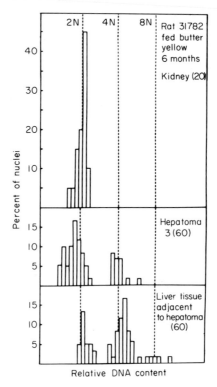

FIG. 8. DNA patterns in a hepatoma are compared with DNA values in kidney and uninvolved liver of the same animal. The tumor is hypodiploid with no evidence for DNA synthesis at the time of removal. Liver-DNA values are normal.

It is not possible to discuss in detail the results of DNA analyses of liver tumors, which began appearing in this experiment after animals were on the DAB diet for about 6 months. Again, kidney nuclei from each animal served as an internal diploid standard by which to compare the relative DNA content of tumors. There was no evidence that this carcinogen affects the morphology or DNA content of kidney cells.

Most of the tumors were diagnosed as hepatomas; a few were classed as cholangiomas. DNA data from three tumors are shown in Fig. 7. The two hepatomas present a scattering of DNA values with more present in the 8N and 16N range than are observed in normal liver. There is no evident mode which dominates either population, nor are the diploid and tetraploid peaks present. On the other hand, the DNA content of the cholangioma appears only slightly different from that of normal liver (Fig. 1). In other hepatomas examined, dominant DNA modes are seen, which occur at the diploid, hypodiploid, or hyperdiploid positions. An example of a hypodiploid tumor is shown in Fig. 8. The lack of intermediate values between 2N and 4N indicates that DNA synthesis had temporarily ceased by the time this tumor was removed. Kidney-DNA values are shown as well as data from liver tissue adjacent to the hepatoma. These two histograms reveal normal patterns. In addition to the finding that each induced liver tumor behaves as an unpredictable individual with respect to its DNA pattern, it is interesting to note that certain nuclear abnormalities accompany the tumors. These alterations consist of large, abnormal-shaped nuclei (Fig. 6c) and hypertrophied nucleoli (Fig. 6d), which suggest a failure of DNA control and heightened nucleolar RNA synthesis.

II. Chemical Synchronization of Cancer Cells *in Vitro*

In order to allow analyses of nucleoproteins and observations of cellular morphology during discrete periods of the cancer cell cycle, techniques have been developed in recent years to synchronize cell populations in culture. We employ an epithelial tumor cell line developed in our laboratory which is termed the CMP cell line. Phase-contrast and electron micrographs of typical CMP cells are shown in Figs. 9 and 10. The synchrony procedure with which our group has the greatest experience is known as the double-thymidine technique (*17, 18, 20*). This work has greatly benefited by the participation of F. Strasser. Briefly, cells are treated with 2.5 mM thymidine for 24 hr, followed by 15 hr in normal medium, and then a second thymidine blockage for 24 hr. DNA synthesis is inhibited with excess thymidine by preventing the formation of deoxycytidine triphosphate from cytidine-5' phosphate (*30*).

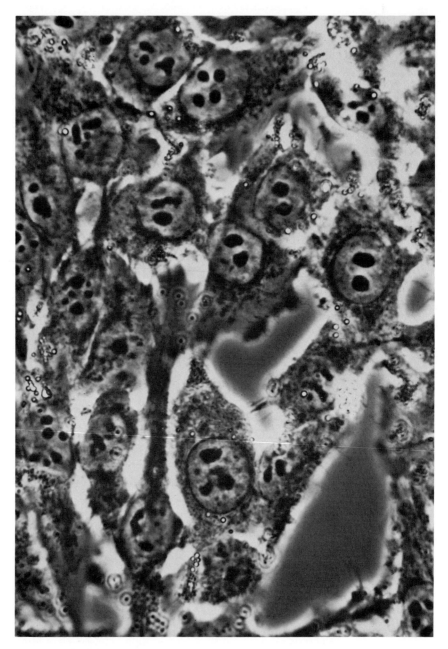

FIG. 9. Phase-contrast photomicrograph of CMP epithelial tumor cells in culture; 3-day culture, passage 100.

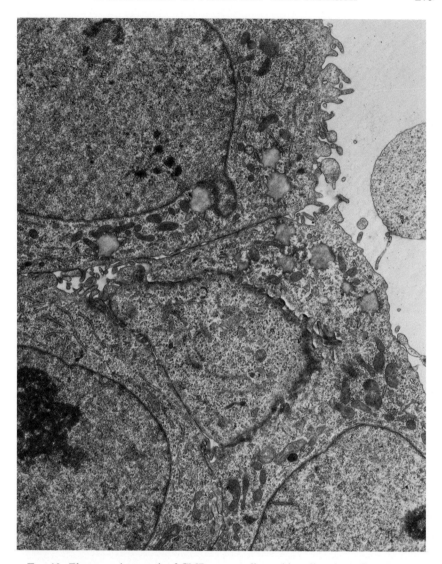

FIG. 10. Electron micrograph of CMP tumor cells used in cell cycle studies. 9735 ×.

There is also a partial blockage of chromosomal RNA synthesis and a complete inhibition of nucleolar RNA synthesis (20). Radioautographs demonstrating this are shown in Fig. 11a and b. Cells are accumulated in the late G_1 phase of the cell cycle and immediately begin DNA synthesis when they are placed in normal medium at the end of the second thymidine block. During the first hour, the radioactive thymidine is diluted by the

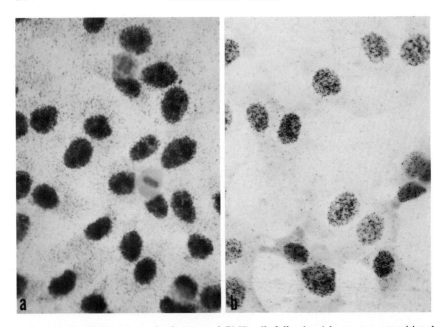

FIG. 11. (a) Radioautograph of untreated CMP cells following 1-hr exposure to tritiated uridine. Labeled RNA appears heavy over nucleoli, moderate over nucleus, and slight over cytoplasm. (b) Cells treated with excess thymidine and labeled as above fail to synthesize RNA in nucleoli and show reduced chromosomal RNA synthesis. [From Kasten *et al.* (*20*) and Kasten and Strasser (*18*).]

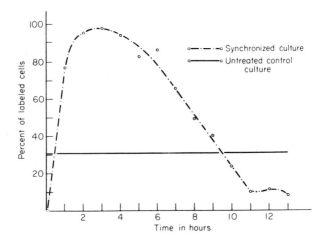

FIG. 12. Curve demonstrates ³H-thymidine labeling index of synchronized cells compared with untreated CMP cells. Synchrony is about 100% near beginning of S phase and is somewhat reduced by the time cells reach the early G_1 phase at 13 hr. [From Kasten and Strasser (*18*).]

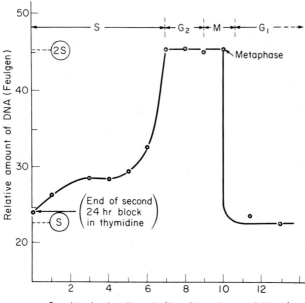

FIG. 13. Feulgen-DNA content of chemically synchronized cells. Data obtained with Barr & Stroud Scanning Microdensitometer. Relative DNA content is close to the basic S value. DNA content during S phase increases nonlinearly. About two-thirds of total DNA is synthesized in the last hour of the S phase. Remainder of curve follows predicted changes.

unused pool of unlabeled thymidine still present in cells from the synchrony pretreatment. DNA synthesis is detected in practically all cells beginning at the second hour, as judged by ^{3}H-thymidine autoradiography (Fig. 12), reflecting the high degree of synchrony in the system. The per cent of labeled cells falls as the cell cycle proceeds into the G_2 and M phases. The rate of DNA synthesis is not uniform during the S phase. By sampling cells at sequential periods, it is possible to show by Feulgen cytophotometry (Fig. 13) that the amount of DNA per nucleus increases gradually, reaches a brief plateau in the mid-S phase, and finally exhibits a very rapid rate of increase to the doubled value in the last hour or two. The nature of the fast replicating DNA of the terminal S phase is unknown. These data were obtained using the Barr & Stroud Scanning Microdensitometer.

By means of time-lapse cinematography (19), it can be demonstrated that a mitotic burst occurs (Fig. 14a–d), which may be as high as 70% at a given time. It is not possible to abstract for this paper all the important film sequences shown in the original synchrony film presented at the conference. These sequences demonstrated the action of excess thymidine in eliciting violent cell membrane activity, heightened vibrations of attachment fibers, increase in cellular and nuclear sizes, and inhibition of mitosis. Other

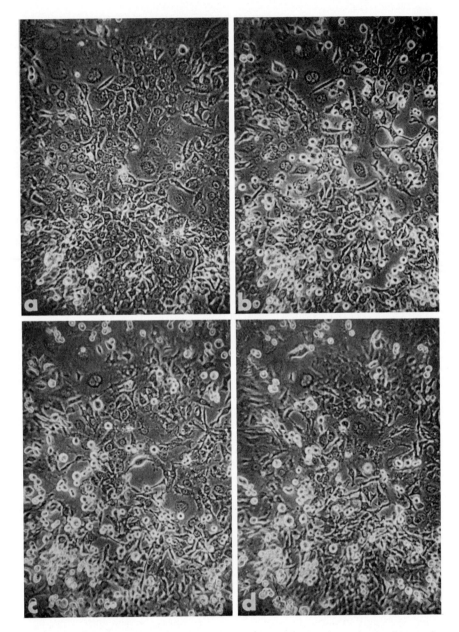

FIG. 14. Abstracts from time-lapse film of synchronized culture shows changes in frequency of dividing cells (halated elements): (a) 10 hr postwash; (b) 12 hr postwash; (c) 14 hr postwash; (d) 16 hr postwash. Mitotic peak in this experiment occurs close to 12 hr.

FIG. 15. View of a portion of time-lapse cinematography laboratory. Four incubated cine units are shown with 16-mm movie cameras above and time control units below. Two units on left side each contain double microscopes to allow simultaneous filming of two fields. A monitor is mounted on second unit to allow closed-circuit TV monitoring on expanded screen. Two other cine units not shown complete the facility.

sequences revealed dynamic changes during the cycle in nucleolar number and morphology at the light and electron microscopic levels. Following mitosis, the early G_1 phase is typified by extremely active membrane movements generally associated with pinocytosis. A photograph of a portion of our time-lapse photography laboratory is shown in Fig. 15. The complete facility includes six time-lapse units equipped for all levels of modular operation. Mitotic counts on fixed and stained preparations likewise reveal a high degree of synchrony (Fig. 16). However, there is a gradual decay with time so that the level of synchrony in the early G_1 phase (14 hr postwash) is approximately 60%, compared with the 95–100% seen at the beginning of the S phase. In its present form, the double-thymidine block technique is a powerful tool, and considerable useful information about the cell cycle has been obtained.

DNA synthetic patterns vary considerably in the S phase (18). Intranucleolar DNA synthesis occurs at two distinctive times near the mid-S phase. The first nucleolar DNA which is synthesized apparently represents template DNA, since nucleolar RNA synthesis is blocked during nucleolar DNA replication (Fig. 17a). Perinucleolar DNA is likewise synthesized at a different time from the intranucleolar DNA. An example of the labeling pattern near the beginning of the S phase is shown in Fig. 17b. The ultra-structural localization of late-replicating DNA is shown by electron radio-autography to be associated with the nuclear membrane and the nucleolus (Fig. 18). This unusual labeling pattern is seen in practically every cell examined from a synchronized population, pulse-labeled for 30 min at the end of the S phase. The results are suggestive of a significant biologic relationship between this DNA fraction and the nuclear membrane which will disappear about an hour later at the onset of mitosis. This finding contradicts a recent report (2) where it is claimed that DNA labeling at the nuclear membrane is associated specifically with the very early S phase.

This synchronized system has been employed to determine the rate of RNA synthesis using tritiated uridine (18). Chromosomal RNA synthesis per nucleus is maximum in the mid-S phase and is preceded by a burst of nucleolar RNA synthesis. The role of the nucleolus is further emphasized by the fact that nucleolar RNA is at least three times more responsible to actinomycin D than is chromosomal RNA (12).

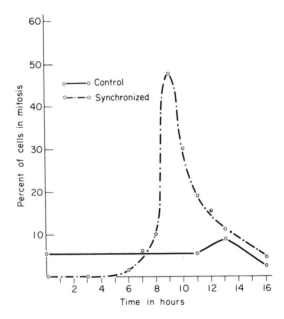

FIG. 16. Mitotic counts of fixed and stained preparations of synchronized cells reveal a mitotic burst of 50% at 9 hr postwash. This percentage is a lower limit since dividing cells are preferentially lost during fixation. Untreated cells have about 5% mitoses.

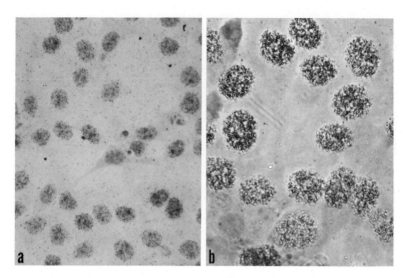

FIG. 17. (a) Radioautograph of synchronized cells labeled with tritiated uridine for 1 hr prior to the mid-S phase. Chromosomal RNA synthesis is seen. Nucleolar RNA synthesis is inhibited due to intranucleolar DNA replication which takes place at this time. (b) Uniform DNA synthetic pattern is seen in this radioautograph of synchronized cells labeled for 1 hr near the beginning of the S phase. [From Kasten and Strasser (*18*).]

In order to attain a higher synchronized G_1-phase population than is possible by the double-thymidine block procedure alone, an important modification was introduced. At the mitotic burst, which occurs usually at 10 hr postwash, large numbers of dividing cells are gently sprayed off the glass surface with calcium- and magnesium-free medium. The tendency of dividing cells to be removed is due to their property of rounding up during mitosis and temporarily separating from the glass surface. Practically a 100% population of viable cells in mitosis may be obtained in large numbers, using the combined chemical and mechanical methods of obtaining synchronized populations. For this reason, the method is termed the "thymidine-spray technique" (*13*). Populations of cells in specific stages of mitosis may be harvested for light microscope, ultrastructural, and biochemical analyses or they may be seeded into new vessels to start a newly synchronized G_1 population. Figure 19a shows untreated CMP cells, which have a normal mitotic rate of 3–4% and a field of dividing cells (Fig. 19b) which were collected and fixed after using the thymidine-spray technique. There is a 100% yield of cells in mitosis, largely metaphase, in this example. This biologic sample proved very useful in combination with phase-interference optics (Nomarski) (Fig. 20). Some of the fibers are seen to run continuously from pole to pole through the intermediate mass of metaphase chromosomes.

III. Virus–Cell Interaction Involving DNA Viruses

Quantitative cytochemistry in combination with fluorescence microscopy and phase-contrast microscopy has yielded considerable information about the nucleic acid changes induced during DNA virus infection or viral neoplastic transformation, according to which host system is infected.

Interaction of DNA viruses with suitable host systems results typically in

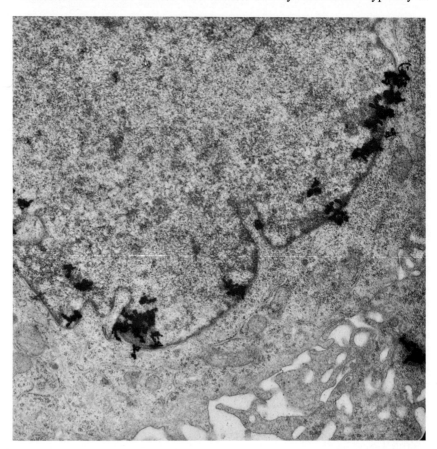

Fig. 18. Typical electron microscope radioautograph of synchronized CMP cell after labeling with tritiated thymidine for 30 min at the end of the S phase. Labeled DNA is restricted to the nuclear membrane. Nucleoli also are simultaneously labeled at this time. [From Kasten and Jacob (16).] 14,750 ×.

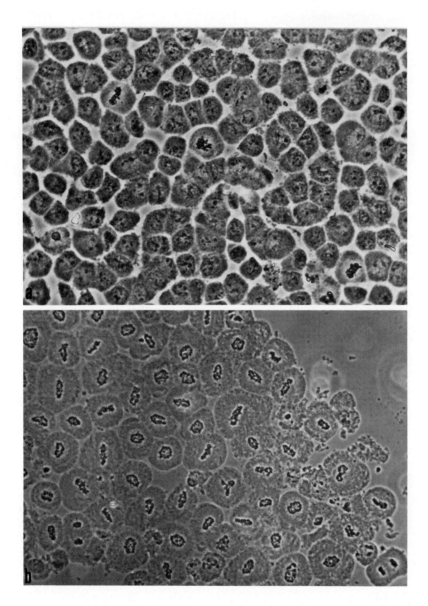

FIG. 19. Fixed preparations of CMP cells stained with aceto-dahlia and photographed with phase optics. (a) Random population. (b) Field of cells in metaphase after collection by thymidine-spray technique.

certain cytopathogenic effects—primarily in the nucleus where the principle
cellular DNA synthesizing system is located. New viral DNA is synthesized
which is associated with involvement of the nucleolus, a reduction of host
DNA, large concentrations of virus DNA, and a rather bizarre nuclear
morphology. Eventually, the infected cell releases large numbers of new
virus particles and dies.

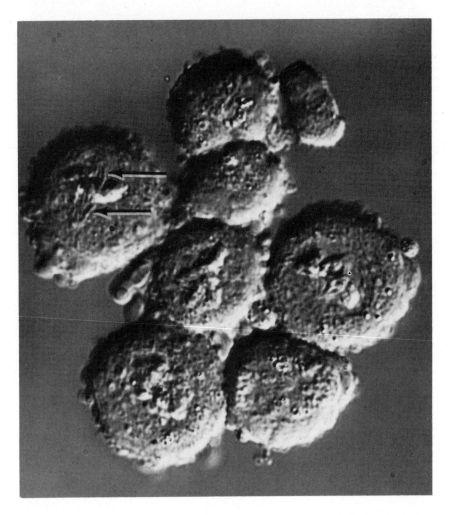

FIG. 20. Group of live tumor cells obtained at mitosis by the thymidine-spray technique
and photographed with the aid of phase-interference (Nomarski) optics. Three-dimensional
features are emphasized. Note presence of spindle fibers. In one case, fiber passes from
pole to pole through the metaphase chromosomes (arrows).

An example of such interaction is the experimental infection of cultured monkey kidney cells with SV40 virus. This simian virus also induces tumors in young hamsters. Both green monkey and baboon kidney cells *in vitro* may be infected to produce typical DNA alterations which are followed by Feulgen staining (22). During the first few days of infection, nuclear DNA becomes aggregated into many coarse granules (Fig. 21a). This is followed by the formation of a central mass of DNA which has a typical homogeneous appearance (Fig. 21b). During the early stages, perinucleolar DNA is activated (Fig. 21b) and abnormal amounts of intranucleolar DNA are seen (Fig. 21c). The nucleus eventually builds up an intensive accumulation of Feulgen-positive material in the form of one or more inclusion bodies (Fig. 21d) which contrast markedly with uninfected nuclei (Fig. 21d). The full course of events described takes about 5–7 days. Similar DNA changes are seen when human cells are infected with various human adenoviruses. In ultrathin sections, such infected nuclei display a bizarre pattern of chromatin, large numbers of incomplete and complete viral particles, and abnormal dense granules within a hypertrophied nucleolus (Fig. 22).

By means of the fluorescent-Feulgen reaction (5–8, 10, 11) the distribution of DNA is seen in fine detail, as in Fig. 23. The nucleus of a cultured salamander lung cell displays coarse, heterochromatic channels interspersed with fine granules and filaments. The nucleolus clearly contains a small amount of DNA. The resemblance of this fluorescence photomicrograph to "Le Grand Charles" is evident. The fluorescence-Feulgen technique is a relatively simple procedure and enables one to follow virus-induced DNA alterations with greater sensitivity and detail than with the standard Feulgen reaction using transmitted light. For example, after 1-day infection, the coarse granules are easily seen (Fig. 21e), as are the emerging inclusions a few days later (Fig. 21f).

In describing the intranuclear-DNA lesions, it is not possible with the Feulgen or the fluorescent-Feulgen techniques to distinguish host DNA from viral DNA. For this purpose, a sequential DNase-fluorescent technique was employed (15) which takes advantage of the fact that viral DNA resists digestion by DNase unless the protein coat is first removed with pepsin (1, 25). The enzymic technique in combination with the fluorescent-Feulgen reaction gives results which are illustrated in Fig. 24. Enzymically untreated populations of infected nuclei (Fig. 24a) show the typical nuclear DNA patterns described earlier. When fixed cells are exposed to DNase, there is first a loosening up of cellular DNA in the nuclear center (Fig. 24b), followed by an increasing loss of DNA within the interior, leaving a narrow rim of DNA at the nuclear membrane (Fig. 24c). Eventually, even this resistant DNA is digested so that the cells appear to contain holes filled with

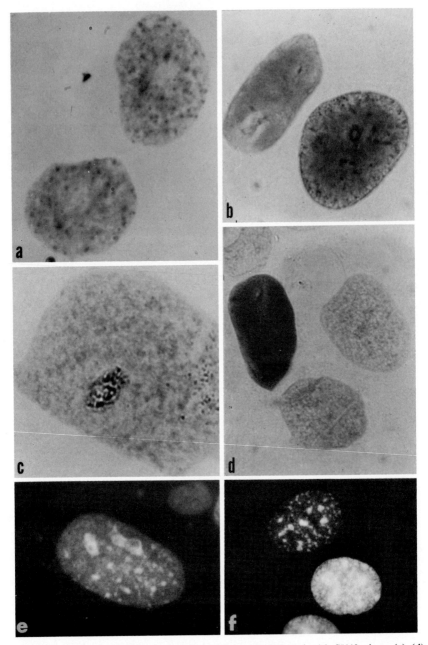

Fig. 21. Cultured monkey and baboon kidney cells infected with SV40 virus: (a)–(d) stained for DNA in Feulgen reaction; (e)–(f) fluorochromed for DNA using the auramine

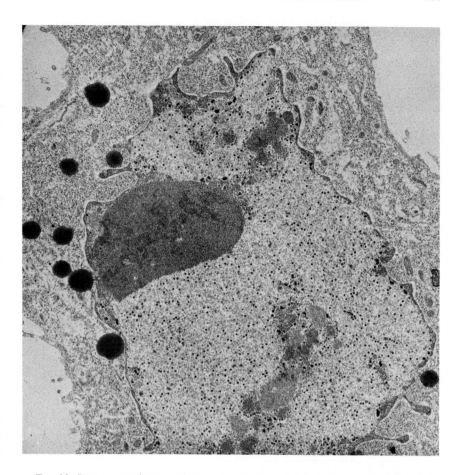

FIG. 22. Low-power electron micrograph of HeLa cell infected with adenovirus 1 for 28 hr. Nucleus is filled with newly synthesized viral particles and abnormal patterns of chromatin. Hypertrophied nucleolus contains unusual dense granules. 7670 ×.

O–SO$_2$ reagent in the fluorescent-Feulgen reaction. (a) 3 days infection. Coarse DNA granules are seen. (b) 5 days infection. Nucleus at left beginning to form inclusion. Nucleus at right in intermediate stage between granule deposition and appearance of inclusion. Perinucleolar DNA very active. (c) 4 days infection. Multiple DNA granules in nucleolus. (d) 5 days infection. Nucleus at left filled with DNA inclusion. Two other nuclei are uninfected. (e) 1 day infection. Abnormal DNA granules present. (f) 3 days infection. Early and late stages of nuclear-DNA alterations. [(a)–(d) From Kasten *et al.* (*22*).]

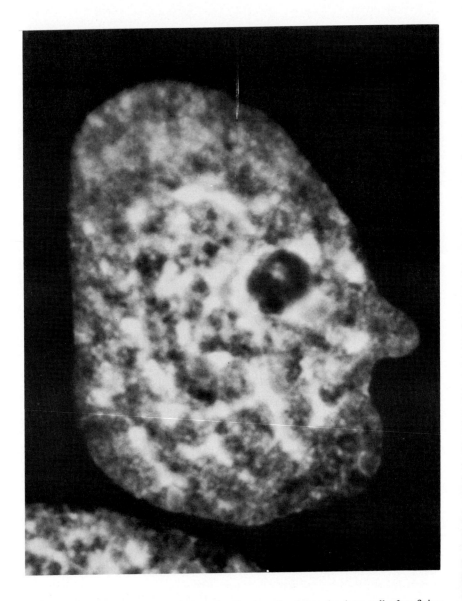

FIG. 23. Fluorescence photomicrograph of cultured salamander lung cell after fixing and staining for DNA by fluorescent-Feulgen reaction (auramine O–SO₂ reagent). Details seen of heterochromatic DNA channels, granules, and filaments. Intranucleolar DNA in "eye" of "Le Grand Charles" is detected.

nucleoli (Fig. 24d). Throughout this digestion and at the end, viral DNA remains unaffected and fluoresces intensely and specifically (Fig. 24b, c, d).

Quantitative DNA measurements (22) of uninfected cells reveals a multimodal population with peaks corresponding to a polyploidy series (Fig. 25). The presence of DNA values bridging the modes suggests that a number of cells are in DNA synthesis. This is supported by mitotic counts which reveal a division rate of 2–3%.

Random measurements of Feulgen-DNA in infected cultures show similar type histograms as the control cultures, with noticeably larger numbers of cells with higher DNA values. When measurements are made on a more selective basis (Fig. 26), it is clear that normal-looking nuclei display the same DNA pattern as in the controls but infected nuclei contain irregular and increasing amounts of DNA. These two populations are nicely separated when the data are plotted on the basis of individual DNA values against nuclear areas (Fig. 27). It is seen that infected nuclei with the same amounts of DNA as the controls are always smaller in size.

When rat fibroblasts are treated with SV40 virus, a completely different sequence of events occurs. Instead of inducing cytopathogenic effects and multiplying, the virus causes a morphologic cell transformation into a tumorigenic population. It is not possible to discuss this problem but I would like to point out that the transformed cells have exactly double the DNA amount as their normal counterparts (28). Feulgen-DNA histograms which illustrate this dramatic difference are shown in Fig. 28.

As a final example of virus–cell interaction, a phase-contrast time-lapse film was originally shown, revealing peculiar cytoplasmic and nuclear phenomena in human diploid fibroblasts infected with cytomegalovirus (CMV)—an unusual DNA virus of the herpes group (29). The virus induces a cellular contraction followed by the appearance of cytoplasmic aggregates at one pole of the nucleus. Intense cytoplasmic activity is generated and appreciated best in the original film. The entire mass of vacuoles and granules migrates to the Golgi zone. In the meantime, the nucleus becomes flattened against the cell membrane and assumes a U-shape where the cytoplasmic mass impinges on it. Peculiar needles or crystals are associated temporarily with the nucleus. All of this occurs within the first 24 hr. Some of these features are shown in Fig. 29a–f.

During the next 24 hr, the nucleus with its associated cytoplasmic lesion always undergoes a slow rotation, much slower and different from the nuclear rotation commonly observed in cultured cells. Small, intranuclear inclusions develop in this period. The nucleolus remains prominent and is frequently

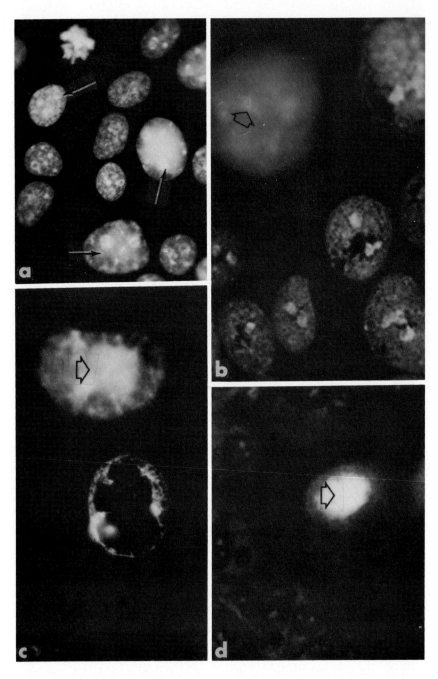

FIG. 24. Monkey kidney cells *in vitro* infected with SV40 2 days and fluorochromed for DNA with fluorescent-Feulgen reaction (auramine O–SO$_2$ reagent). (a) No DNase pre-

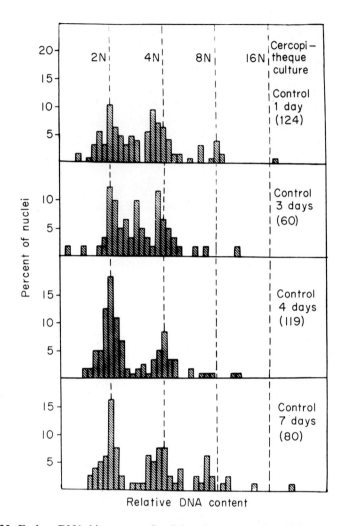

FIG. 25. Feulgen-DNA histograms of uninfected green monkey kidney cells *in vitro*. [From Kasten *et al.* (*22*).]

treatment. Note normal nuclei and others in different stages of infection (arrows). (b)–(d) Predigested with DNase for varying times before staining. Note loosening up of DNA within nuclei of (b); further removal of cellular DNA and presence of resistant DNA fraction at nuclear membrane in (c); and complete removal of host DNA from cells at left side of (d). Viral DNA (arrows) present in inclusions of (b)–(d) are not digested and become easily detected. [From Kasten and Gerber (*14*).]

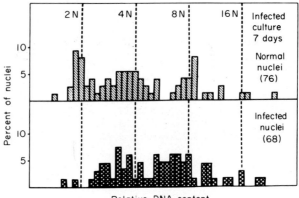

FIG. 26. Feulgen-DNA histograms from a monkey kidney culture infected 7 days with SV40 virus. Distributions of two selected cell populations are shown. Infected nuclei contain irregular and larger amounts of DNA than normal-looking nuclei of the same culture. [From Kasten *et al.* (*22*).]

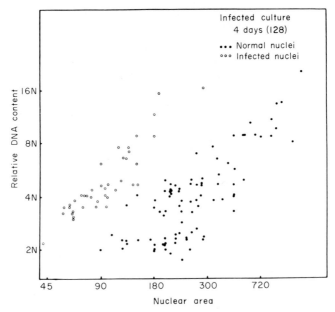

FIG. 27. Feulgen-DNA data from two cell populations of SV40-infected culture. Infected nuclei are smaller than normal ones with same DNA content. [From Kasten *et al.* (*22*).]

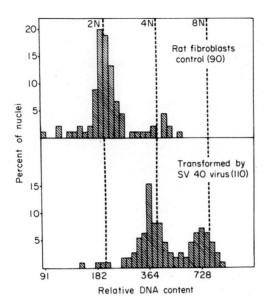

FIG. 28. Feulgen-DNA histograms of normal and viral-transformed cells. [From Vendrely *et al.* (*28*).]

hypertrophied. Phase-contrast photomicrographs and film abstracts of these phenomena are shown in Fig. 30a–f.

In the final stage, a new cytoplasmic inclusion appears in the center of the Golgi-centered lesion. The entire cell undergoes violent activity while in a contracted state and finally explodes at 2–4 days very dramatically, leaving a shrunken, nonliving cellular remnant. A series of film abstracts which illustrate the final stage is presented in Fig. 31a–d.

This phase-contrast cinematographic investigation by itself sheds no light on the cytochemical events associated with CMV infection. Other studies (*26*) showed that the cytoplasmic lesion contains lipid, carbohydrate, protein, and some RNA. The nuclear inclusions are Feulgen-positive, as is the cyto-plasmic inclusion which develops later. In electron microscopy studies (*27*) it is shown that intranuclear viral particles develop an extra viral coat membrane at the nuclear membrane enroute to the cytoplasm. Lysosomes apparently are stimulated by the virus infection and play an important role in the formation of the cytoplasmic lesion. Many viral particles are found within lysosomes in the end stage. It is not clear why this particular DNA virus is so exceptional in inducing the formation of a cytoplasmic DNA inclusion in addition to the typical nuclear inclusions induced by this and other DNA viruses.

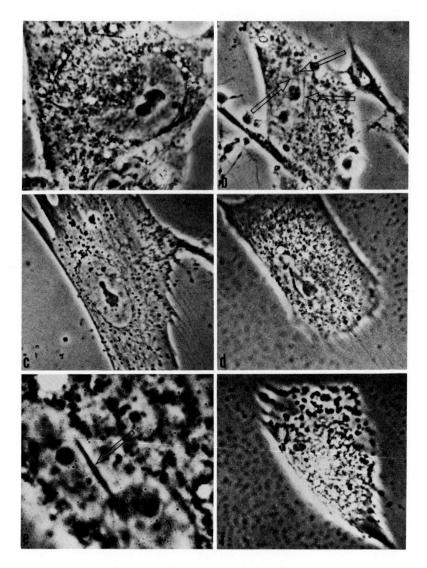

FIG. 29. Phase-contrast photomicrographs of human diploid fibroblasts (W138). All
fields are of cells infected with cytomegalovirus (CMV) except (c); (c)–(f) are film abstracts
of same cell. (a) 77 hr infection. Compressed U-shaped nucleus. (b) 21 hr infection. Multiple
needles associated with nucleus at arrows. (c) Control cell. (d) 22 hr infection. Note compressed cell and needle in nucleus (arrow). (e) 31 hr infection. High magnification of nucleus
shows two nucleoli and needle at same optical plane. (f) 84 hr infection. Appearance of
contracted cell just prior to explosion and death. Violent membrane activity shown by
phase-dense structures directed upward. (Cells and virus supplied to author by H. T. Wright,
Jr. and R. M. McAllister.)

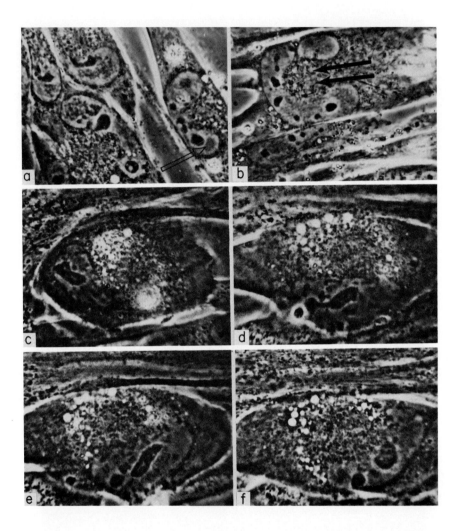

FIG. 30. Same material as in Fig. 29; (c)–(f) are film abstracts of same cell. (a) and (b) 77 hr infection. Note prominent nucleoli, nuclear inclusions (arrows), U-shaped nuclei, and cytoplasmic lesions in Golgi zones (double arrow). (c) 20 hr infection. Contracted cell with vacuoles marking edge of cytoplasmic lesion. (d) 31 hr infection. Nuclear shift, nucleolus in process of cleavage, small nuclear inclusions. (e) 32 hr infection. Further nuclear rotation, nucleolus fully separated. (f) 39 hr infection. Continued unidirectional nuclear rotation.

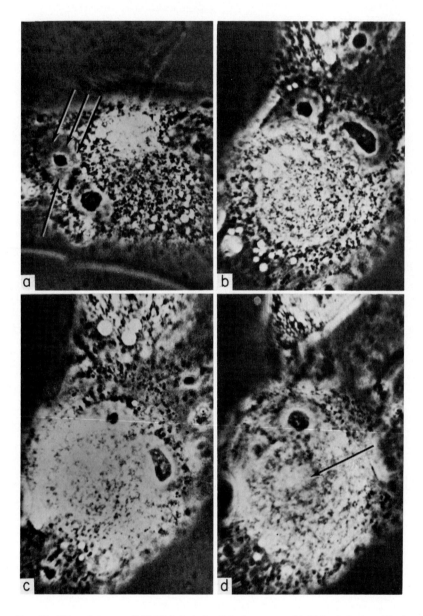

FIG. 31. Film abstracts of CMV-infected cell in late stage. (a) 88 hr infection. Nucleus with prominent nucleoli and inclusions (arrows) surrounded by cytoplasmic granules and vacuoles. (b) 99 hr infection. Nuclear rotation with violent cytoplasmic activity. (c) 103 hr infection. Cell boiling. (d) Large cytoplasmic inclusion (arrow), cell close to explosive point.

REFERENCES

1. Anderson, E. S., Armstrong, J. A., and Niven, J. S. F., Fluorescence microscopy: Observations of virus growth with aminoacridines. *Symp. Soc. Gen. Microbiol.* **9**, 224–255 (1959).
2. Comings, D., The rationale for an ordered arrangement of chromatin in the interphase nucleus. *Am. J. Human Genet.* **20**, 440–460 (1968).
3. Daoust, R., and Molnar, F., Cellular populations and mitotic activity in rat liver parenchyma during azo dye carcinogenesis. *Cancer Res.* **24**, 1898–1909 (1964).
4. Grundmann, E., Quantitative cytochemistry of carcinogenesis in the rat liver. *Acta, Unio. Intern. Contra Cancrum* **19**, 571–575 (1963).
5. Kasten, F. H., Schiff-type reagents in cytochemistry. 1. Theoretical and practical considerations. *Histochemie* **1**, 466–509 (1959).
6. Kasten, F. H., Burton, V., and Glover, P., Fluorescent Schiff-type reagents for cyto-chemical detection of polyaldehyde moieties in sections and smears. *Nature* **184**, 1797 (1959).
7. Kasten, F. H., The chemistry of Schiff's reagent. *Intern. Rev. Cytol.* **10**, 1–100, (1960).
8. Kasten, F. H., Schiff-type reagents in cytochemistry. 3. General applications. *Acta Histochem.* Suppl. **3**, 240–247 (1963).
9. Kasten, F. H., The Feulgen reaction—an enigma in cytochemistry. *Acta Histochem.* **17**, 88–99 (1964).
10. Kasten, F. H., Schiff-type reagents. *In* "Laboratory Techniques in Biology and Medicine" (V. M. Emmel and E. V. Cowdry, eds.), pp. 390–392. Williams & Wilkins, Baltimore, Maryland, 1964.
11. Kasten, F. H., Histochemical methods in the study of nucleic acids. *In* "Encyclopedia of Biochemistry" (R. Williams and E. M. Lansford, Jr., eds.), pp. 408–410. Reinhold, New York, 1967.
12. Kasten, F. H., Chromosomal and nucleolar RNA synthetic rates and inhibitory patterns after actinomycin D pulses in the synchronized cell cycle. *J. Cell Biol.* **35**, 64A (1967).
13. Kasten, F. H., Use of a thymidine-spray technique to yield highly synchronized, isolated cell populations in mitosis. *J. Cell Biol.* **35**, 153A (1967).
14. Kasten, F. H., and Gerber, P., Cytochemical demonstration of viral DNA in infected cells. *Proc. 2nd Intern. Congr. Histochem. Cytochem., Frankfurt, Germany,* 1964 p. 199.
15. Kasten, F. H., and Gerber, P., Unpublished data (1964).
16. Kasten, F. H., and Jacob J., Unpublished data (1967).
17. Kasten, F. H., and Strasser, F. F., Amino acid incorporation patterns during the cell cycle of synchronized human tumor cells. *J. Natl. Cancer Inst.* **23**, 353–368 (1966).
18. Kasten, F. H., and Strasser, F. F., Nucleic acid synthetic patterns in synchronized mammalian cells. *Nature* **211**, 135–140 (1966).
19. Kasten, F. H., and Strasser, F. F., Dynamic activities of the synchronized mammalian cell cycle: A time-lapse study. *J. Cell Biol.* **35**, 150A (1967).
20. Kasten, F. H., Strasser, F. F., and Turner, M., Nucleolar and cytoplasmic ribonucleic acid inhibition by excess thymidine. *Nature* **207**, 161–164 (1965).
21. Kasten, F. H., Vendrely, C., and Riviere, M., DNA changes in rat liver during butter yellow carcinogenesis. *Folia Histochem. Cytochem.* **1**, Suppl. 1, 169–170 (1963).
22. Kasten, F. H., Vendrely, C., Tournier, P., and Wicker, R., DNA lesions induced in nuclei and nucleoli of monkey kidney cells in tissue culture by Simian vacuolating virus (SV40). *J. Cellular Comp. Physiol.* **66**, 33–48 (1965).

23. Koulish, S., and Lessler, M. A., Cytological aspects of normal and tumorous liver. *Cancer Res.* **22**, 1188–1196 (1962).
24. Maini, M. M., and Stich, H. F., Chromosomes of tumor cells. 2. Effect of various liver carcinogens on mitosis of hepatic cells. *J. Natl. Cancer Inst.* **26**, 1413–1427 (1961).
25. Mayor, H. D., The nucleic acids of viruses as revealed by their reactions with fluorochrome acridine orange. *Intern. Rev. Exptl. Pathol.* **2**, 1–45 (1900).
26. McAllister, R. M., Straw, R. M., Filbert, J. E., and Goodheart, C. R., Human cytomegalovirus. Cytochemical observations of intracellular lesion development correlated with viral synthesis and release. *Virology* **19**, 521–531 (1963).
27. McGavran, M. H., and Smith, M. G., Ultrastructural, cytochemical and microchemical observations of cytomegalovirus (salivary gland virus) infection of human cells in tissue culture. *Exptl. Mol. Pathol.* **4**, 1–10 (1965).
28. Vendrely, C., Tournier, P., Wicker, R., Grange, M. T., and Kasten, F. H., Teneur en ADN en rapport avec le nombre chromosomique et les phases du cycle de generation de fibroblastes de rat transformés par les virus polyoma et SV40. *Bull. Assoc. Franc. Etude Cancer* **51**, 447–454 (1964).
29. Wright, H. T., Jr., Kasten, F. H., and McAllister, R. M., Human cytomegalovirus. Observations of intracellular lesion development as revealed by phase-contrast, time-lapse cinematography. *Proc. Soc. Exptl. Biol. Med.* **127**, 1031–1036 (1968).
30. Xeros, N., Deoxyriboside control and synchronization of mitosis. *Nature* **194**, 682–683 (1962).

SPECIAL APPLICATIONS OF TWO-WAVELENGTH CYTOPHOTOMETRY IN BIOLOGIC SYSTEMS

Ellen M. Rasch* and Robert W. Rasch

DEPARTMENT OF BIOLOGY, MARQUETTE UNIVERSITY, MILWAUKEE, WISCONSIN; DEPARTMENT OF PHYSIOLOGY, MARQUETTE SCHOOL OF MEDICINE, INC., MILWAUKEE, WISCONSIN

I. Photometric Relationships in the Two-Wavelength Method

Measurement of the projected area of a chromophore is an inherent property of two-wavelength cytophotometry, a fact explicitly recognized by Ornstein in his ingenious derivation of the method (*39*), but a relationship thus far not utilized by workers in the field of quantitative cytochemistry.

A. GENERAL CONSIDERATIONS

Usual methods of measuring irregular areas in cytologic material by image projection and planimetry are both tedious and time-consuming. Even the determination of areas from regular geometric forms by direct measurement of dimensions with an ocular micrometer is an irksome task. Fortunately, the two-wavelength method offers a simple and accurate method of measuring area from both regularly and irregularly shaped objects when the material of interest is stained with an appropriate chromophore. In addition to usual computations for total amount of dye per object (*33, 39, 40, 51*), values for the projected area of chromophore and its extinction corrected for distributional error can be calculated independently and used separately, or in conjunction with each other, for interpretation of quantitative cytologic studies.

* U. S. Public Health Service, National Institutes of Health Research Career Development Awardee (1-K3 GM 3455). Supported by grants from the U. S. Public Health Service (GM 10503 and GM 14644) and the National Science Foundation (GB 7393).

In brief, the quantity $BL_1/(2 - Q)$ is the projected area, or A, of the regions of an object containing chromophore. The quantity $2 \log 1/(Q - 1)$ is the extinction of chromophore corrected for distributional error, or $E_2{}^*$, expressed in terms of the wavelength of greater absorption. As defined by Patau (40), $Q = L_2/L_1$, where $L = (1 - T)$ and subscripts for transmittances are used to refer to two wavelengths chosen so that their extinction coefficients are in a 2:1 ratio, with λ_2 having the greater absorption. B is the area of an optic field delimited to contain the object for measurement. Values for projected area may be expressed in relative arbitrary units, or, if the actual dimensions of B are known, values computed for projected areas will be in similar terms, for example, μ^2.

B. THEORETICAL CONSIDERATIONS

It is instructive to consider Ornstein's original approach to deriving the two-wavelength method because of the novel model he proposed for light absorption (39). He considered the probability of photon capture by chromophore to be σ/A, where σ is the cross-sectional area of the chromophore and A is the area containing the chromophore. For a single-chromophore molecule the probability of photons not being captured is given by $1 - \sigma/A$, and for n molecules of chromophore it is $(1 - \sigma/A)^n$, so that T^*, which is the transmittance of a randomly distributed chromophore, is given by

$$T^* = (1 - \sigma/A)^n \tag{1}$$

This relationship can be simplified if $\sigma \ll A$, because under these conditions $\ln T^* = n \ln(1 - \sigma/A)$ and $\ln(1 - \sigma/A) = -\sigma/A$, when σ/A is small. Therefore $\ln T^* = -n\sigma/A$ or

$$\log \frac{1}{T^*} = \frac{n\sigma}{A \ln 10} = E^* \tag{2}$$

where E^* is the extinction of a randomly distributed population of chromophore molecules. Alternatively, if $n\sigma \ll A$, a condition that can sometimes be realized,

$$T^* = \left(1 - \frac{\sigma}{A}\right)^n = 1 - \frac{\sigma n}{A} + \frac{n(n-1)}{2!}\left(\frac{\sigma}{A}\right)^2 \cdots$$

and

$$T^* = 1 - \frac{n\sigma}{A}$$

or

$$(1 - T^*) = L^* = \frac{n\sigma}{A} \tag{3}$$

Equation (2) is useful because it indicates that the projected area containing a randomly distributed chromophore is the parameter pertinent to cytophotometry, i.e., volume is not directly relevant to a measurement of extinction. By the term "projected area" we mean that area containing chromophore molecules which is cast as a shadow by parallel rays of light traversing an object and impinging on a photocathode, as is approximated in a properly aligned cytophotometer with low condenser numeric aperture (39, 40). For regular objects, the above equation can be expressed in terms of concentration and thickness. With irregular objects, however, it is usually the projected chromophore area per se which is relevant.

Equation (3) is of interest, not only because it represents the model for Patau's independent derivation of the two-wavelength method (40), but also, as will be shown later, because it can be quite useful when very few chromophore molecules are present in a photometric field.

In the case of nonrandom distribution of chromophore, where packages of chromophore are distributed within an area that lacks chromophore, for example, the DNA-Feulgen dye complex in chromosomes of a metaphase plate, one of the relationships proposed by Ornstein (39) was that

$$T = F + (1 - F)T^* \tag{4}$$

where T is the transmittance of the total optical field; F is that fraction of the area of the optical field not occupied by chromophore; and T^* is the transmittance of randomly distributed chromophore, as defined above. Any area lacking chromophore has a transmittance of 1, because there is no opportunity for photon loss or capture when both chromophore and light scattering are absent. When T is measured over an optical area, B, then $B(1 - F) = A$.

Solving Eq. (4) for T^*:

$$T^* = \frac{T - F}{1 - F} \quad \text{and} \quad \log \frac{1}{T^*} = \log \frac{1 - F}{T - F} \tag{5}$$

If the same object is measured at two different wavelengths, chosen so that the extinctions of a-homogeneously dispersed sample of the chromophore will be in the ratio of 2:1, then $\sigma_{\lambda_2} = 2\sigma_{\lambda_1}$. For an object with inhomogeneous distribution of chromophore, the following relations, from Eqs. (2) and (5) will hold:

at λ_1

$$E_1{}^* = \frac{n\sigma_1}{A \ln 10} \quad \text{and} \quad E_1{}^* = \log \frac{1 - E}{T_1 - F} \tag{6}$$

at λ_2

$$E_2{}^* = \frac{n\sigma_2}{A \ln 10} \quad \text{and} \quad E_2{}^* = \log \frac{1 - F}{T_2 - F} \tag{7}$$

If A is independent of wavelength, with $\sigma_2 \doteq 2\sigma_1$, then

$$n = \frac{A \ln 10}{2\sigma_1} \log \frac{1 - F}{T_2 - F} = \frac{A \ln 10}{\sigma_1} \log \frac{1 - F}{T_1 - F} \tag{8}$$

and

$$\frac{1 - F}{T_2 - F} = \frac{(1 - F)^2}{(T_1 - F)^2} \tag{9}$$

Define: $L_1 = (1 - T_1)$; $L_2 = (1 - T_2)$; and $Q = L_2/L_1$. By substitution, solve for

$$(1 - F) = L_1/(2 - Q) \tag{10}$$

and

$$A = B(1 - F) = BL_1/(2 - Q) \tag{11}$$

Alternatively, when $n\sigma \ll A$, using Eq. (3) and Eq. (4) with appropriate substitutions of L for T, since

$$L^* = \frac{n\sigma}{A}, L^* = \frac{L}{(1 - F)}$$

and $B(1 - F) = A$, then

$$n = \frac{LB}{\sigma} \tag{12}$$

Other useful identities can be recognized in the above equations. A number of these are summarized in Table I. When using Table I, methods based on common wavelengths and extinction coefficients will be in the same units and should be directly comparable.

TABLE I

Photometric Identities of Two-Wavelength and Two-Area Cytophotometry

Transmissions corrected for distributional error

At λ_1 $T_1^* = Q - 1$

At λ_2 $T_2^* = (Q - 1)^2$

Extinctions corrected for distributional error

At λ_1 $E_1^* = \log 1/(Q - 1)$

At λ_2 $E_2^* = 2 \log 1/(Q - 1)$

Measures of area

$A = B L_1/(2 - Q)$ (two-wavelength method)

$A = B L_1/L_1^{\cdot}$ (two-area method)

Measures of amount in terms of σ_1 and σ_2

$$M = \frac{(\ln 10)BL_1\, 2 \log 1/(Q - 1)}{\sigma_2\, (2 - Q)}$$

At λ_2

Pateau (*40*) $M = \dfrac{2\, BL_1\, C}{\sigma_2}$, where $C = \dfrac{1}{2 - Q} \ln \dfrac{1}{(Q - 1)}$

Swift and Rasch (*51*) $M = \dfrac{(\ln 10)\, BL_1\, D}{\sigma_2}$, where $D = \dfrac{2}{2 - Q} \log \dfrac{1}{(Q - 1)}$

At λ_1

Pateau (*40*) $M = \dfrac{BL_1 C}{\sigma_1}$ $M = \dfrac{(\ln 10)BL_1 D}{2\, \sigma_1}$

Garcia (*15*) $M = \dfrac{(\ln 10)BL_1 E_1^*}{\sigma_1 L_1^{\cdot}}$ (two-area method)

Measures of amount in terms of k_1 and k_2

At λ_2

$$M = \frac{2\, BL_1 C}{k_2 \ln 10} \qquad M = \frac{BL_2\, E_2^*}{k_2\, L_2^{\cdot}}$$

Swift and Rasch (*51*) $M = \dfrac{BL_1 D}{k_2}$

At λ_1

$$M = \frac{BL_1 D}{2\, k_1} \qquad M = \frac{BL_1\, C}{k_1 \ln 10}$$

$$M = \frac{BL_1 E_1^*}{k_1\, L_1^{\cdot}}$$

(only when $n\sigma \ll A$) $M = \dfrac{L_1 B}{k_1 \ln 10}$

II. Comments on Materials and Methods

A. Tissue Preparation

In the following presentation of results and observations a variety of plant and animal materials have been used as photometric test objects. Cells from lily anthers or from onion root meristems and rat liver were selected for study because they are familiar cytologic objects of ready availability. Giant polytene chromosomes from the salivary gland cells of dipteran larvae and the very small hemocytes of such animals not only provided material of topical interest in the field of developmental biology, but also are test objects which span the range of visible light cytophotometry, with respect both to the sizes and the staining intensities usually encountered.

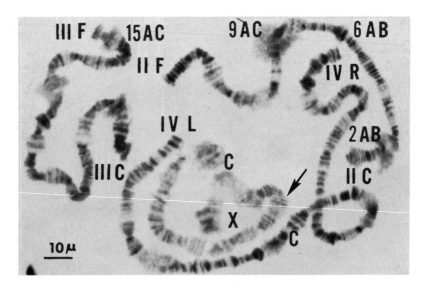

FIG. 1. Polytene salivary gland chromosomes of *Sciara coprophila* from female larva at late fourth instar. Centromere end (C) and free end (F) shown for Chromosomes II, III, and X; right arm (R) and left arm (L) shown for Chromosome IV, which has a submedian centromere. Expanded regions of Chromosome II designated 2AB, 6AB, or 9AC are sites of prominent DNA puffs which characteristically form during this stage of larval development. Duality of the synapsed homologs which comprise these giant chromosomes is evident in the small, asynapsed, and slightly twisted region of the X chromosome marked by the arrow. Gland dissected in buffered insect saline (3) and fixed briefly in ethanol–acetic acid (3:1) prior to smearing in 45% acetic acid, freezing with dry ice for removal of coverslip, and postfixation for 30 min in 10% neutral formalin. Stained with the Feulgen reaction for DNA after 10 min of hydrolysis in 1 N HCl at 60°C.

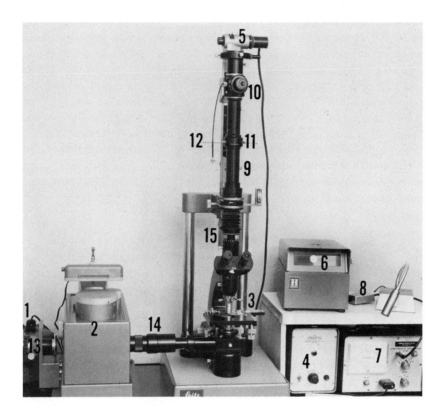

FIG. 2. Visible light cytophotometer, using a ribbon filament source with line voltage stabilizer (1), a Leitz linear mirror monochromator (2), and an Ortholux microscope (3). A Farrand, line-operated high-voltage supply (4) with a 1P21 electron multiplier phototube (5) is coupled to a Honeywell galvanometer (6) through an RCA ultrasensitive DC micro-ammeter (7) via an impedence matching circuit (8). The system allows final readout in three figures on the illuminated scale of the galvanometer, as well as full use of the six sensitivity ranges (0.01 × to 1000 ×) of the microammeter. By easily varied combinations of objectives and oculars, the image of the object for measurement is enlarged optically 100 to 1000 times and projected with an intermediate lens system (9) onto the plane of a field diaphragm which is viewed through a side tube (10) containing a focusing 10 × ocular with a micrometer scale. Field sample size is selected either with a manually operated iris diaphragm (11) or a movable slide (12) with six different apertures of fixed sizes. Use of a condenser with n.a. 0.4 and insertion of limiting diaphragms at the source (13), at the exit slit of the monochromator (14), and in the microscope ocular (15) reduce instrumental flare to less than 0.3%. Reproducibility of measurements made on a single test object such as a fish or chick erythrocyte nucleus is within 2%. The coefficient of variation for measurements of different populations of 50 such nuclei consistently is within 3% to 5%.

Specific details of routine tissue handling, fixation procedures, etc., are indicated in appropriate figure or table legends. All preparations, whether pieces of tissue sectioned in Paraplast after dehydration in *tert*-butyl alcohol, or cell nuclei initially fixed in ethanol–acetic acid (3:1) before dispersal and squashing in 45% acetic acid, were hydrolyzed in 1 N HCl at 60°C, rinsed briefly in cold 1 N HCl, and then stained for 2 hr in Schiff reagent made according to Lillie (28). Three rinses of 10 min each in sulfite water, followed by 10 min of washing in running tap water, preceded dehydration in graduated changes of ethanol, clearing in xylene, and mounting of tissues in refractive index liquid (R.P. Cargille Laboratories, Inc., Cedar Grove, New Jersey) or in D. P. X. (ESBE, Toronto, Canada).

Designations of particular regions of salivary gland chromosomes of *Sciara coprophila* (Lintner) (see Fig. 1) are according to the recently revised maps for this species given by Gabrusewycz-Garcia (12).*

B. PHOTOMETRY

Data of the present study were obtained with a simple cytophotometer using Leitz optic components, shown in Fig. 2. Specific features of this instrument are indicated in the figure legend. Appropriate procedures and tests were applied (44) to validate sensitivity, accuracy, and reproducibility in the performance of this cytophotometer, following recommendations discussed by Swift and Rasch (51), Mendelsohn (32), Garcia (13), and Garcia and Iorio (18). Uniformity of illumination of the measuring field was achieved by the Koehler principle through projection of the image of the exit slit into the condenser diaphragm, using an iris diaphragm at the lens in front of the exit slit to serve as an adjustable stop for the illuminated field. A Leitz linear mirror monochromator with a prism of F_2 glass was set with both entrance and exit slits at 0.1 mm to provide a nominal bandwidth of approximately 3 mμ for the range of wavelengths of interest. With a Leitz H 20 × quartz objective (n.a. 0.4) used as a condenser in this system, satisfactory performance was achieved to minimize instrument flare and to fulfill the important criteria for monochromasy and uniformity of illumination discussed by Patau (40) and Patau and Swift (41).

Determination of the area of a particular cytophotometric measuring field is another important potential source of error in either the two-wavelength or the two-area method of microphotometry (14, 18, 40, 51). For the work reported here, several different limiting apertures of fixed sizes were used, the largest of which was calibrated directly with the aid of a stage micro-

---

*Populations of *S. coprophila* presently maintained in our laboratory originate from stocks with the wavy wing marker for the X' chromosome which were made available to us in 1962 through the kind courtesy of Dr. Helen V. Crouse.

meter. Areas of smaller apertures were determined photometrically, using galvanometer deflections and a calibrated standard reference area in the manner suggested by Garcia and Iorio (*18*).

III. Photometric Identities in Two-Wavelength and Two-Area Cytophotometry

Different authors in writing about two-wavelength cytophotometry have used various letter designations and subscripts to denote particular identities and relationships among functions of transmittances, "one-minus-trans-mittances," pairs of wavelengths, and correction factors for distribution error (*11, 14, 15, 32, 35, 39, 40, 51*). The intent here is not to add yet another system of nomenclature to the literature, but to provide a consistent set of definitions with appropriate designations, arranged to show functional inter-relationships among the several photometric parameters inherent in both the two-wavelength method and the two-area method of cytophotometry—relationships of considerable potential interest for quantitative studies of biologic materials. We hope further to provide recognizable identities that will prove useful for required computations and, as will be discussed later, which will be useful for interpreting the values themselves.

A. DATA TRANSFORMATION

Specific applications of the relationships summarized in Table I are shown in Tables II, III, IV, and V, in which various identities of the two-area and the two-wavelength methods are derived from actual photometric data and used to determine three parameters:

(1) Chromophore extinction, i.e., proportional to concentration of stain corrected for distributional error, as if measured at λ_2.

(2) Relative projected chromophore area, e.g., relative cross-sectional nuclear area.

(3) Total relative amount of chromophore in the photometric field, which in the cases illustrated here is the relative amount of DNA-Feulgen per nucleus or per chromosome segment.

Particular use of the term *relative* in cytophotometry is deserving of commentary and specific clarification at this point. When the term *relative* is used with reference to amount of chromophore, for example, it means the number of molecules multiplied by the chemical extinction coefficient at the wavelength used, or $M \equiv nk$. Since the true or actual value of k is rarely, if ever, actually determined in cytophotometric studies, a commonly accepted expression of the amount M, defined as *relative amount of chromophore*, is achieved by arbitrarily setting all measurements to a single wavelength and giving k a value of 1.0. Thus a particular numeric designation or value

TABLE II

PHOTOMETRIC DATA FROM TWO-AREA AND TWO-WAVELENGTH MEASUREMENTS ON FEULGEN-STAINED CHROMOSOME SEGMENTS OF *Sciara coprophila*[a]

Two-area method[b] ($\lambda = 560$ mμ)

Chromosome region	Field diameter (μ)	Galvanometer deflection With object (I'')	Without object (I_0)	$T\cdot$	Field diameter (μ)	Galvanometer deflection With object (I_n)	Without object (I_0)	T
II 9 AC	3.6	18.0	47.4	0.380	14.6	39.1	73.1	0.535
II 2 AB	3.6	15.1	48.0	0.315	10.9	24.0	39.0	0.615

Two-wavelength method[c]

Chromosome region	Field diameter (μ)	$\lambda_1 = 502$ mμ Galvanometer deflection With object (I_n)	Without object (I_0)	T_1	$\lambda_2 = 560$ mμ Galvanometer deflection With object (I_n)	Without object (I_0)	T_2
II 9 AC	14.6	35.0	49.2	0.711	39.0	73.1	0.534
II 2 AB	10.9	20.4	27.0	0.756	24.0	39.1	0.614

[a] Smear preparation of giant salivary gland chromosomes from animal sacrificed at pupal molt. Gland fixed for 5 min in ethanol–acetic acid (3:1) prior to spreading in 45% acetic acid. Standard Feulgen reaction after 10 min of hydrolysis in 1 N NCl at 60°C.

[b] $T\cdot$, transmittance through chromophore sample only at λ_x in photometric field $B\cdot$. T, transmittance at λ_x in photometric field B, which includes chromophore plus surrounding rim of nonabsorbing background area.

[c] Two wavelengths selected so $E(\lambda_2) = 2 E(\lambda_1)$ on sample of chromophore homogeneously dispersed within measuring field. Here the

TABLE III

PHOTOMETRIC IDENTITIES DERIVED FROM DATA OF TABLE II

Entry	Photometric identities and equivalents		Chromosome regions	
			II 9 AC	II 2 AB
	Two-area method ($\lambda = 560$ mμ)			
1	Area of sample through chromophore	$B^\bullet$	10.2 μ^2	10.2 μ^2
2	Area of photometric field with clear surround	B	167.4 μ^2	93.3 μ^2
3		$L^\bullet = 1 - T^\bullet$	0.620	0.685
4		$L = 1 - T$	0.465	0.385
5	Relative projected chromophore area	$(1 - F) = L/L^\bullet$	0.750	0.562
6	Chromophore area	$A = B(1 - F)$	125.6 μ^2	52.4 μ^2
7	Extinction corrected for distributional error	$E^* = \log 1/T^\bullet$	0.420	0.502
8	Relative amount of chromophore or after Garcia (*15*)	$M = E^*A \lambda_{560/m\mu}$ $AE^* = BL/L^\bullet \log 1/T^\bullet$	52.7	26.3
	Two-wavelength method ($\lambda_1 = 502$ mμ and $\lambda_2 = 560$ mμ)			
9		$L_1 = 1 - T_1$	0.289	0.244
10		$L_2 = 1 - T_2$	0.466	0.386
11		$Q = L_2/L_1$	1.612	0.582
12	Corrected transmittance at λ_1	$T_1^* = Q - 1$	0.612	0.582
13	Corrected transmittance at λ_2	$T_2^* = (Q - 1)^2$	0.374	0.339
14	Corrected extinction at λ_1	$E_1^* = \log 1/T_1^*$	0.213	0.235
15	Corrected extinction at λ_2	$E_2^* = 2 E_1^*$	0.426	0.470
16	Relative projected chromophore area	$(1 - F) = L_1/(2 - Q)$	0.745	0.584
17	Area of chromophore	$A = B(1 - F)$	124.7 μ^2	54.5 μ^2
18	Relative amount of chromophore (DNA-Feulgen per segment)			
	18-1:	$M = nk_2 = AE_2^*$	53.1 k_2	25.6 k_2
	18-2: [Patau (*40*)][a]	$\gamma = L_1CB$	61.2[b]	29.5[b]
			(53.2)	(25.6)
	18-3: [Swift and Rasch (*51*)][c]	$M = L_1DB$	53.2[c]	25.6[c]
	18-4: [after Ornstein (*39*)]	$Q = B(1 - F) \log \dfrac{(1 - F)}{(T_2 - F)}$	53.2	25.6

[a] Values for C obtained from Table 1 of Patau (*40*). Values for (L_1 C) can also be read directly from tables of T_1 and T_2 given by Mendelsohn (*33*).

[b] When Patau's correction factor C is used, the amount of chromophore is expressed in terms of $\sigma_1 = k_1 \ln 10$. Correction factor D expresses it in terms of k_2, the chemical extinction coefficient. Values adjusted for these differences in terms are shown in parentheses.

[c] Values for D obtained from Table 5 of Swift and Rasch (*51*), where $D = 0.868$ C.

TABLE IV

PHOTOMETRIC DATA FROM TWO-WAVELENGTH MEASUREMENTS ON FEULGEN-STAINED LILY NUCLEI[a]

| | $\lambda_1 = 498$ mμ | | | | $\lambda_2 = 560$ mμ | | | |
| | Galvanometer deflection | | | | Galvanometer deflection | | | |
Cell type	With nucleus (I_n)	Without nucleus (I_0)	T_1	$\bar{T}_1$	With nucleus (I_n)	Without nucleus (I_0)	T_2	$\bar{T}_2$
(1) Nuclear diameter								
(2) Field diameter								
Whole nucleus leptotene microsporocyte								
(1) 14.4 μ	30.0	46.1	0.651	0.652	36.2	73.9	0.490	0.490
(2) 17.2 μ	30.1	46.2	0.652		36.2	74.0	0.489	
Whole nucleus anther wall								
(1) 13.3 μ	20.0	39.5	0.506	0.507	20.1	63.0	0.319	0.319
(2) 14.7 μ	20.1	39.6	0.508		20.0	62.7	0.319	
Cut nucleus anther wall								
(1) 13.3 μ	35.0	46.1	0.759	0.760	46.0	73.0	0.630	0.630
(2) 18.2 μ	35.1	46.2	0.760		45.9	73.0	0.629	

[a] Wall parenchyma and microsporocyte nuclei from anther of *Lilium davidii*, fixed in Navashin's fluid and sectioned at 25 μ. Standard Feulgen reaction after 15 min of hydrolysis in 1 N NCl at 60°C. Wavelengths selected for measurement so that $E(\lambda_2) = 2E(\lambda_1)$ for homogeneous samples of same preparation. Ratio $E \lambda_{560m\mu} : E \lambda_{498m\mu} = 1.987 \pm 0.01$, $n = 20$.

TABLE V

PHOTOMETRIC IDENTITIES AND PROCEDURES FOR COMPUTING NUCLEAR AREA, CORRECTED
EXTINCTION, AND RELATIVE AMOUNTS OF DNA-FEULGEN PER NUCLEUS, USING TWO-
WAVELENGTH MEASUREMENTS FROM DATA OF TABLE IV

Entry	Photometric identities and equivalents		Whole leptotene nucleus	Cell type whole wall nucleus	Cut wall nucleus
1	Area of photometric field	B	260 μ^2	170 μ^2	260 μ^2
2		$L_2 = 1 - T_2$	0.510	0.681	0.370
3		$L_1 = 1 - T_1$	0.348	0.493	0.240
4		$Q = L_2/L_1$	1.466	1.381	1.542
5	L_1 corrected for distributional error	$L_1^* = 2 - Q$	0.534	0.619	0.458
6	Transmittance λ_1, corrected for distributional error	$T_1^* = Q - 1$	0.466	0.381	0.542
7	Extinction λ_1, corrected for distributional error	$E_1^* = \log 1/T^*$	0.332	0.419	0.266
8	Extinction λ_2, corrected for distributional error	$E_2^* = 2 E_1^*$ also: $E_2^* = 2 \log (1/Q - 1)$	0.664	0.838	0.532
9	Relative projected chromophore area	$(1 - F) = L_1/(2 - Q)$ also: $(1 - F) = L_1/L_1^*$	0.652	0.796	0.524
10	Projected chromophore area	$A = B(1 - F)$	169.5 μ^2	135.3 μ^2	136.2 μ^2
11	Nuclear area from direct micrometer measurement	$A = \pi r^2$	162.9 μ^2	138.9 μ^2	138.9 μ^2
12	Relative amount of chromophore (DNA-Feulgen per nucleus) 12-1:	$M = nk_2 = AE_2^*$	112.6k_2	113.4k_2	72.5k_2
	12-2: [Patau (40)][a]	$\gamma = L_1CB$	129.4[a] (112.4)	130.6[a] (113.5)	83.4[a] (72.4)
	12-3: [Swift and Rasch (51)][b]	$M = L_1DB$	112.4[b]	113.5[b]	72.4[b]
	12-4: [after Ornstein (39)]	$Q = B(1 - F)\log \dfrac{(1 - F)}{(T_2 - F)}$	112.4	113.5	72.4

[a] Values for correction factor C obtained from Table 1 of Patau (40), which is derived from his formulations using $\sigma_1 = k_1 \ln 10$. Values shown in parentheses were corrected to common chemical extinction coefficient (k_2).

[b] Values for correction factor D obtained from Table 5 of Swift and Rasch (51), in which $D = 0.868C$, which conversion translates the expression by Patau of the same relationship to common logs and in terms of k_2.

TABLE VI

VALUES FOR CORRECTED EXTINCTION AT λ_2 OR E_2^* ($\times 10^3$) FOR DIFFERENT VALUES OF Q^a

Q	+ 0	0.01	0.02	0.03	0.04	0.05	0.06	0.07	0.08	0.09
1.9	91.5	81.9	72.4	63	53.7	44.6	35.5	26.5	17.5	8.7
1.8	194	183	172	162	151	141	131	121	111	101
1.7	310	298	285	273	262	250	238	227	216	205
1.6	444	429	415	401	388	374	361	348	335	322
1.5	602	585	568	551	535	519	504	488	473	458
1.4	796	774	754	733	713	694	674	656	638	620
1.3	1046	1017	990	963	937	912	887	864	840	818
1.2	1398	1356	1315	1276	1240	1204	1170	1137	1106	1075
1.1	2000	1917	1842	1772	1708	1648	1592	1539	1489	1442
1.0	—	4000	3398	3046	2796	2602	2444	2310	2194	2092

^{a} Computed from $E_2^* = 2 \log 1/(Q - 1)$, where $Q = L_2/L_1$.

computed for M is not in terms of an absolute amount of chromophore, but is in actuality an expression of a *relative amount of dye*. Such *relative* values of M are comparable to other measurements at the same wavelength for those objects which have essentially similar absorption curves. In practice, with Feulgen cytophotometry, these conditions for directly comparable measurements of *relative amounts of dye* are usually met only by slides of similarly fixed tissues, from a given acid hydrolysis series, which have been stained simultaneously with a particular lot of Schiff reagent. Considerations in general similar to those above apply to the use of the term *relative* with regard to projected chromophore area, which is simply that fraction of the total area of the optic field occupied by chromophore molecules. Where the area of the optic field, B, in actual square microns, or other specific units, is not stated, the term *relative projected chromophore area* again denotes appropriate qualification for numeric values obtained.

Typical, raw photometric data are displayed in Tables II and IV. Specific transformation of these data into useful photometric identities is shown in Tables III and V, the entries of which are arranged to display possible alternative schemes for computing each of the three parameters listed above. Steps used in these calculations are summarized below. Computations of corrected extinctions at λ_2 are made unnecessary by reference to Table VI, which gives values for E_2^* as a function of Q. Further, as was shown in Table I, since $E_1^* = E_2^*/2$, Table VI can also be used to derive corrected extinctions at λ_1 in those situations where the latter is a desired parameter.

B. Steps in the Calculations

Procedures for photometric determination of chromophore area, corrected extinction, and total amount of chromophore presume selection of a pair of wavelengths, λ_1 and λ_2, so that $k_{\lambda_2} = 2k_{\lambda_1}$, where $k = \sigma/\ln 10$, and involve the following steps in the calculations:

(1) Choose object of interest and circumscribe with iris diaphragm or appropriate aperture of fixed size.

(2) Determine transmittance at λ_2: $I_n/I_0 = T_2$.

(3) Determine transmittance at λ_1: $I_n/I_0 = T_1$.

(4) Determine area of measuring field, B, using a calibrated standard reference area, or directly by micrometer readings of field diameter.

(5) Compute $L_1 = (1 - T_1)$ and $L_2 = (1 - T_2)$.

(6) With L_2 and L_1, compute $Q = L_2/L_1$.

(7) Compute $(1 - F) = L_1/(2 - Q)$.

(8) Compute area, $A = B(1 - F)$.

(9) Compute E_2^* as $2 \log 1/(Q - 1)$, or read E_2^* directly from values of Q using Table VI.

(10) Compute total relative amount of chromophore: $M = E_2^*B(1 - F)$.

C. Comparison of Two-Wavelength and Two-Area Determinations

In measuring with the two-area method, it is assumed that a transmittance sampled through a thin, flat plate of chromophore is representative of the whole sample (15, 19). A similar presumption was made by Patau in deriving a correction factor for distributional error in a flat plate measured in an area with chromophore-free surroundings (40). Therefore, the value for corrected transmission (T^*) from a two-wavelength determination should agree with the measurement of transmission in a plug sample through chromophore ($T^{\cdot}$) done by the two-area method at the same wavelength. That this is indeed the case here at $\lambda_{560\ m\mu}$ is shown by comparing values of $T^{\cdot}$ from Table II with those for T_2^* in entry 13 of Table III. The entity T^* is useful, therefore, for evaluating measurements on new or unknown material, for it provides an immediate check on the implicit assumption of a uniform plate of chromophore as required by the two-area method (15). Also in Table III, note that values for E_2^* from the two-wavelength method agree with those obtained for $E^{\cdot}$ with the two-area method (entries 15 and 7, respectively). Note the identity $Q = 1 + T_1^*$ (entries 11 and 12, respectively). It follows then that for corrected transmittance at the alternative wavelength, $T_2^* = (Q - 1)^2$. Also in Table III, the derivation of projected chromophore area, $A = B(1 - F)$, from measurements by the two-wavelength method (entry 17) yields values directly equivalent to those for the same identity but derived from photometric data obtained by the two-area method of measurement (entry 6). Finally, agreement within 2% obtains for computed estimates of the net amount of DNA-Feulgen per chromosome segment when separate sets of data on the same object are compared using the two-wavelength and the two-area methods of measuring (entries 18 and 8, Table III). This level of agreement is well within the photometric error assessed for the particular cytophotometer used (44), and approaches the level of performance commonly claimed for scanning procedures (1, 7, 8, 29, 37, 46). Also, it approximates the 1% estimate for wavelength error anticipated by Patau in his original derivation of the two-wavelength method (40). No assessment or claim is implied here, however, of error estimates involving the factor of proportionality between the amount of dye bound and the amount of natural substance indicated by it. This problem, a topic of considerable, renewed interest, has been discussed by Patau and Swift (41), Srinivasachar and Patau (48), and more recently by Garcia (16, 17, 17a), Gledhill (20, 20a, b), Mayall (29, 30), and Mayall and Mendelsohn (30, 31a).

Four alternative procedures for computing relative amounts of chromophore from two-wavelength measurements are shown in Table III (entry 18) and Table V (entry 12), along with appropriate expressions of the particular photometric terms involved and pertinent citations to the previous use of these formulations in the literature. Slight discrepancies ($\approx 1\%$) in the actual numbers generated by each of these procedures will be encountered in practice due to rounding and reading errors, e.g., from use of a slide rule for computations, or from interpolation of values taken from tables for C (40), D (51), or L_1C (33). The data presented here on several different biologic objects clearly document the direct equivalence of these several expressions, the relationships among which are also evident from consideration of the photometric identities involved.

Readers interested in further discussion of the problem of distributional error and two-wavelength cytophotometry will find instructive the recent summary by Garcia (14) in which original formulations according to Ornstein (39) or to Patau (40) and the alternative solution by Mendelsohn (34) are compared to demonstrate the mathematic equivalence of the particular terms involved.

IV. Determination of Chromophore Area by Two-Wavelength Cytophotometry

Our current primary interest in quantitative cytochemical analysis by two-wavelength microphotometry has been the problem of determining the area (A) occupied by absorbing material, or the quantity $B(1 - F)$, as described above, and as discussed by others using alternative approaches to the same problem (1, 2, 4, 7, 8, 15, 20, 26, 43, 45, 46).

A. EXPERIMENTAL VALIDATION

Some degree of empiric validation of the proposed method for photometric determination of area occupied by chromophore is provided by the data and transformations shown in Table V. In comparing the values for area, where $A = \pi r^2$, obtained by direct micrometer measurement (entry 11) with those derived for the same identity, where $A = B(1 - F)$, computed from photometric data (entry 10), agreement of values within less than 5% was obtained.

The results of further explorations to assess the validity of such an approach to determining area in materials of biologic interest are given in Figs. 3A, 4A, and 5A. In the cases of liver parenchymal nuclei and liliaceous anther

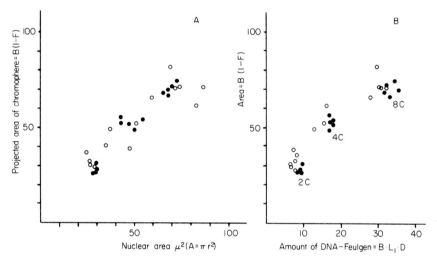

FIG. 3. (A) Determination of nuclear areas in rat liver cells by two-wavelength cyto-photometry correlated with values for areas computed from direct micrometer measurement of nuclear diameters. The linear relationship shown by both cut and uncut nuclei demonstrates that the photometric method is independent of volume; $r = 0.92$; slope $= 0.847 \pm 0.07$, which is not significantly different from 1.0. Solid circles represent values for whole nuclei; open circles values for cut nuclei. The random distribution of area estimates for cut nuclei both above and below those for whole nuclei shows that the photo-metric method is measuring the projected area of the nucleus-containing chromophore. Navashin-fixed tissue sections 8, 12, and 18 μ thick were mounted on a single slide and hydrolyzed for 15 min in 1 N NCl at 60°C prior to staining for 2 hr in Schiff reagent. Preparation mounted in refractive index oil n_D 1.564 for measurement. ($\lambda_2 = 560$ mμ; $\lambda_1 = 498$ mμ.) (B) Projected chromophore area, $B(1 - F)$, plotted against relative amount of DNA-Feulgen per nucleus, computed as $M = BL_1D$, for the same population of cells shown in (A). Solid circles represent values for whole nuclei; open circles, values for cut nuclei. Scattering of DNA values for cut nuclei is in marked contrast to the close clustering of values within expected DNA classes shown by whole nuclei.

nuclei, use of sectioned material provided an opportunity to select on the one hand whole nuclei covering a range of area from about 20 μ^2 to about 200 μ^2, and to compare the values so obtained with similar determinations on doubly cut nuclear sectors which had only about half of the volume of the whole nuclei, but which displayed about the same nuclear cross-sectional area when they were viewed microscopically. In each case, a direct measurement of nuclear diameter (or of major and minor axes) was made with an ocular micrometer at a magnification of 1600 and the nuclear area then computed as $A = \pi r^2$. Finally, to encompass an even wider range of chromophore areas, 20 to 1000 μ^2, a range likely to be encountered in diverse biologic

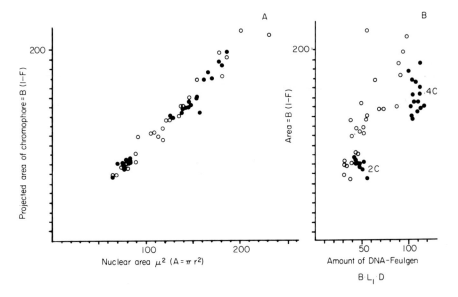

Fig. 4. (A) Two-wavelength photometric determinations of nuclear areas in lily correlated with values for areas computed from direct micrometer measurement of nuclear diameters. Solid circles represent values for whole nuclei; open circles, values for cut nuclei; $r = 0.98$; slope $= 0.996 \pm 0.026$. Wall and tapetal cells from anther of *Lilium davidii* fixed in ethanol-acetic acid (3:1), embedded in Paraplast, and sectioned at 15 and 35 μ. Hydrolysis for 12 min in 1 N NCl at 60°C before staining with Feulgen reaction for DNA. ($\lambda_2 = 560$ mμ; $\lambda_1 = 498$ mμ.) (B) Projected chromophore area, $B(1 - F)$, plotted against relative amount of DNA-Feulgen per nucleus, computed as $M = BL_1D$ for the same population of cells shown in (A). Solid circles represent whole nuclei; open circles, cut nuclei. DNA-Feulgen values for cut nuclei fall significantly to the left of those for whole nuclei with comparable cross-sectional areas. DNA values for whole nuclei cluster into two well-defined classes, here designated 2C and 4C. Many of the cut nuclei with DNA values near the 2C level (between 40 and 50 units) show nuclear areas which are consonant with those found for whole 4C nuclei, and thus clearly reflect measurement only of a median sector from those nuclei chosen as doubly nicked.

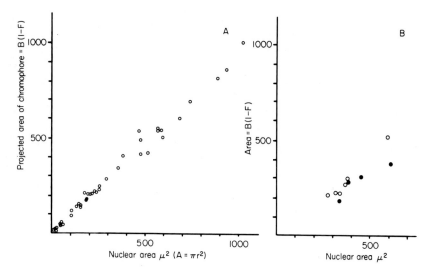

FIG. 5. (A) Two-wavelength photometric determination of nuclear areas in onion interphase cells correlated with values for areas computed from direct micrometer measurements of nuclear diameters; $r = 0.99$; slope $= 0.912 \pm 0.016$. Departure from a slope of 1.0 in this case may be attributable in part to the presence of prominent nucleoli which in many thoroughly flattened nuclei of a squash preparation no longer show rims of chromatin overlying nucleolar material and thus would not register as part of the projected chromophore area. Feulgen-stained, onion root meristem smeared in 45% acetic acid, frozen on dry ice for coverslip removal, and remounted in refractive index liquid n_D 1.570. ($\lambda_2 = 560$ mμ; $\lambda_1 = 498$ mμ.) (B) Projected chromophore area, $B(1 - F)$, plotted against nuclear areas estimated from direct micrometer measurement of major and minor axes of nuclei at midprophase (open circles) or at prometaphase (solid circles). Preparation as in (A).

materials, a squash preparation was made of Feulgen-stained onion root tips which had been grown for 72 hr in 0.1% colchicine. Through distortion of normal karyokinesis by inhibition of spindle function in actively dividing cells of these root meristems, this treatment provided test material containing small micronuclei, many without nucleoli, well-flattened chromosomes in either prometaphase or metaphase configurations, and a population of diploid, tetraploid, and octaploid interphase nuclei. Pertinent statistics are given for each group of data in the appropriate figure legends.

The curves of Figs. 3A, 4A, and 5A show straight-line relationships, with intercepts at 0, between nuclear area determined by micrometer readings of nuclear diameters and nuclear area determined photometrically as the projected chromophore area, $B(1 - F)$. Over an extensive range of areas in biologic test materials, the parameter $B(1 - F)$ represents with small error the *actual area* projected by an absorbing chromophore, and is *not* a volume-dependent function, an essential relationship recognized by Ornstein in his original formulation of the method (*39*). This important point is further demonstrated by the strict linearity obtained with regard to *areas* determined from either whole or cut nuclei (Figs. 3A and 4A), but the marked discrepancies with regard to distributions of the *amounts* of DNA-Feulgen per nucleus when the same two populations are compared (Figs. 3B and 4B). The clustering of DNA-Feulgen values for the whole liver nuclei, for example, fits expectations for 2C, 4C, and 8C DNA classes (*50*), while DNA-Feulgen values for nuclei which were selected as doubly cut, median sectors either fall below the 2C level or scatter between the 2C and 4C or the 4C and 8C levels.

B. INTERPRETATION OF DEVIANT RESULTS

Another relevant point, illustrated in Fig. 5B, is that areas computed as $B(1 - F)$ may sometimes show significant deviation from area values estimated from πr^2 on the same nuclei by direct micrometer measurements. In our experience such deviations occur when chromatin condensation, as in the case of onion prometaphase nuclei, or the presence of large nucleoli, as in the case of squashed polyploid onion interphase nuclei, provide an appreciable nonabsorbing area enclosed within the usually accepted limits of the nuclear boundary (Fig. 5B). In the case of whole, Feulgen-stained liver nuclei, however, unstained nucleoli with their adherent rims of chromatin would cast a shadow and effectively register their own discrete hollow area as part of the projected chromophore area and thus contribute much less error.

The linear regressions obtained for interphase nuclei of liver, lily, and onion clearly demonstrate both the validity and utility of assessing the quantity $B(1 - F)$ from two-wavelength cytophotometry to determine areas of appropriately stained, irregularly shaped objects of biologic interest. Recent advances in computer analysis of microscopic images (*36*) suggest that transformation of photometric data into an area function may provide a useful discriminatory parameter in some types of programs for automated cytoscanning.

V. Variables Influencing Two-Wavelength Cytophotometry

The determination of projected chromophore area, $B(1 - F)$, and extinction of chromophore corrected for distributional error, E^*, allow specific evaluation of some variables which may influence two-wavelength measurements.

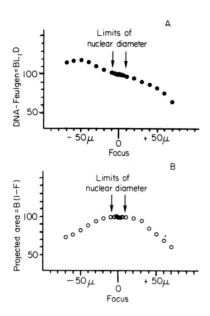

FIG. 6. (A) Effect of objective focus on two-wavelength estimates of the relative amount of DNA-Feulgen staining in an onion interphase nucleus. DNA values with the nuclear image in sharp focus (0) are set equal to 100, and values obtained at varying levels above (+) or below (−) focus have been adjusted accordingly. Squash preparation of Feulgen-stained, onion root meristem; mounted in refractive index liquid n_D 1.550. ($\lambda_2 = 560$ mμ; $\lambda_1 = 498$ mμ.) (B) Effect of objective focus on two-wavelength estimates of relative projected area of chromophore. Values with nuclear image in sharp focus (0) are set equal to 100 for comparison with values obtained at different levels above (+) or below (−) focus. Preparation as in (A).

A. EFFECT OF FOCUS AND DETERMINATION OF AREA

As discussed by Mendelsohn in his presentation of data on a test system of dye droplets (*32*), the two-wavelength method should be expected to give results which are independent of focus, provided all light passing through the object is included in the measured photometric field. As shown in Fig. 6, a series of two-wavelength measurements were made on a Feulgen-stained onion interphase nucleus, varying the levels of objective focus on the object plane from 0, which refers to the nucleus in sharp focus, to +70 μ (above) or −70 μ (below) focus on the nuclear outline. Intermediate settings were obtained by progressively turning the fine focus control of the microscope upwards through a series of calibrated markings and then by resetting to sharp focus (= 0) and turning the fine focus control downward through a similar series of calibrated distance intervals. Diameter of the nucleus used as a test object for Fig. 6 was approximately 18.5 μ, so that at focus +20 or -20 μ the image viewed microscopically was only a soft shadow of absorbance, but not a clearly defined nuclear outline. As shown by the values obtained, however, even such gross change in objective focus yielded estimates of relative amount of chromophore, which were within 6% of the mean value obtained by five replicate measurements at the 0 or sharp focus setting (Fig. 6A). The two-wavelength method used with properly corrected optics (47) thus exhibits "visibility" beyond the apparent level of focus of the objective, a point discussed also by Ornstein, who noted that chromophore packets in microscopic photometric fields may only *appear* to depart from situations to which the Beer-Lambert laws may be properly applied (39, 43). Of course, as emphasized also by Mendelsohn (32), appreciable alterations in condenser focus with regard to object plane would violate the requirement for a nearly parallel beam of monochromatic light of normal incidence and uniform intensity in the plane of the microscopic image of the object. Such distortion of proper measuring conditions would markedly alter values obtained by the two-wavelength method.

In the data of Fig. 6A, deviations of values for relative total due content obtained at extreme settings of over or under focus reflect in part absorbance of chromophore extending beyond the photometric field. This relationship is demonstrated in Fig. 6B by the significant lowering of values computed for the projected area of chromophore, when photometric measurements from focus settings beyond the limit of the nuclear diameter are considered, i.e. beyond $+20$ or -20 μ.

On the other hand, an apparent increase of total dye content at settings of less than 20 μ *below* focus, but an apparent decrease in similar values at settings of more than 20 μ *above* focus are attributable, at least in part, to refractive differences between the object (n_D about 1.570) and its mounting medium (refractive index liquid n_D 1.550). As illustrated diagrammatically by Schillaber (47; his Figs. 199–203), displacement of the Becke line toward the object of higher refractive index, in this case the nucleus, when above focus would direct light into the area of absorbing chromophore, effectively reducing its measured absorbance by spurious increase in transmittance. Or, on focusing below the object, the converse obtains which would result in refraction of light out of the photometric field, thus affecting a decrease in the amount of light transmitted to the phototube, a spurious enhancing of the measured absorbance. These expectations are consistent with the plot of data in Fig. 6A and show empirically some of the residuals and rarely encountered foibles of two-wavelength methodology.

The same data also, of course, demonstrate consistent and reproducible estimates both of total dye content and of nuclear area under normal measuring conditions of proper objective focus and reasonable matching of object refractive index.

B. EFFECT OF WAVELENGTH AND DETERMINATION OF AREA

In derivation of formulas for computing net amounts of bound dye by the two-wavelength method, it was assumed by Patau (40) and explicitly stated by Ornstein (39) that the area of absorbing chromophore is a constant proportion of the total area of the microscopic field sampled, i.e., for the case of a photometric field of unknown distribution, consisting of two or more discrete but uniformly absorbing regions of chromophore, "the value of F calculated for any pair of wavelengths for which $n = 2$ will be constant" (39). The quantity $(1 - F)$ of Ornstein should therefore be independent of wavelength. This point was illustrated by Ornstein with curves computed for values of F where transmittances of a photometric field would be expected to vary with the different distributions of chromophore presumed for given model system (39; curve D, his Fig. 1).

In the case of small, irregularly shaped diploid nuclei, such as insect hemocytes, or particularly in the case of polytene chromosome segments spread thin enough to allow selection of specific regions for photometric analysis, there is sufficient dispersion of material at the edges of the object to cause concern that at shorter wavelengths, i.e., at lower values of k, thinner portions of an object might become "invisible" to the phototube, thus violating an inherent assumption of the two-wavelength method. We felt compelled, therefore, to assess the validity of this basic assumption experimentally on specific materials of interest for cytologic study, ecompassing a range of wavelengths useful for visible light photometry of the chromophores produced by Schiff reagent as used in the Feulgen reaction for DNA.

To evaluate the influence of wavelength extremes and the selection of alternative pairs of wavelengths on estimates of relative chromophore area, a number of absorption curves were run on whole nuclei and on segments of flattened salivary gland chromosomes of *Sciara*. Values for T (transmittance), L or $(1 - T)$, and E (extinction) were obtained at 5-mμ intervals from 470 to 590 mμ from curves run through the test object only (homogeneous sample) and from curves of the same object including a rim of surrounding, but nonabsorbing background area (inhomogeneous sample). In effect, a series of measurements by the two-area method of Garcia (15) was made over a range of wavelengths with widely differing values of k. A dozen pairs of wavelengths were then selected graphically so that $E(\lambda_2) = 2E(\lambda_1)$ as determined from 10 to 12 replicate curves on the homogeneous test samples.

In a previous context, it was shown that the exact function for $(1 - F)$ can be computed from $L_1/(2 - Q)$. Values so derived from different pairs of wavelengths over the range of 470 to 590 mμ are shown by the open symbols in Fig. 7. Objects which were tested included a whole, polytene salivary gland nucleus from a *Sciara* larva at late fourth instar and two different segments of salivary gland chromosomes which show disruption of normal band structure associated with the formation of specific "puffs" which characteristically occur prior to pupation in *Sciara*. The error introduced by using alternative pairs of wavelengths was within 3%, which is consistent with the magnitude of general instrumental error for the cytophotometer used (*44*). Errors of reproducibility in repeated determinations

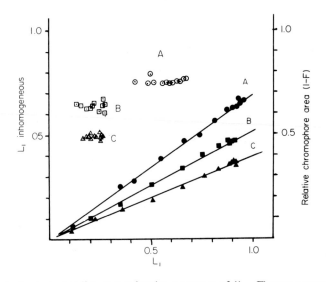

Fig. 7. A simple and sufficient test for the constancy of $(1 - F)$ over a range of wavelength pairs. Open symbols: values of $(1 - F)$, computed from $L_1/(2 - Q)$, as a function of L_1 over a range of wavelengths from 470 to 590 mμ for three different types of dye distribution: A, B, C. Solid symbols: values for a simple test of the constancy of $(1 - F)$. The ordinate is a measurement over the object including a rim of surrounding, but nonabsorbing background, while the abscissa is a measurement taken through the object only at the same wavelength. When such pairs of measurements are obtained over a range of wavelengths, the slope of the curve they describe is a function of $(1 - F)$. If the plot is linear, this is a sufficient test for the constancy of $(1 - F)$, and the more laborious calculation of $(1 - F) = L_1/(2 - Q)$ is avoided when testing a number of different wavelength pairs. A, Whole, Feulgen-stained, salivary gland nucleus of *Sciara* fourth-instar larva $(1 - F) = 0.76 \pm 0.02$, $n = 12$. B, Flattened chromosome segment III 15 AC, free end with an expanded RNA bulb, from Feulgen-stained salivary gland smear of *Sciara* at pupal molt: $(1 - F) = 0.64 \pm 0.01$, $n = 11$. C, Chromosome region II 2 AB, site of a prominent DNA puff, from Feulgen-stained salivary gland smear of *Sciara* prepupa: $(1 - F) = 0.48 \pm 0.01$, $n = 12$. (Values shown for computed areas do not include correction for different optical magnifications or sizes of photometric fields and hence are not directly comparable with each other or with values shown in Table III.) Slide preparation as in Fig. 1.

on a single test object also were within 3%. These data, obtained for a specific biologic system to be analyzed by two-wavelength cytophotometry, show that estimates of relative area of the absorbing chromophore are indeed independent of wavelength within the range tested, so long as the wavelength pairs selected for measuring are related as $k_{\lambda_2} = 2k_{\lambda_1}$.

While instructive for evaluating certain difficult photometric objects, the above testing procedure and required computations are demanding of instrument performance, time, and operator patience. As shown by the three curves with solid symbols in Fig. 7, we have found it particularly convenient to plot (L inhomogeneous) vs. (L homogeneous) to determine possible sources of error in estimates of $(1 - F)$. The slope of the line obtained in such a plot is a function of the quantity $(1 - F)$, and if a strictly linear relationship is found, this should be a sufficient test for constancy of $(1 - F)$ within the range of wavelengths of interest. On the other hand, if values do not display a linear relationship, either further validation is required to demonstrate adherence to the presumption that chromophore area is independent of wavelength, or measurements using such deviant wavelength pairs must be viewed with due caution.

C. Effect of Chromophore Fading and Determination of Area

The utility of considering separate parameters inherent in two-wavelength cytophotometry as discrete entities is further shown by evaluating the problem of chromophore fading. In recent years several workers have reported loss of color in Feulgen-stained preparations following prolonged storage, exposure to light, or after mounting slides in certain media (10, 18, 22–24, 49). In one case, statistic analysis of relative DNA-Feulgen values from populations of nuclei remeasured at 1, 2, or $3\frac{1}{2}$ year intervals indicated a direct proportionality between decolorization of the Feulgen-DNA chromophore and the nuclear area or exposed surface of the various cell types sampled (18). Evaporation of refractive index liquid used for mounting the preparations was suggested as a major factor in the fading observed (18, 49).

Squash preparations of dipteran salivary gland nuclei provide a useful photometric test material in which individual chromosomes and even particular regions of bands or puffs can be readily identified and precisely relocated for measurement. Such material affords a unique opportunity for direct assessment of possible changes in absorption curve characteristics, degree of object flattening, stain decoloration, etc. The values given in Table VII for several particular regions of Chromosome II of *Sciara* are from measurements made in September of 1966 on a preparation mounted in D. P. X., and after slide storage in the dark, remeasurement of the same

TABLE VII

EFFECT OF PROLONGED STORAGE ON FADING OF THE DNA-FEULGEN CHROMOPHORE[a,b]

Chromo-some region	Measure-ment date	L_2	L_1	Q	$B(1-F)$ (μ^2)	E_2*	M	Change in relative dye content
II 1 AC	(1966)	0.243	0.160	1.520	58.1	0.568	33.0	−25.4%
	(1968)	0.213	0.132	1.613	57.8	0.425	24.6	
II 2 AB	(1966)	0.428	0.271	1.579	59.6	0.475	28.3	−25.8%
	(1968)	0.378	0.228	1.658	57.6	0.364	21.0	
II 9 AC	(1966)	0.466	0.298	1.564	116.0	0.497	57.6	−15.1%
	(1968)	0.429	0.264	1.625	119.9	0.408	48.9	
II 14 C	(1966)	0.451	0.295	1.529	26.1	0.553	14.4	−31.9%
	(1968)	0.385	0.234	1.645	25.8	0.381	9.8	

[a] Values from measurements repeated over a 2-year interval on squash preparation of salivary gland chromosomes of *Sciara coprophila*, stained with Feulgen for DNA, and mounted in D.P.X.

[b] Ratios for $E\lambda_{560m\mu}/E\lambda_{560m\mu}$ on homogeneously dispersed samples of chromophore:

Date	Mean	S.E.	n
September 19, 1966	1.987	± 0.01	20
September 30, 1968	2.005	± 1.01	12

material in September of 1968. A reduction of roughly 25% was found in the total amount of DNA-Feulgen chromophore for each of the chromosome regions tested after this prolonged storage period. Our data agree closely with the degree of dye decoloration reported by Garcia and Iorio (*18*) for Feulgen-stained smears of human blood nuclei after 2 or 3 years of slide storage. Reassessment of absorption curve characteristics in 1968 confirmed a 2:1 ratio for extinctions determined on homogeneously dispersed samples of chromophore at the same two wavelengths used for measurement in 1966 (see footnote *b* in Table VII). Presumably then, no marked changes had occurred in the dye–substrate complex which might be attributed to alteration and/or selective loss of one or more configurational isomers of pararosaniline, an issue discussed by Hardonk and van Duijn in considering deviant shapes of absorption curves for the Feulgen dye complex in model systems (*21*). As shown in Table VII when sets of data for individual chromosome regions were used to compute projected chromophore area, or $B(1-F)$, values derived from 1968 measurements agreed within 5% of those obtained in 1966, which may be taken to demonstrate insignificant alteration in object geometry during storage, either from coverslip pressure or from evaporation of the mounting medium. On the other hand, inspection of values for Q and computation of E^* showed highly significant changes, averaging about a 25% decrease, which clearly accounts for the net loss of DNA-Feulgen staining observed in this experiment. Finally, comparisons of corrected extinction and projected chromophore area for the large, expanded puff region (II 9 AC) and those for the smaller region of dense bands (II 14 C) showed somewhat less dye decoloration for the large puff. In this material, at least, fading of the chromophore is not directly correlated with magnitude of exposed surface area; in fact, a reverse trend is shown.

A similar type of analysis was applied to evaluate the effect of acetic acid on fading of the DNA-Feulgen chromophore. Several metaphase configurations from a freshly squashed onion root meristem, which had been stained *in toto*, were measured in 45% acetic acid. The preparation was then stored overnight at 4°C in a moist chamber, and the same cells measured again the following morning. Values for projected chromophore area showed increases of 5% to 10%, probably due to flattening from coverslip pressure in a wet mount. Values for E_2^*, however, showed appreciable decreases, ranging between 15% and 25%. Estimates of total amount of chromophore showed corresponding reductions of about 20% again largely attributable to decoloration or loss of DNA-bound dye.

In agreement with Garcia and Iorio (*18*), we recommend caution in measuring old, stained preparations and cannot condone their use as "standards" by which recently made slides are evaluated. These conclusions do not ignore the fact that useful information can still be obtained from slides which, for some reason, have been stored for prolonged intervals. At least in preparations mounted in D. P. X., comparative ratios of relative amounts of DNA-Feulgen for several chromosome regions were essentially unchanged by the intervening storage period (Table VII), although the net amount of chromophore per se was significantly reduced in this test preparation.

VI. A Simple Method of Approximation for Measurement of Objects of High Transmittance

In his original formulation of the two-wavelength method Patau (*40*) considered the possibility of using a single wavelength for measurement, at which stained objects would be so highly transparent that all dye molecules would be exposed to practically the same flux of photons, i.e., that there would be no effective shadowing. In such a case, the number of chromophore packages would be essentially proportional to the amount of light absorbed out of a given light flux. Patau, however, discounted as useless for practical purposes his Eq. (1), $\gamma = KBL$, considering that it would be reasonably accurate only for such high transmissions that light loss could not be measured with any precision, and that shifting of wavelengths for measurement in the direction of increasing absorption, i.e., increasing k, would lead to progressive deviation in the amount *estimated* for chromophore and the *true* amount of dye in the photometric field.

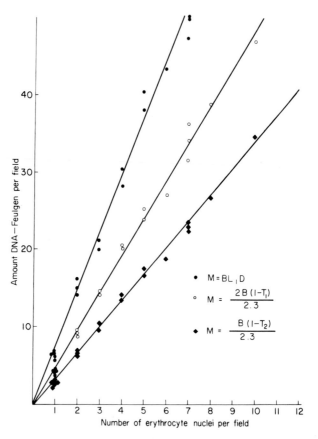

FIG. 8. Amount of DNA-Feulgen per photometric field plotted against the number of chick erythrocyte nuclei included in each field. Dye amounts computed with full two-wavelength correction (●); by simple approximation using appropriate correction factor for λ_1 (○); by simple approximation using appropriate correction factor for λ_2 (◆). A linear function is shown by each method of computation, demonstrating a capacity to count numbers of nuclei per field, despite departures of about 30% at λ_1 and about 50% at λ_2 from estimates of the total amount of DNA-Feulgen per field determined by conventional two-wavelength cytophotometry. Air-dried smear of chicken blood, postfixed in methanol–acetic acid–formalin (*14*, *15*). Hydrolysis for 60 min in 5 N HCl at room temperature (*6*, *9*), prior to staining for 2 hr in Schiff reagent. ($\lambda_2 = 560$ mμ; $\lambda_1 = 505$ mμ.)

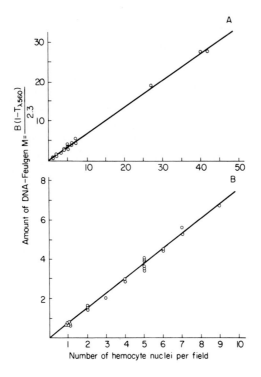

FIG. 9. Correlation of estimated amounts of DNA-Feulgen staining per photometric field with numbers of *Drosophila* hemocyte nuclei per field. Amounts of dye complex computed for $\lambda560$ mμ, using simple correction for objects of high transmittance. (See text.) The cluster of values between 1 and 10 nuclei in A has been expanded in B to indicate the kind of variability actually encountered in 24 different sampling fields. Air-dried smear of hemolymph of *Drosophila melanogaster*, from male larva at late third instar, postfixed in 10% neutral formalin, washed for 30 min in running tap water, and stained with Feulgen reaction for DNA after 40 min of hydrolysis in 5 N HCl at room temperature (6, 9).

Equations (3) and (12) of this chapter presume the ideal photometric conditions outlined above, from which the last photometric identity of Table I is derived, namely, $M = L_1 B / k_1 \ln 10$. Keeping in mind the practical limitations envisioned by Patau for using such a simple approximation to the problem of distributional error (40) and the cautions expressed by Mendelsohn (35), we have applied this method to several biologic test systems. As shown in Fig. 8, a linear relationship was obtained between the numbers of chicken erythrocyte nuclei per measuring field and the amount of chromophore estimated for such a field, although the relative amount of DNA-Feulgen in the photometric field, as expected, is appreciably under-estimated by computations involving measurement only at λ_1 and even more depressed by measurement only at λ_2, in comparison with amounts determined by two-wavelength correction. Fowl erythrocytes contain about 2.3 pg (picograms) DNA per nucleus (*38, 54, 55*) and do not truly represent a transparent type of object. They provide, however, an example of material which can be readily measured by either the two-wavelength or the two-area method as well as by this simple, one-wavelength method of approximation. They thus offer some basis for comparing data obtained by the latter procedure with those found by more conventional methods.

Unlike avian blood cells, however, the very small, pale nuclei of many dipteran hemocytes contain 0.5 pg or less of DNA per cell (*5*) and after standard Feulgen-staining procedures may often give such high values of Q, because of low absorbance at λ_1, that measurements by the usual

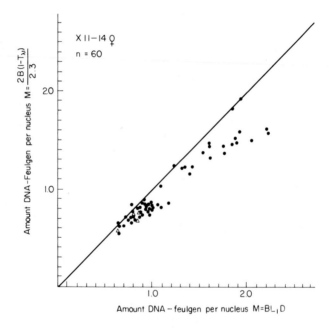

FIG. 10. Estimates of DNA-Feulgen amounts in small objects of high transmittance by simple approximation for correction of distributional error. Values computed using measurements only at λ_1 are plotted against those obtained using full two-wavelength correction. Although this function rapidly departs from the expected slope as the amount of staining material increases, it is apparent that the function is still linear over a surprisingly wide range, making possible the use of simple, experimentally determined correction factors for distributional error when large numbers of measurements are to be done within predetermined limits of dye content. Values shown are for 60 hemocyte nuclei of *Sciara coprophila* from a female larva sacrificed within 1 hr of molt to fourth instar; slide preparation as in Fig. 9. Solid circles represent interphase nuclei; open symbols, telophase nuclei. ($\lambda_2 = 560_{m\mu}$; $\lambda_1 = 502_{m\mu}$.)

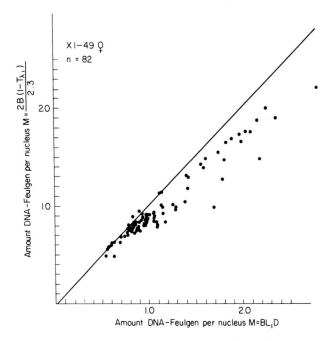

FIG. 11. A simple approximation for correction of distributional error with small objects of high transmittance. DNA-Feulgen values estimated by simple, one-wavelength measurement are plotted against those obtained using conventional, two-wavelength correction. The function described approaches linearity for nuclei with lowest dye contents. In spite of departures from the expected slope as the amount of staining material increases, such curves provide a basis for empirical adjustment either of individual values or of the means determined on a large population of small pale nuclei measured in this simple way. Values are shown for 82 hemocyte nuclei from a female *Sciara* larva sacrificed at late fourth instar, after development of pigmentation in adult eye anlage; slide preparation as in Fig. 9. ($\lambda_2 = 560$ mμ; $\lambda_1 = 502$ mμ.)

two-wavelength method are made only with due caution and trepidation. The minute size of these nuclei also precludes their accurate measurement by the two-area method, which would at least allow use of wavelengths at or near the absorption maximum of the Feulgen chromophore (*15, 19*). Consequently, we have applied the method of simple approximation by one-wavelength measurement to hemocytes of *Drosophila*, using 560 mμ to estimate net amounts of Feulgen dye in photometric fields containing varying number of nuclei (Fig. 9). A linear function was found for the range of 1 to 40 nuclei, indicating a direct proportionality between the amount of DNA chromophore estimated and the number of nuclei included within a given photometric field. Although essentially confirming Patau's original and practical expectations for this method (*40*), the findings reported here offer

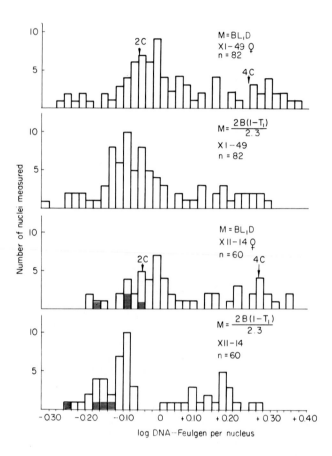

FIG. 12. Comparison of two-wavelength correction with a simple, one-wavelength correction for estimating amounts of DNA in small nuclei of high transmittance. Frequency histograms of DNA-Feulgen values for two populations of individual hemocyte nuclei of *Sciara* larvae at fourth instar. Open bars represent interphase nuclei; shaded bars, nuclei at telophase. Although a bimodality of DNA values is shown by either method of computation, note that the modes determined with two-wavelength correction for distributional error are consistently displaced to the left when relative DNA values for the same nuclei are computed from measurements made only at one wavelength. Nuclear subpopulations, here designated 2C and 4C, were established by probit analysis (*42*). When expressed in arbitrary photometric units using two-wavelength correction for distributional error, 2C = 0.87 ± 0.02, $n = 37$; 4C = 1.70 ± 0.06, $n = 19$; and 2C telophase nuclei = 0.80 ± 0.05, $n = 4$, for animal XII-14. For animal XI-49: 2C = 0.86 ± 0.03, $n = 60$; 4C = 1.76 ± 0.07, $n = 22$. Slide preparation as in Fig. 9.

a reasonable, empirical validation for potential use of this method to count small objects stained by a specific chromophore which shows appropriate stoichiometric relationships within a range of useful wavelengths. It is reasonable to suppose, for example, that the aggregate area of mitochondria within individual cells might be estimated by this approach, using the selective reduction of appropriate tetrazolium salts to provide a site-specific stain which would cast a shadow within a nearly transparent measuring field.

Results from further assessment of this approach to the problems of measuring small, pale objects are shown in the data displays of Figs. 10, 11, and 12. Estimates of DNA-Feulgen amounts are shown for individual nuclei from hemocytes of two different larval stages of *Sciara coprophila*. Amounts computed by the conventional two-wavelength method, where $M = BL_1D$, are plotted against those obtained by the method of one-wavelength, simple approximation, where $M = 2BL_1/2.3$ (Figs. 10 and 11). Deviations of values estimated by the latter method were not as great as might have been anticipated in view of the often appreciable influence of small measuring errors in determining $(1 - \text{transmittance})$ at the wavelength of lower extinction coefficient, i.e., L_1 at λ_1. In considering such a "blunt, head-on approach to cytophotometry" Mendelsohn (*35*) estimated an error of 40% or more, "particularly if the mean absorbance of the nucleus is high." The possibility was acknowledged, however, of decreasing distributional error in this simple method by the use of more transparent nuclei, achieved either by change of measuring wavelength or of stain intensities. Each of the pairs of histograms in Fig. 12 clearly shows a bimodality of values estimated for DNA content of *Sciara* hemocyte nuclei. In this figure, distributions of individual DNA-Feulgen values are displayed in the same arbitrary photometric units to encourage direct comparison of modal values for each of the methods of computation. Values indicated as representing means for 2C and 4C nuclear DNA classes were established from the two-wavelength determinations by probit analysis (*42*). Displacement of values to the left of these means serves to emphasize the small, but consistent underestimates of dye amounts which result from our one-wavelength method of simple approximation. This difference is quite apparent when comparing values for pairs of telophase nuclei (shaded bars in Fig. 12). Nonetheless, as shown in Figs. 10 and 11, reasonably good, direct linear correlations with two-wavelength measurements can be obtained by simple approximation, particularly for nuclei of the 2C or diploid class. Increasingly larger deviations from a strictly linear relationship were encountered for nuclei of the 4C or tetraploid class, as would be expected from their greater DNA content, enhanced staining and absorbance, and hence decreased transparency in the photometric field.

Although these comments on application of Patau's idealized method (*40*) are not intended to encourage its use when the objects of special interest can be readily and more accurately measured by the usual two-wavelength or two-area procedures, particular circumstances of object size and staining intensities may sometimes warrant use of this simple approach. Deviations from ideal photometric conditions can be evaluated, and with appropriate assessment of systematic bias, meaningful inferences can be drawn from systems which otherwise might not permit quantitative cytochemical study.

VII. Concluding Remarks

Readers interested in other applications of the method of two-wavelength microspectophotometry to biologic objects are well advised to refer to the illuminating discussions of model systems and other theoretical considerations in the original papers on this method by Ornstein (39) and Patau (40), to the summary of methodology given by Garcia (14), and to recent comparisons with photographic colorimetry (25, 52, 53). Specific details of measuring technique, validation of presumptions, potential sources of error, and various practical considerations of cytophotometric instrumentation are also to be found in papers by Patau and Swift (41), Leuchtenberger (27), Mendelsohn (32, 35), Pollister and Ornstein (43), Swift and Rasch (51), van Duijn et al. (11), Garcia (13), Garcia and Iorio (19), and Mayall (30), Mayall and Mendelsohn (31a), and in the general text on quantitative cytochemistry edited by Wied (56).

References

1. Atkin, N. B., and Richards, B. M., DNA in human tumors as measured by microspectrophotometry of Feulgen stain: A comparison of tumors arising from different sites. Brit. J. Cancer 10, 769–786 (1956).
2. Bachmann, K., and Cowden, R. R., Quantitative cytophotometric studies on polyploid liver cell nuclei of frog and rat. Chromosoma 17, 181–193 (1965).
3. Becker, H. J., Die Puffs der Speicheldrüsenchromosomen von Drosophila melanogaster. I. Mitteilung. Beobachtungen zum Verhalten des Puffmusters im Normalstamm und bei zwei Mutanten, giant und giant-lethal-larvae. Chromosoma 10, 654–678 (1959).
4. Carlson, L., Caspersson, T., Foley, G. E., Kudynowski, J., Lomakka, G., Simonsson, E., and Sören, L., The application of quantitative cytochemical techniques to the study of individual mammalian chromosomes. Exptl. Cell Res. 31, 589–594 (1963).
5. Daneholt, B., and Edström, J. E., The content of deoxyribonucleic acid in individual polytene chromosomes of Chironomus tentans. Cytogenetics (Basel) 6, 350–356 (1967)
6. De Cosse, J. J., and Aiello, N., Feulgen hydrolysis: Effect of acid and temperature. J. Histochem. Cytochem. 14, 601–604 (1966).
7. Deeley, E. M., An integrating microdensitometer for biological cells. J. Sci. Instr. 32, 263–267 (1955).
8. Deeley, E. M., Scanning apparatus for the quantitative estimation of deoxyribonucleic acid content. Biochem. Pharmacol. 4, 104–112 (1960).

9. Deitch, A. D., Wagner, D., and Richart, R. M., Conditions influencing the intensity of the Feulgen reaction. *J. Histochem. Cytochem.* **16**, 371–379 (1968).
10. De la Torre, L., and Salisbury, G. W., Fading of Feulgen-stained bovine spermatozoa. *J. Histochem. Cytochem.* **10**, 39–41 (1962).
11. Duijn, P. van, den Tonkelaar, E. M., and Hardonk, M. J., An improved apparatus for quantitative cytochemical studies and its use in an experimental test for the two-wavelength method. *J. Histochem. Cytochem.* **10**, 473–481 (1962).
12. Gabrusewycz-Garcia, N., Cytological and radioautographic studies in *Sciara coprophila* salivary gland chromosomes. *Chromosoma* **15**, 312–344 (1964).
13. Garcia, A. M., Studies on DNA in leucocytes and related cells of mammals. I. On microspectrophotometric errors and statistical models. *Histochemic* **3**, 170–177 (1962).
14. Garcia, A. M., Studies on DNA in leucocytes and related cells of mammals. II. On the Feulgen reaction and two-wavelength microspectrophotometry. *Histochemie* **3**, 178–194 (1962).
15. Garcia, A. M., A one-wavelength, two-area method in microspectrophotometry for pure amplitude objects. *J. Histochem. Cytochem.* **13**, 161–167 (1965).
16. Garcia, A. M., On Feulgen-deoxyribonucleic acid haploid and diploid values. *J. Histochem. Cytochem.* **15**, 778 (1967).
17. Garcia, A. M., Feulgen-deoxyribonucleic acid values after nuclear swelling. *J. Histochem. Cytochem.* **16**, 509 (1968).
17a. Garcia, A. M., Stoichiometry of dye binding versus degree of chromatin coiling. *In* "Introduction to Quantitative Cytochemistry-II" (G. L. Wied and G. F. Bahr, eds.), p. 53. Academic Press, New York, 1970.
18. Garcia, A. M., and Iorio, R., Potential sources of error in two-wavelength cytophotometry. *In* "Introduction to Quantitative Cytochemistry" (G. L. Wied, ed.), pp. 215–237. Academic Press, New York, 1966.
19. Garcia, A. M., and Iorio, R., A one-wavelength, two-area method in cytophotometry for cells in smears or prints. *In* "Introduction to Quantitative Cytochemistry" (G. L. Wied, ed.), pp. 239–245. Academic Press, New York, 1966.
20. Gledhill, B. L., Studies on the DNA content, dry mass, and optical area of bull spermatozoal heads during epididymal maturation. *Acta Vet. Scand.* **7**, 131–142 (1966).
20a. Gledhill, B. L., Changes in nuclear stainability associated with spermateliosis, spermatozoal maturation and male infertility. *In* "Introduction to Quantitative Cytochemistry-II" (G. L. Wied and G. F. Bahr, eds.), p. 125. Academic Press, New York, 1970.
20b. Gottlieb, P., A photometric study of the Feulgen dye-content and area of human leukocyte nuclei. M. S. Thesis, New York University, New York, 1968.
21. Hardonk, M. J., and van Duijn, P., Studies on the Feulgen reaction with histochemical model systems. *J. Histochem. Cytochem.* **12**, 758–767 (1964).
22. Kasten, F. H., Stability of the Feulgen-deoxyribonucleic acid absorption curve *in situ* with variation in nuclear protein content and other factors. *J. Histochem. Cytochem.* **4**, 462–470 (1956).
23. Kasten, F. H., Relative stability of cellular DNA, apurinic acid, and the DNA-dye complex in the Feulgen reaction. *Proc. 3rd Intern. Cong. Histochem. Cytochem. New York, 1968*, Summary Reports, p. 126. Springer, Berlin, 1968.
24. Kasten, F. H., Kiefer, G., and Sandritter, W., Bleaching of Feulgen-stained nuclei and alteration of absorption curve after continuous exposure to visible light in a cytophotometer. *J. Histochem. Cytochem.* **10**, 547–555 (1962).
25. Kelly, J. W., Color film cytophotometry. *In* "Introduction to Quantitative Cytochemistry" (G. L. Wied, ed.), pp. 247–280. Academic Press, New York, 1966.

26. Korson, R., A microspectrophotometric study of red cell nuclei during pyknosis, *J. Exptl. Med.* **93**, 121–128 (1951).
27. Leuchtenberger, C., Quantitative determination of DNA in cells by Feulgen microspectrophotometry. *Gen. Cytochem. Methods* **1**, 219–278 (1958).
28. Lillie, R. D., Simplification of the manufacture of Schiff reagent for use in histochemical procedures. *Stain Technol.* **25**, 163–165 (1951).
29. Mayall, B. H., The detection of small differences in DNA content with microspectrophotometric techniques. *J. Cell Biol.* **31**, 74A (1966).
30. Mayall, B. H., Deoxyribonucleic acid cytophotometry of stained human leukocytes. I. Differences among cell types. *J. Histochem. Cytochem.* **17**, 249–257 (1969).
31. Mayall, B. H., and Mendelsohn, M. L., Chromatin and chromosome compaction and the stoichiometry of DNA staining. *J. Cell Biol.* **35**, 88–89A (1967).
31a. Mayall, B. H., and Mendelsohn, M. L., Errors in absorption cytophotometry: Some theoretical and practical considerations. *In* "Introduction to Quantitative Cytochemistry-II" (G. L. Wied and G. F. Bahr, eds.). Academic Press, New York, 1970.
32. Mendelsohn, M. L., The two-wavelength method of microspectrophotometry. I. A microspectrophotometer and tests on model systems. *J. Biophys. Biochem. Cytol.* **4**, 407–414 (1958).
33. Mendelsohn, M. L., The two-wavelength method of microspectrophotometry. II. A set of tables to facilitate calculations. *J. Biophys. Biochem. Cytol.* **4**, 415–424 (1958).
34. Mendelsohn, M. L., The two-wavelength method of microspectrophotometry. IV. A new solution. *J. Biophys. Biochem. Cytol.* **11**, 509–513 (1961).
35. Mendelsohn, M. L., Absorption cytophotometry: Comparative methodology for heterogeneous objects, and the two-wavelength method. *In* "Introduction to Quantitative Cytochemistry" (G. L. Wied, ed.), pp. 201–214. Academic Press, New York, 1966.
36. Mendelsohn, M. L., Mayall, B. H., Prewitt, J. M. S., Bostrom, R. C., and Holcomb, W. G., Digital transformation and computer analysis of microscopic images. *In* "Advances in Optical and Electron Microscopy" (V. E. Cosslett and R. Barer, eds.), pp. 77–150. Academic Press, New York, 1968.
37. Mendelsohn, M. L., and Richards, B. M., A comparison of scanning and two-wavelength photometry. *J. Biophys. Biochem. Cytol.* **5**, 707–709 (1958).
38. Mirsky, A. E., and Ris, H., The desoxyribonucleic acid content of animal cells and its evolutionary significance. *J. Gen. Physiol.* **34**, 451–462 (1951).
39. Ornstein, L., The distributional error in microspectrophotometry. *Lab. Invest.* **1**, 250–262 (1952).
40. Patau, K., Absorption microphotometry of irregular-shaped objects. *Chromosoma* **5**, 341–362 (1962).
41. Patau, K., and Swift, H., The DNA-content (Feulgen) of nuclei during mitosis in a root tip of onion. *Chromosoma* **6**, 149–169 (1953).
42. Pettit, B. J., Rasch, R. W., and Rasch, E. M., DNA synthesis in the giant salivary chromosomes of *Drosophila virilis* prior to pupation. *J. Cell Physiol.* **69**, 273–280 (1967).
43. Pollister, A. W., and Ornstein, L., The photometric chemical analysis of cells. *In* "Analytical Cytology" (R. C. Mellors, ed.), 2nd ed., pp. 431–518. McGraw–Hill, New York, 1959.
44. Rasch, E. M., Darnell, R. M., Kallman, K. D., and Abramoff, P., Cytophotometric evidence for triploidy in hybrids of the gynogenetic fish, *Poecilia formosa. J. Exptl. Zool.* **160**, 155–170 (1965).

45. Rasch, R. W., Rasch, E. M., and Woodard, J. W., Heterogeneity of nuclear populations in root meristems. *Caryologia* **20**, 87–100 (1967).
46. Richards, R. M., Walker, P. M. B., and Deeley, E. M., Changes in nuclear DNA in normal and ascites tumor cells. *Ann. N.U. Acad. Sci.* **63**, 831–846 (1956).
47. Schillaber, C. P., "Photomicrography in Theory and Practice" Wiley, New York, 1944.
48. Srinivasachar, D., and Patau, K., Proportionality between nuclear DNA-content and Feulgen-dye content. *Exptl. Cell Res.* **17**, 286–298 (1959).
49. Swartz, F. J., and Nagy, E. R., Feulgen stain stability in relation to three mounting media and exposure to light. *Stain Technol.* **38**, 179–185 (1962).
50. Swift, H. H., The desoxyribose nucleic acid content of animal nuclei. *Physiol. Zool.* **23**, 169–198 (1950).
51. Swift, H., and Rasch, E., Microphotometry with visible light. *Phys. Tech. Biol. Res.* **3**, 353–400 (1965).
52. Tonkelaar, E. M., den, and van Duijn, P., Photographic colorimetry as a quantitative cytochemical method. I. Principles and practice of the method. *Histochemie* **4**, 1–9 (1964).
53. Tonkelaar, E. M. den, and van Duijn, P., Photographic colorimetry as a quantitative cytochemical method. II. Determination of relative amounts of DNA in cell nuclei. *Histochemie* **4**, 10–15 (1964).
54. Tonkelaar, E. M. den, and van Duijn, P., Photographic colorimetry as a quantitative cytochemical method. III. Determination of the absolute amount of DNA in cell nuclei. *Histochemie* **4**, 16–19 (1964).
55. Vendrely, R., The desoxyribonucleic acid content of the nucleus. *In* "The Nucleic Acids" (E. Chargaff and J. N. Davidson, eds.), Vol. 2, pp. 155–188. Academic Press, New York, 1955.
56. Wied, G. L., ed., "Introduction to Quantitative Cytochemistry." Academic Press, New York, 1966.

TWO-WAVELENGTH CYTOPHOTOMETRY OF SCIARA SALIVARY GLAND CHROMOSOMES

Ellen M. Rasch*

DEPARTMENT OF BIOLOGY, MARQUETTE UNIVERSITY, MILWAUKEE, WISCONSIN

I. Introduction

The stability and fate of the nucleoprotein materials which accumulate at specific sites of dipteran giant chromosomes during certain stages of larval development are important issues in studies of the biologic significance of the so-called "puffing" process (*7, 17, 38, 40, 79*). A particularly interesting case is that of the "anomalous" or "extra" DNA shown by salivary gland chromosomes of *Rhynchosciara* or *Sciara* (*13, 43a, 53, 53a, 62, 72, 78*). Unlike the localized alterations in chromosome structure and biosynthetic capacity described for several dipteran species by Beermann and co-workers (*6–8, 16, 55*) or by Swift (*78, 79*), and others (*5, 9, 36, 50, 66, 73, 76*), many of the prominent puff loci in salivary gland chromosomes of sciarid larvae are recognized *not* by enhanced levels of RNA basophilia or rapid uptake of tritiated RNA precursors, but by localized accumulations of deoxyribonucleoprotein materials (*53, 71, 72*) and by vigorous incorporation of tritiated thymidine (*23, 27, 78*). The possibility that such DNA puffs represent sites of selective and excessive gene replication has been of particular interest to us and prompted the present cytophotometric analysis of puff formation in polytene chromosomes of *Sciara coprophila*. Preliminary results have been reported elsewhere (*59, 60, 62*).

* U.S. Public Health Service, National Institutes of Health Research Career Development Awardee (1-K3 GM 3455). Supported by grants from the U.S. Public Health Service (GM 10503 and GM 14644).

II. Resumé of the Biological Problem

Since the classic papers of Beermann (6) and Pavan and Breuer (13, 52) in the early 1950's recognizing the functional significance of chromosomal "puffs," the dipteran salivary gland has provided a useful model system for studies of differential gene activation and the molecular events of development. (For recent reviews, see 8, 38, 40, 53a, 75, 79.)

The salivary gland of dipteran larvae offers several unique features for studies in developmental cytology. It is a purely somatic type of tissue, devoid of the complications of cell division throughout larval life (10, 18). As a consequence of a several thousand fold amplification in size of its cells during larval development, salivary gland nuclei possess giant, polytene chromosomes which allow correlation of visible changes in chromosome structure with alteration in physiologic state of the tissue (2, 7, 78). During specific stages of larval life, these chromosomes evidence characteristic patterns of localized swellings or expansions of the chromosome axis and disruption of the normal, cross-banded appearance of these regions. In several species of *Chironomus* and *Drosophila*, such regions, called *puffs* by Beermann (6, 7), and other workers, not only show net accumulations of stainable ribonucleoprotein materials (5, 9, 16, 50, 55, 66, 76, 78), but they also are active sites of RNA biosynthesis, showing rapid uptake and incorporation of tritiated uridine or cytidine (55, 66, 73, 78), an activity which is inhibited by treatment with actinomycin D (8, 17, 36, 55). In addition to showing puff sites like those just described, salivary gland chromosomes from larvae of sciarid flies, a more primitive dipteran group, also have prominent puffed regions which after puffing show enhanced ultraviolet absorption (72), increased stainability with the Feulgen reaction for DNA (53, 60, 62, 78), and during puff formation are sites of selective, heavy labeling when glands are incubated with tritiated thymidine (15, 22, 23, 43, 43a, 53a, 61, 78).

Despite the many descriptive studies of various metabolic aspects of the puffing phenomenon in dipteran chromosomes, there have been relatively few quantitative studies detailing cytochemical concomitants of the process. Among these are the elegant microchemical studies by electrophoresis to determine base ratios for isolated segments of *Chironomus* chromosomes by Edström and Beermann (21); the careful comparison by scanning microphotometry of DNA amounts in individual, specific chromosome bands of *Chironomus* hybrids by Keyl (32); the Feulgen cytophotometry of *Chironomus* puffs by Sebeleva et al. (76); and the demonstration by ultraviolet microdensitometry of a disproportionate accumulation of DNA and certain protein fractions in specific puff sites of *Rhynchosciara* by Rudkin and Corlette (72) and Rudkin (71). A similar phenomenon of localized, differential DNA accretion during puff formation at several specific sites of the salivary gland chromosomes of *Sciara* was also found by Feulgen cytophotometry by Rasch (*59*, cited in *78*), as recently confirmed by Keyl and Crouse (*19b, 33*). The examples in *Sciaridae* of marked disproportionality in DNA levels at some puff sites are in striking contrast to the stability of deoxynucleohistone levels found by Gorovsky and Woodard (*29*) during the process of RNA puff expansion in *Drosophila virilis*, and by Keyl during the formation of puffs in the midge, *Glyptotendipes* (*31*). Recently, Keyl and Hägele (*34*) have reported microphotometric measurements on nuclei of *Chironomus melanotus* in which salivary gland chromosomes sometimes show an unusual proliferation and ejection of heterochromatin material from kinetochore regions.

Finally to be noted is the growing body of evidence from oocyte systems for selective gene amplification, namely, the differential replication of genes for ribosomal RNA production during amphibian oogenesis (*14, 25, 25a, 67*). This work is directly relevant to the problem of DNA puffs in dipteran chromosomes outlined above, and perhaps to those cases in other cell systems where a close association of Feulgen-positive bodies or DNA granules with nucleolar materials has been reported (*3, 4, 24, 39, 42, 46, 54, 68*).

The varied implications (*70, 75, 79*) of the several studies briefly mentioned above emphasize the need for additional, quantitative information on the behavior, stability, and intracellular disposition of the nucleoprotein materials accreted at sites of puff formation in dipteran chromosomes.

III. Materials and Methods

A. THE ORGANISM AND TISSUE PREPARATION

We have used giant polytene salivary gland chromosomes from both female and male larvae of *Sciara coprophila* (Lintner), reared throughout their life cycle at 17°C. Our stocks, now in their sixtieth generation of single-pair matings, were originally obtained through the courtesy of Dr. Helen Crouse, and carry the wavy wing marker for the X′ chromosome, which identifies female-producing mothers in this monogenetic species. [See Metz (*45*) and Crouse (*19*) for reviews on cytogenetic eccentricities of this interesting species.]

Since the appearance of puffs in *Sciara* salivary gland chromosomes is highly correlated with stage of larval development (*23, 59, 60, 62*), our assessment of salivary gland differentiation during larval life has been based not only on time in days since hatching of eggs, but also on several, specific, morphologic criteria of development such as size and pigmentation of the larval head capsule after each molt to a successive instar and during fourth instar by the appearance, growth, and pigmentation of the imaginal eye

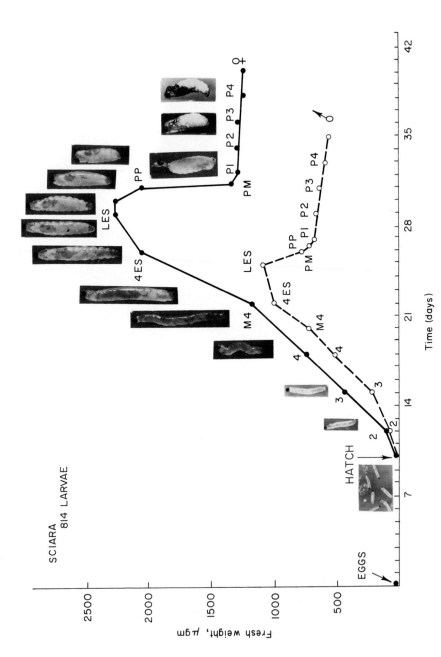

FIG. 1. Schedule of developmental stages for male and female larvae of *Sciara coprophila* reared at 17°C throughout their life cycle and fed a mixture of brewer's yeast, mushroom powder, and ground straw. Representative photographs of female larvae or pupae are shown.

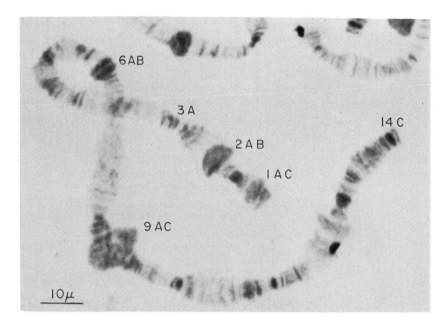

FIG. 2. Chromosome II of *Sciara coprophila*, from salivary gland of a female at pupal molt. Sites of DNA puff formation in Regions 2 AB, 6 AB, and 9 AC are shown. Feulgen stain.

anlage, here called the eyespots or ES (see Fig. 1). Within the fourth-instar period, larval stages are designated as just prior to appearance of eyespots (OES), early eyespots (EES), mid-eyespots (MES), or late eyespots (LES). Additional developmental criteria include the change in color of the gastric caeca, patterns and extent of deposition in the larval fat body, and the swallowing of air which initiates evacuation of the gastrointestinal tract to mark the beginning of the prepupal stage (PP). Lateral displacement of imaginal eyespots and detachment of the larval head capsule in prepupae precedes the shedding of larval cuticle, or pupal molt (PM). The latter stage can be defined ±30 min by noting pigmentary changes in the abdominal spiracles of the pupa. Age of pupae is then timed, in hours or in days, from this well demarcated clock point, and for convenience, stages for pupae within the first 24 hr after molt are called pupa 1; those after an additional 48 hr are called pupa 2, etc. (Fig. 1). Salivary glands in *Sciara*, unlike *Drosophila* (10), persist through formation of the puparium and undergo histolysis only during the pupal 2 stage (59).

Feulgen-stained "squash" preparations were made of salivary glands from animals carefully staged according to the above criteria. Glands, dissected into cold, buffered insect saline (5), were fixed for 3–5 min in ethanol–acetic acid (3 : 1), smeared in a drop of 45% acetic acid, and frozen on dry ice for removal of coverslips. Postfixation for 30 min in 10% neutral formalin was followed by washing for 30 min in running tap water and storage of slides at 4°C in 70% ethanol until all preparations from sibling animals sampled at different stages of development could be accumulated and processed simultaneously through hydrolysis for 10 min in 1 N HCl at 60°C, and staining for 2 hr in Schiff reagent freshly prepared according to Lillie

TABLE I

METHODOLOGICAL COMPARISON OF ESTIMATES OF DNA-FEULGEN CONTENT FOR SPECIFIC REGIONS OF *Sciara* SALIVARY GLAND CHROMOSOME II[a]

Regions of Chromosome II	Estimated DNA-Feulgen content			
	Two-wavelength method		Two-area method	Scanning method
1 AC	22.1		24.2	22.9
	22.9		23.8	
2 AB	27.5	23.1	28.0	
	24.0	26.5	26.5	24.1
	23.6	24.2	25.2	
3 A	8.8		—	13.7
6 AB	15.8		—	17.2
	17.6			
9 AC	48.6	52.5	49.6	
	44.2	47.2	48.3	51.1
	46.5	52.7	43.2	
14 C	11.7		11.4	
	12.1		11.6	10.1
	11.4		11.2	

[a] Measurements are from the chromosome sample shown in Fig. 2. All values expressed in terms of λ 560 mμ, the single wavelength used for measuring with both the two-area and the scanning methods. For two-wavelength determinations: $\lambda_2 = 560$mμ; $\lambda_1 = 502$ mμ.

(41). In several cases, a squash preparation was made of one member of a gland pair and a whole mount was made of the other member to allow cytophotometry on whole nuclei from specific regions of the gland for which puffing stages of the chromosomes were known. In addition, two types of standard tissue—cryostat sections of rat liver and air-dried films of chicken blood—were processed each day with each set of *Sciara* preparations, both to provide a check on possible variations in tissue handling and to serve as "standards" of known DNA content (65).

Designations in tables and figures of particular regions of salivary gland chromosomes of *S. coprophila* are according to the recently revised maps for this species given by Gabrusewycz-Garcia (*23*). The specific regions of Chromosome II used for intensive cytophotometric analysis are illustrated in Fig. 2 and are representative of glands sampled at pupal molt.

B. CYTOPHOTOMETRIC CONSIDERATIONS

The instrumentation and photometric procedures as outlined previously (*63*) were used to analyze details accompanying the formation of several prominent puff sites on Chromosomes II and III of *Sciara*. Total amounts of Feulgen dye, taken as estimates of net DNA content, were measured by two-wavelength cytophotometry (*48, 51*) at each presumptive, active, or regressing puff site. Data obtained for regions involved in puffing were compared with similar measurements on certain other, nonpuffing regions of the same chromosomes.

As illustrated in Table I, measurements were taken as sets on individual chromosomes. Selection of well-spread chromosomes for analysis was based on the presence in acceptable condition for photometry of at least four of the desired regions, one of which must be the free end of Chromosome II, i.e., Region II 14 C. The reliability of estimates with the two-wavelength method was checked in a number of cases by repeated measurements on different days of the same chromosomal regions and by using the two-area procedure of Garcia (*26*), or the scanning technique of Deeley with a Barr and Stroud Integrating Microdensitometer (*20*). Essentially similar estimates of (1) total amount of DNA-Feulgen per chromosome segment and of (2) ratios between amounts from two different segments were obtained with each of the tested procedures (Table I).

M = amount of dye at λ_2 corrected for distributional error

$$M = \frac{BL_1 \, D}{k_2} \quad \text{or} \quad M = \frac{B \, (I-F) \cdot E_2^{\star}}{k_2} \qquad (1)$$

where: B = area of measurement

$$(I-F) = \text{area of chromophore} = \frac{L_1}{(2-Q)} \qquad (2)$$

$$E_2^{\star} = \text{corrected extinction} = 2 \, \log \frac{1}{1-Q} \qquad (3)$$

$$L_2 = (I - \text{Transmission at } \lambda_2)$$

$$L_1 = (I - \text{Transmission at } \lambda_1)$$

$$Q = \frac{L_2}{L_1}$$

$$D = \frac{2}{2-Q} \, \log \frac{1}{Q-1}$$

FIG. 3. Summary of equations for photometric entities.

Consideration of the technique of two-wavelength photometry shows that two major variables are being assessed simultaneously. These are (1) the area of the absorbing chromophore, expressed by the quantity $B(1 - F)$, as this term was used by Ornstein (48); and (2) the optical density, or extinction of the chromophore corrected for distributional error. As shown by the relationships summarized in Fig. 3 and discussed in an earlier chapter (63), the exact functions of parameters for area, dye concentration, and relative dye content can be readily derived from measurements of transmissions obtained by two-wavelength cytophotometry, making possible quantitative evaluation of changes in size of an expanding puff region, changes in relative DNA concentration of that region, and assessment of the contribution of each of these parameters to changes in the net amount of DNA characteristic of the particular region under study.

Values for DNA-Feulgen amounts given in Table I or in Figs. 4 and 5 are expressed in relative, arbitrary photometric units. Presuming a DNA content of 2.3 pg (picograms) per fowl erythrocyte nucleus (47, 81) or a

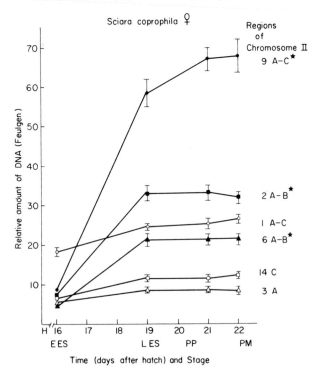

FIG. 4. Amounts of DNA-Feulgen in specific regions of Chromosome II from female larvae of *Sciara coprophila* prior to pupation. Starred regions are sites of DNA puff formation. A total of 420 measurements on individual chromosome regions is shown.

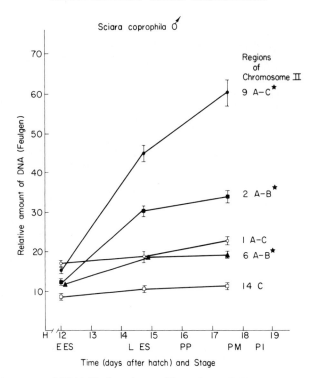

FIG. 5. Amounts of DNA-Feulgen in specific regions of Chromosome II from male larvae of *Sciara coprophila* prior to pupation. Starred regions are sites of DNA puff formation. A total of 150 measurements on individual chromosome regions is shown.

DNA content of about 6.0 pg per diploid rat liver nucleus (*80*), each photometric unit here is equivalent to roughly 0.82–0.86 pg of DNA. Data for chromophore areas in Fig. 5, also expressed in arbitrary units, have not been adjusted for the optical magnification used in measuring and hence are not directly expressed in terms of square microns.

Finally, in addition to possible variables in tissue processing such as fixation and hydrolysis times, which could markedly influence comparative, cytophotometric studies of DNA-Feulgen levels, other potential sources of systematic error should be noted. The possibility of an appreciable proportionality error, about 10% has been recently discussed in terms of chromatin compaction by several workers to account for discrepancies found when comparing amounts of DNA-Feulgen in several types of human or avian blood cells (*27, 30, 43a, 44, 64*), or during the maturation of bull spermatozoa (*28*). Also the possibility of differences in the lability of different types of deoxyribonucleoprotein complexes to the acid hydrolysis of the Feulgen reaction cannot be discounted, an issue recently emphasized for certain types of embryonic and neoplastic tissue systems (*1, 11*). These variables have not yet been evaluated for the *Sciara* chromosomes. Inferences on the biologic significance of variations in DNA-Feulgen content drawn in the present study are thus based on presumptions of an appropriate stoichiometric relation between observed maximal extinction and DNA concentration in the objects measured.

IV. Observations and Results

A. DNA-Feulgen Content of Specific Chromosome Regions during Formation of Puffs

As illustrated in Fig. 2, during late larval life and at pupation some regions of *Sciara* Chromosome II, Regions 2 AB, 6 AB, and 9 AC, show marked local distension of the chromosome axis and an apparent accumulation of DNA-Feulgen stainable material. Other sites of the same chromosome, the centromere-containing end, Region 1 AC, the free end, Region 14 C, and an intercalary site which is adjacent to a DNA puff, Region 3 A, evidence only normal cross-banding.

The data plotted in Figs. 4 and 5 show net amounts of DNA-Feulgen estimated for these six different sites of Chromosome II before, during, and after maximum puff expansion. Stages of larval development are given in terms of morphologic criteria as well as in terms of days since egg hatching. Figure 4, for female larvae, and Fig. 5, for male larvae, comprise almost 400 individual determinations of specific chromosome regions. They show that nonpuffing regions essentially double in net DNA content during this period of a last chromosomal replication cycle in larvae with eyespots, but that the puffing sites not only double their initial DNA content, but continue to accumulate Feulgen-positive material to levels some two to four times above those anticipated from the amount of DNA present in the region before the initiation of puffing. Further, the puff sites evidence an entirely different relative rate of DNA increase and duration of synthesis from that shown by other, nonpuffing regions of the same chromosome. Note the marked differences in slopes of the lines for puff sites vs. nonpuff sites between days 16 and 19 for females, or between days 12 and 15 for males. Also, the largest puff of Chromosome II, Region 9 AC, continues to show net accumulation of DNA throughout the prepupal period, including the last 2 days of larval life, by which time the other DNA puff sites of the same chromosome, II 2 AB and II 6 AB, as well as the nonpuffing sites along the chromosome seem essentially to have shut down DNA synthesis. Thus, characteristic differences in time of maximal puffing and in duration of

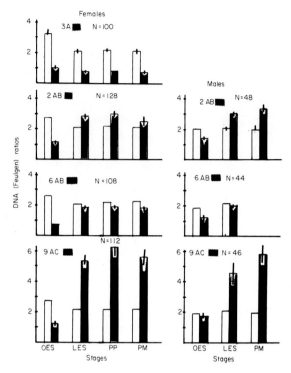

FIG. 6. Changes in amounts of DNA at puffing and nonpuffing sites of Chromosome II of *Sciara* salivary nuclei. Data expressed as mean ratios (± standard errors) of DNA-Feulgen content at indicated site (shaded bar) relative to DNA content in Region II 14 C, the nonpuffing, free end of this chromosome. Unshaded bars represent mean rations of DNA content in the centromere-containing region, II 1 AC, compared to that of the free end, II 14 C. Different stages of larval development are designated OES (before development of imaginal eye spots), LES (3 to 4 days after appearance of eye spots), PP (prepupae), and PM (within 30 min after pupal molt). These stages encompass a 4-day period for male larvae, and a 7-day period for female larvae, when animals are reared at 17°C.

synthesis, as well as in relative rate of synthesis and net amount of DNA accumulated are manifest by each of the regions studied and perhaps are intrinsic properties of these sites. Our observations on changes in net DNA content and the differences in relative rates of DNA synthesis inferred are completely consistent with autoradiographic findings reported for similar stages of larval development in *S. coprophila* (*15, 23, 60, 61*). The data of Fig. 5 for male larvae show essentially the same patterns of changes found for female larvae (Fig. 4), differing only in details of their respective time schedules, as would be predicted from the more rapid rate of development from initiation of eyespot to the time of pupal molt characteristically shown by male larvae (Fig. 1).

The magnitude of disproportionality in accumulation of DNA at sites of puff formation in *Sciara* chromosomes is further emphasized by the summary of data shown in Fig. 6, in which elevations in DNA content at each of the puffing segments of Chromosome II have been related to changes in the amount of DNA in the nonpuffing, free end of that same chromosome. By using a single region of the chromosome as an internal reference "standard" (*31, 72*) to assess the relative magnitude of changes in other regions of the same chromosome, differences in DNA-Feulgen levels due solely to DNA increases during a last round of chromonemal replication would be accounted for and manifest as a stable 1:1 ratio for the region tested before

TABLE II

AMOUNTS OF EXTRA DNA ACCUMULATED AT INDIVIDUAL DNA PUFF SITES OR IN ADJACENT REGIONS OF CHROMOSOME II FROM SALIVARY GLAND NUCLEI OF *Sciara coprophila* DURING LATE FOURTH LARVAL INSTAR[a]

Chromosome II region	Ratio of change in DNA amount after puff/before puff (pupal molt/no eyespots)					
	Female larvae			Male larvae		
	Mean	S.E.	N	Mean	S.E.	N
Nonpuffing sites						
14 C	1.00			1.00		
1 AC	0.81 ± 0.09		50	0.98 ± 0.13		24
3 A	0.80 ± 0.05		50	—	—	—
Puffing sites						
2 AB	2.10 ± 0.36		42	2.14 ± 0.16		24
6 AB	2.41 ± 0.23		36	1.63 ± 0.27		28
9 AC	4.48 ± 0.51		46	3.26 ± 0.41		24

[a] Values, computed from data of Fig. 6, are expressed as ratios of the relative increase in DNA levels at puff sites, exclusive of coincident DNA synthesis at nonpuffing site 14 C, which was set equal to 1. Total number of measurements involved in each comparison shown by N. (See text for additional discussion.)

and after puffing and the free end of the chromosome before and after puffing. Clearly, from the differences in ratios shown in Table II for puff sites compared to nonpuff sites after the formation of puffs, there is approximately a twofold increase of "extra" DNA in Region 2 AB and about a fourfold increase of "extra" DNA in Region 9 AC, *above and beyond* levels anticipated from a single cycle of DNA replication along the whole chromosome. Furthermore the ratios given in Table II are not appreciably altered

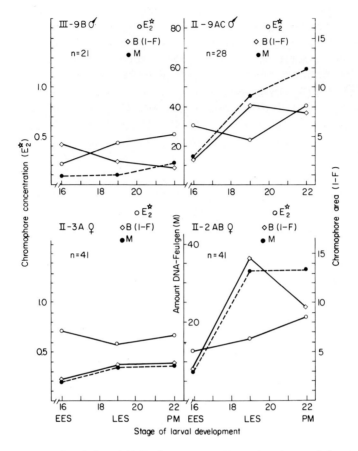

FIG. 7. Changes in relative DNA-Feulgen content, M; projected area of chromophore, $B(1 - F)$; and corrected extinction, E_2^*, for puffing and nonpuffing regions of salivary gland Chromosomes II and III of *Sciara coprophila* prior to pupation. Values are expressed in relative arbitrary photometric units, and represent the average of 6–14 two-wavelength determinations on specific regions from different chromosomes at each of the stages of larval development shown.

if the centromere end of the chromosome, Region 1 AC, is used as the internal standard of comparison. For example, in male larvae the ratio of DNA amounts at sites II 2 AB/II 1 AC, after puffing compared to levels in these regions before puffing, is 2.02 $\pm$ 0.17, which is not significantly different from 2.14 $\pm$ 0.16, the ratio obtained when site II 14 C is used as the basis for comparison. These indications of one, two, or more additional cycles of DNA replications, but only at sites of the DNA puffs, imply separate and multiple initiation points for DNA synthesis along the axis of these giant chromosomes (*56, 57*) and the discriminatory operation of some regulatory mechanism to limit or shut down DNA synthesis at these points, albeit at a much later time for each of the DNA puff sites than for other regions of the same chromosome. Support for the latter suggestion comes from the striking parallels in patterns of puffing response shown by each of the DNA puff sites, whether in male or female larvae and despite obvious differences in their developmental schedules, which presumably reflect other differences in the internal milieu of these two kinds of animals with similar chromosomes.

The amounts of "extra" DNA accumulated at puff sites are appreciable, comprising in some cases as much as 15 to 20% of the total DNA of Chromosome II. In comparisons with DNA-Feulgen measurements made on diploid rat liver nuclei as a known standard of about 6 pg of DNA (*80*), the puffed regions II 2 AB and II 9 AC together may contain about 50 to 60 pg of DNA. In several cases where whole nuclei containing puffs were measured from one member of a pair of glands and their DNA-Feulgen contents compared with the amounts estimated for major puff sites of Chromosomes II and III from similar cells, the puffs accounted for some 10–12% of the total DNA-Feulgen of the nucleus, or roughly about 150–180 pg DNA. In cells from animals at pupal molt, for example, as shown in Fig. 2, the amount of DNA at a single puff site such as II 2 AB may be equivalent to the DNA content of an octaploid rat liver nucleus, or roughly 20 to 24 pg of DNA, while the DNA level at site II 9 AC may be more than 16 times that found for a fowl erythrocyte nucleus or about 36–38 pg of DNA. It seems unlikely that differences of these magnitudes could be accounted for in terms of proportionality errors associated with chromatin compaction (*27, 43b, 44, 64*) or alterations in the acid lability of various deoxyribonucleoprotein moieties (*1, 11*), as mentioned in an earlier section. Indeed, if these factors are operative here, then the actual changes in DNA amounts at puff sites may be even greater than indicated.

B. Changes in Area and Dye Concentration of Specific Chromosome Regions during Formation of Puffs

As illustrated in Fig. 7, when variables for the total amount of DNA-Feulgen, or M, relative area of the Feulgen chromophore, or $B(1 - F)$, and corrected extinction or relative dye concentration, E_2^*, are plotted against increasing time determined from larval staging, highly significant differences are apparent in the behavior of various chromosomal components during the period of puff formation. Values shown are taken from matched sets of measurements at each of the three stages of larval development indicated.

As described earlier, Region II 2 AB, site of a prominent DNA puff, evidences a striking increase in relative amount of DNA-Feulgen, reaching a plateau by the late eyespot stage of development. Average concentration of the Feulgen chromophore also shows a net increase throughout this time period. Puff area, however, reaches a maximum at late eyespot, and then shows a marked net decrease, suggesting that some component, possibly a proteinaceous material, leaves this chromosomal site. On the other hand, the adjacent, nonpuffing segment of Chromosome II, Region 3 A, shows during the same period of larval development an approximate doubling of DNA-Feulgen content and a parallel doubling in area of the chromosome segment, but no remarkable change in dye concentration, save for the suggestion of a slight decrease in corrected extinction during the doubling of chromosome area at late eyespot.

When the large, submedian DNA puff site of Chromosome II, Region 9 AC, is analyzed in the same manner, the pattern of changes noted for companion puff 2 AB are more dramatically shown, both in terms of the magnitude of increase in total amount of DNA-Feulgen material accumulated and in the initial, marked increase in area of this chromosome segment by the late eyespot stage of development. The latter increase, however, is followed by only a slight decline in area of Feulgen-staining material by the time of pupal molt, which suggests that regression of puff 2 AB starts at an earlier time than does the regression of puff 9 AC, or that each of these two sites have as different a pattern of regression as they do of expansion. In the case of II 9 AC, unlike II 2 AB, localized chromosomal expansion during late eyespot, the time of maximal puffing activity, is accompanied by a net decrease of Feulgen extinction, as if net accumulation of DNA material were not quite keeping apace with the rate of accumulation of other components involved in puff expansion and manifest as increases in area of chromophore at puff sites.

Finally, a segment of Chromosome III, Region 9 B, which often exhibits marked accumulation of RNA fluorescence in preparations stained with coriphosphine (60a), was analyzed as an example of puffing in *Sciara* which does not involve localized DNA accumulation. This site shows an approximate doubling of total DNA during the time interval studied, but a net decrease in puff area coupled with a significant increase in Feulgen concentration, which suggests regression of an RNA-protein type of puff, in sharp distinction to the pattern of changes shown by site II 9 AC which is a DNA-histone type of puff from the same series of chromosome measurements.

V. Discussion

The results presented illustrate several points of interest relevant to the biology of the puffing phenomenon and demonstrate the utility and power of two-wavelength cytophotometry in studies of developmental cytology. Our findings support the notion that certain genetic loci in *Sciara* chromosomes undergo excessive replication, accumulating some two to four times the amount of DNA accounted for by DNA synthesis in adjacent, non-puffing regions during the last replication cycle of these giant, polytene chromosomes. Further, the formation of puffs and the rate of DNA synthesis at puff sites seem intimately correlated with specific stages of larval development, suggesting the operation of some type of hormonal mediation or programming of the changes observed (*19a, 37, 69, 74*).

Special application of the two-wavelength method to determine changes in puff area and dye concentration show that net DNA accumulation may occur during periods of maximal puff expansion, for example, at site II 2 AB, although in some cases, the rate of puff expansion may outpace net DNA synthesis, for example, at site II 9 AC. As suggested by the decrease in area occupied by Feulgen chromophore in these two chromosomal segments by the time of pupal molt, alterations in DNA/protein ratios may be characteristic features of puff expansion and regression, with some chromosomal components, possibly a specific protein, leaving the puffed region after cessation of DNA synthesis at that site. Movement of DNA from the puff site, which would also be consistent with observed decreases in the area occupied by Feulgen chromophore, seems highly unlikely from our data, for the concomitant increase in dye concentration which occurs during puff regression in Region II 9 AC at pupal molt compensates for the slight decrease in area at this site to yield a net amount of DNA appreciably greater than that found previously at the late-eyespot stage of development. Finally, the behavior of an RNA-type of puff, Region III 9 B, and that of a non-puffing, but replicating region, II 3 A, are distinctly different from one another and from the pattern of changes in puff area, Feulgen concentration, and net DNA levels shown by those sites on the same chromosomes which exhibit selective synthesis of "extra" DNA.

In the polytene chromosomes of *Sciara*, synthesis of DNA in approximately geometric increases at local sites along the chromosomes occurs both in a manner and at a time which seem independent of the DNA synthesis at other puff sites or of simple, regular chromosomal replication. If these sites are considered to represent individual growing points for specific templates or "replicons" (*56, 57*), then the present data on *Sciara*, as well as that of Keyl (*32*) and Keyl and Pelling (*35*) on hybrids of *Chironomus*, would say that each of these regions or "replicons" is able to synthesize DNA independently of the rest of the chromosome, or at least to manifest a selective biosynthetic response to those regulatory mechanisms which initiate, maintain, or shut off DNA synthesis in the system.

VI. Concluding Remarks

The well-known hypothesis of DNA constancy simply states that the amount of DNA per nucleus is parallel with the number of sets of genes (*12*). There is a wealth of excellent data from a number of laboratories concerned with diverse plant and animal tissue systems which support this hypothesis. (For reviews, see *47, 77, 81, 82*.) Yet, as was pointed out by Pollister, Swift, and Alfert in the early 1950's, "A 'DNA-gene' is constant and it apparently does self-duplicate, but this very constancy leaves us still on the horns of the old differentiation dilemma" (*58*). So, we ask today, as was asked about 20 years ago: What is the equivalence of information in the nuclei of different tissues? Do biologic systems allow the possibility of change in the informational content of nuclei during cellular specialization? Is there substantial and valid evidence of disproportional changes in DNA levels associated with cellular differentiation? The findings of the present study add support to the growing body of evidence from a number of different tissue systems (*3, 4, 13, 25a, 39, 42, 46, 54, 68, 70*) which provide an affirmative answer to these questions. They add new relevance to the suggestion made a number of years ago by Painter (*49*), who was considering the way that germ cells of some organisms contain "extra" DNA which is eliminated from cells of the soma: "the ability of a nucleus to synthesize proteins depends on the amount of a special kind of DNA." Whether all cases of selective DNA synthesis, or differential gene amplification (*14, 25*), are associated with a common, similar cellular function such as ribosome biosynthesis (*67*), or in certain cases, with some distinctive and more highly specialized aspect of metabolism remains for future study.

REFERENCES

1. Agrell, I., and Bergqvist, H-A., Cytochemical evidence for varied DNA complexes in the nuclei of undifferentiated cells. *J. Cell. Biol.* **15**, 604–606 (1962).
2. Alfert, M., Composition and structure of giant chromosomes. *Inter. Rev. Cytol.* **3**, 131–176 (1954).
3. Bauer, H., Die wachsenden Oocytenkerne einiger Insekten in ihrem Verhalten zur Nuklealfärbung. *Z. Zellforsch. Mikroskop. Anat.* **18**, 254–298 (1933).
4. Bayreuther, K., Die Oogenese der *Tipuliden*. *Chromosoma* **7**, 508–557 (1956).
5. Becker, H. J., Die Puffs der Speicheldrüsenchromosomen von *Drosophila melanogaster*. II. Mitteilung. Die Auslösung der Puffbildung, ihre Spezifität und ihre Beziehung zur Funktion der Ringdrüse. *Chromosoma* **13**, 341–384 (1962).
6. Beermann, W., Chromosomenkonstanz und spezifische Modifikationen der Chromosomenstruktur in der Entwicklung und Organdifferenzierung von *Chironomus tentans*. *Chromosoma* **5**, 139–198 (1952).
7. Beermann, W., Riesenchromosomen. *Protoplasmatologia* **6D**, 1–161 (1962).

8. Beermann, W., Control of differentiation at the chromosomal level. *J. Exptl. Zool.* **157**, 49–62 (1964).

9. Berendes, H. D., The induction of changes in chromosomal activity in different polytene types of cell in *Drosophila hydei. Develop. Biol.* **11**, 371–384 (1965).

10. Bodenstein, D., The postembryonic development of *Drosophila. In* "Biology of Drosophila" (M. Demerec, ed.), pp. 275–367, Wiley, New York, 1950.

11. Bohm, N., and Sandritter, W., Feulgen hydrolysis of normal cells and mouse ascites tumor cells. *J. Cell Biol.* **28**, 1–7 (1966).

12. Boivin, A., Vendrely, R., and Vendrely, C., L'acide désoxyribonucléique du noyau cellulaire dépositaire des caractères héréditarires; arguments d'ordre analytique. *Compt. Rend.* **226**, 1061–1063 (1948).

13. Breuer, M. E., and Pavan, C., Behavior of polytene chromosomes of *Rhynchosciara angelae* at different stages of larval development. *Chromosoma* **7**, 371–386 (1955).

14. Brown, D. D., and Dawid, I. B., Specific gene amplification in oocytes. *Science* **160**, 272–280 (1968).

15. Cannon, G. B., Puff development and DNA synthesis in *Sciara* salivary gland chromosomes in tissue culture. *J. Cellular Comp. Physiol.* **65**, 163–182 (1965).

16. Clever, U., Genakitivitaten in den Riesenchromosomen von *Chironomus tentans* und ihre Beziehung zur Entwicklung. *Chromosoma* **13**, 385–436 (1962).

17. Clever, U., Puffing in giant chromosomes of diptera and the mechanism of its control. *In* "The Nucleohistones" (J. Bonner and P. O. Ts'o, eds.), pp. 317–334. Holden-Day, San Francisco, California, 1964.

18. Cooper, K. W., Concerning the origin of the polytene chromosomes of *Diptera. Proc. Natl. Acad. Sci. U.S.* **24**, 452–458 (1938).

19. Crouse, H. V., Translocations in Sciara: Their bearing on chromosome behavior and sex determination. *Missouri, Univ., Agr. Exptl. Sta.* 75, *Bull.* 379, (19 1–43).

19a. Crouse, H. V., The role of ecdysone in DNA-puff formation and DNA synthesis in the polytene chromosomes of *Sciara coprophila. Proc. Natl. Acad. Sci. U.S.* **61**, 971–978 (1968).

19b. Crouse, H. V., and Keyl, H.-G., Extra replications in the "DNA-puffs" of *Sciara coprophila. Chromosoma* **25**, 357–364 (1968).

20. Deeley, E. M., An integrating microdensitometer for biological cells. *J. Sci. Instr.* **32**, 263–267 (1955).

21. Edström, J.-E., and Beermann, W., The base composition of nucleic acids in chromosomes, puffs, nucleoli, and cytoplasm of *Chironomus* salivary gland cells. *J. Cell Biol.* **14**, 371–380 (1962).

22. Ficq, A., and Pavan, C., Autoradiography of polytene chromosomes of *Rhynchosciara angelae* at different stages of larval development. *Nature* **180**, 983–984 (1957).

23. Gabrusewycz-Garcia, N., Cytological and autoradiographic studies in *Sciara coprophila* salivary gland chromosomes. *Chromosoma* **15**, 312–344 (1964).

24. Gabrusewycz-Garcia, N., and Kleinfeld, R. G., A study of the nucleolar material in *Sciara coprophila. J. Cell Biol.* **29**, 347–359 (1966).

25. Gall, J. G., Differential synthesis of the genes for ribosomal RNA during amphibian oogenesis. *Proc. Natl. Acad. Sci. U.S.* **60**, 553–560 (1968).

25a. Gall, J. G., and Pardue, M. L., Formation and detection of RNA–DNA hybrid molecules in cytological preparations. *Proc. Natl. Acad. Sci. U.S.* **63**, 378–383 (1969).

26. Garcia, A. M., A one-wavelength, two-area method in microspectrophotometry for pure amplitude objects. *J. Histochem. Cytochem.* **13**, 161–167 (1965).

27. Garcia, A. M., Feulgen-deoxyribonucleic acid values after nuclear swelling. *J. Histochem.* **16**, 509 (1968).

28. Gledhill, B. L., Gledhill, M. P., Rigler, R., Jr., and Ringertz, N. R., Changes in deoxyribonucleoprotein during spermiogenesis in the bull. *Exptl. Cell Res.* **41**, 652–665 (1966).
29. Gorovsky, M., and Woodard, J., Histone content of chromosomal loci active and inactive in RNA synthesis. *J. Cell Biol.* **33**, 723–728 (1967).
30. Hale, A. J., The leucocyte as a possible exception to the theory of deoxyribonucleic acid constancy. *J. Pathol. Bacteriol.* **85**, 311–326 (1963).
31. Keyl, H.-G., DNS-Konstanz im Heterochromatin von *Glyptotendipes. Exptl. Cell Res.* **30**, 245–247 (1963).
32. Keyl, H.-G., A demonstrable local and geometric increase in the chromosomal DNA of *Chironomus. Experientia* **21**, 191–193 (1965).
33. Keyl, H.-G., and Crouse, H. V., Ways of local increase of DNA in polytene chromosomes. *Symp. Organ. Genet. Mater., 6th Ann. Meeting, Am. Soc. Cell. Biol., Houston, Texas,* unpublished (1966).
34. Keyl, H.-G., and Hägele, K., Heterochromatinproliferation an den Speicheldrüsenchromosomen von *Chironomus melanotus. Chromosoma* **19**, 223–230 (1966).
35. Keyl, H.-G., and Pelling, C., Differentielle DNS-Replikation in den Speicheldrüsenchromosomen von *Chironomus thummi. Chromosoma* **14**, 347–359 (1963).
36. Kidnadze, I. I., Functional changes of giant chromosomes under conditions of inhibited RNA synthesis. *Cytologia (USSR)* **7**, 311–318 (1965).
37. Krishnakumaran, A., Berry, S. J., Oberlander, H., and Schneiderman, H. A., Nucleic acid synthesis during insect development. II. Control of DNA synthesis in the *Cecropia* silkworm and other *Saturniid* moths. *J. Insect. Physiol.* **13**, 1–57 (1967).
38. Kroeger, H., and Lezzi, M., Regulation of gene action in insect development. *Ann. Rev. Entomol.* **11**, 1–22 (1966).
39. Kunz, W., Lampenbürstenchromosomen und multiple Nukleolen bei *Orthopteren. Chromosoma* **21**, 446–462 (1967).
40. Laufer, H., Developmental interactions in the dipteran salivary gland. *In Vitro* **3**, 93–103 (1968).
41. Lillie, R. D., Simplification of the manufacture of Schiff reagent for use in histochemical procedures. *Stain Technol.* **25**, 163–165 (1951).
42. Lima-de-Faria, A., and Moses, M. J., Ultrastructure and cytochemistry of metabolic DNA in *Tipula. J. Cell Biol.* **30**, 177–192 (1966).
43. Mattingley, E. M., Variations in metabolic activity during the larval development of *Rhynchosciara. J. Cell. Biol.* **31**, 74A (1966).
43a. Mattingly, E., and Parker, C., Nucleic acid synthesis during larval development of *Rhynchosciara. J. Insect Physiol.* **14**, 1077–1083 (1968).
43b. Mayall, B. H., Deoxyribonucleic acid cytophotometry of stained human leukocytes. I. Differences among cell types. *J. Histochem. Cytochem.* **17**, 249–257 (1969).
44. Mayall, B. H., and Mendelsohn, M. L., Chromatin and chromosome compaction and the stoichiometry of DNA staining. *J. Cell Biol.* **35**, 88A–89A (1967).
45. Metz, C. W., Chromosome behavior, inheritance, and sex determination in *Sciara. Am. Naturalist* **72**, 485–520 (1938).
46. Miller, O. L., Jr., The fine structure of lampbrush chromosomes. *Natl. Cancer Inst. Monograph* **18**, 79–99 (1965).
47. Mirsky, A. E., and Ris, H., The desoxyribonucleic acid content of animal cells and its evolutionary significance. *J. Gen. Physiol.* **34**, 451–462 (1951).
48. Ornstein, L., The distributional error in microspectrophotometry. *Lab. Invest.* **1**, 250–262 (1952).
49. Painter, T. S., The elimination of DNA from soma cells. *Proc. Natl. Acad. Sci. U.S.* **45**, 897–902 (1959).

50. Panitz, R., Hormonkontrollierte Genaktivate in den Riesenchromosomen von *Acricotopus lucidus. Biol. Zentr.* **83**, 197–230 (1964).

51. Patau, K., Absorption microphotometry of irregular-shaped objects. *Chromosoma* **5**, 341–362 (1952).

52. Pavan, C., and Breuer, M. E., Polytene chromosomes in different tissues of *Rhynchosciara. J. Heredity* **43**, 151–157 (1952).

53. Pavan, C., and Breuer, M. E., Differences in nucleic acids content of the loci in polytene chromosomes of *Rhynchosçiara angelae* according to tissues and larval stages. *In* "Symposium on Cell Secretion" (G. Schreiber, ed.), pp. 90–99, Belo Horizonte, Brazil, 1955.

53a. Pavan, C., and da Cunha, A. B., Chromosomal activities in *Rhynchosciara* and other Sciaridae. *Ann. Rev. Genet.* **3**, 425–450 (1969).

54. Peacock, W. J., Chromosome replication. *Natl. Cancer Inst. Monograph* **18**, 101–131 (1965).

55. Pelling, C., Ribonucleinsaure Synthese der Riesenchromosomen. Autoradiographische Untersuchungen on *Chironomus. Chromosoma* **15**, 71–122 (1964).

56. Plaut, W., On the replicative organization of DNA in the polytene chromosome of *Drosophila melanogaster. J. Mol. Biol.* **7**, 632–635 (1963).

57. Plaut, W., Nash, D., and Fanning, T., Ordered replication of DNA in polytene chromosomes of *Drosophila melanogaster. J. Mol. Biol.* **16**, 85–93 (1966).

58. Pollister, A. W., Swift, H. H., and Alfert, M., Studies on the desoxypentose nucleic acid content of animal nuclei. *J. Cellular Comp. Physiol.* **38**, 101–120 (1951).

59. Rasch, E. M., Developmental changes in patterns of DNA synthesis by polytene chromosomes of *Sciara coprophila. J. Cell Biol.* **31**, 91A (1966).

60. Rasch, E. M., Nucleoprotein metabolism during larval development in *Sciara. Proc. 3rd Intern. Congr. Histochem. Cytochem., New York,* 1968, Summary Reports, pp. 215–216, Springer, Berlin, 1968.

60a. Rasch, E. M., and Barr, H. J., Unpublished observations (1967).

61. Rasch, E. M., and Pettit, B. J., Nucleoprotein metabolism in salivary gland chromosomes of *Sciara* during pupation. *J. Cell Biol.* **35**, 110A (1967).

62. Rasch, E. M., and Rasch, R. W., Extra DNA synthesis at specific sites of *Sciara* salivary gland chromosomes. *J. Histochem. Cytochem.* **15**, 793 (1967).

63. Rasch, E. M., and Rasch, R. W., Special applications of two-wavelength cytophotometry in biologic systems. *In* "Introduction to Quantitative Cytochemistry-II" (G. L. Wied and G. F. Bahr, eds.) p. 357. Academic Press, New York, 1970.

64. Ringertz, N. R., and Bolund, L., Nucleoprotein changes in hen erythrocyte nuclei undergoing reactivation. *Proc. 3rd Intern. Cong. Histochem. Cytochem. New York,* 1968, Summary Reports, pp. 221–222, Springer, Berlin, 1968.

65. Ris, H., and Mirsky, A. E., Quantitative cytochemical determination of desoxyribonucleic acid with the Feulgen nucleal reaction. *J. Gen. Physiol.* **33**, 125–146 (1949).

66. Ritossa, F. M., Behavior of RNA and DNA synthesis at the puff level in salivary gland chromosomes of *Drosophila. Exptl. Cell Res.* **36**, 515–523 (1964).

67. Ritossa, F. M., Atwood, K. C., Lindsley, D. L., and Spiegelman, S., On the chromosomal distribution of DNA complementary to ribosomal and soluble RNA. *Natl. Cancer Inst. Monograph* **23**, 449–472 (1966).

68. Roberts, B., DNA granule synthesis in the giant food pad nuclei of *Sarcophaga bullata. J. Cell Biol.* **39**, 112A (1968).

69. Rodman, T. C., Relationship of developmental stage to initiation of replication in polytene nuclei. *Chromosoma* **23**, 271–287 (1968).

70. Roels, H., "Metabolic" DNA: A cytochemical study. *Intern. Rev.Cytol.* **19**, 1–34 (1966).

71. Rudkin, G. T., The proteins of polytene chromosomes. *In* "The Nucleohistones" (J. Bonner and P. O. Ts'o, eds.), pp. 184–192. Holden-Day, San Francisco, California, 1964.
72. Rudkin, G. T., and Corlette, S. L., Disproportionate synthesis of DNA in a polytene chromosome region. *Proc. Natl. Acad. Sci. U.S.* **43**, 964–968 (1957).
73. Rudkin, G. T., and Woods, P., Incorporation of H^3-cytidine and H^3-thymidine into giant chromosomes of *Drosophila* during puff formation. *Proc. Natl. Acad. Sci. U.S.* **45**, 997–1003 (1959).
74. Schneiderman, H., and Gilbert, L., Control of growth and development in insects. *Science* **143**, 325–333 (1964).
75. Schultz, J., Genes, differentiation, and animal development. *Brookhaven Symp. Biol.* **18**, 116–147 (1965).
76. Sebeleva, T. E., Sherudilo, A. I., and Kidnadze, I. I., Quantitative estimation of DNA in puff formation in *Chironomus dorsalis*. *Genetika* **2**, 102–105 (1965).
77. Swift, H., Quantitative aspects of nuclear nucleoproteins. *Intern. Rev. Cytol.* **2**, 1–76 (1953).
78. Swift, H., Nucleic acids and cell morphology in dipteran salivary glands, *In* "The Molecular Control of Cellular Activity" (J. Allen, ed.), pp. 73–125. McGraw-Hill, New York, 1962.
79. Swift, H., Molecular morphology of the chromosome. *In Vitro* **1**, 26–49 (1965).
80. Thompson, R. Y., Heagy, F. D., Hutchinson, W. C., and Davidson, J. N., The desoxyribonucleic acid content of the rat cell nucleus and its use in expressing the results of tissue analysis in particular reference to the composition of liver tissue. *Biochem. J.* **53**, 460–474 (1953).
81. Vendrely, R., The desoxyribonucleic acid content of the nucleus. *In* "The Nucleic Acids" (E. Chargaff and J. N. Davidson, eds.), Vol. 2, pp. 155–188. Academic Press, New York, 1955.
82. Vendrely, R., and Vendrely, C., The results of cytophotometry in the study of the deoxyribonucleic acid (DNA) content of the nucleus. *Intern. Rev. Cytol.* **5**, 171–197 (1956).

DNA CYTOPHOTOMETRY OF SALIVARY GLAND NUCLEI AND OTHER TISSUE SYSTEMS IN DIPTERAN LARVAE

Ellen M. Rasch*

DEPARTMENT OF BIOLOGY, MARQUETTE UNIVERSITY, MILWAUKEE, WISCONSIN

I. Introduction

In view of considerable contemporary interest by developmental biologists in the use of various tissues of insect larvae as models for studying the molecular events of cellular differentiation (*5, 7, 11, 52, 54, 75, 76, 110, 122*), we have undertaken a systematic comparison of the quantitative behavior of DNA at particular stages of development in several different, dipteran species. The suggestion by Rudkin (*98–99b*) of underreplication of DNA by heterochromatic segments of *Drosophila* chromosomes during tissue differentiation, and the indications of excessive DNA synthesis within particular regions of the giant salivary gland chromosomes of *Rhynchosciara* or *Sciara* at certain stages of tissue specialization (*19, 25a, 37, 61, 73, 75, 75a, 99, 100, 103, 109*) are fraught with implications relevant to the hypothesis of DNA constancy (*18, 66, 106*) and to current notions on the regulation of selective gene amplication for these and other developing cell systems (*20, 39, 74, 75a, 90, 96*).

Quantitative cytophotometry provides certain unique advantages for the study of such biologic problems. Although there are several reports on estimated DNA content per nucleus for salivary gland cells of *Drosophila melanogaster* (*21, 56, 70, 71*), of *Drosophila hydei* (*27, 28*), and more recently for *Chironomus* (*55*), and *Sciara* (*43*), the data given are based on microchemical determinations adjusted for total number of cells from pooled

* U.S. Public Health Service, National Institutes of Health Research Career Development Awardee (1-K3 GM 3455). Supported by grants from the U.S. Public Health Service (GM 10503 and GM 14644). The collaboration of Dr. R. W. Rasch in the statistical analyses of this study is gratefully acknowledged.

samples of from 36 to 100 pairs of glands, and thus, by necessity, values represent average or modal DNA levels for the cell type. They do not reflect, for example, the proximal–distal gradient in cell size and degree of nuclear polytenization which was described many years ago by Bodenstein for *Drosophila virilis* (*16*, *17*), by Cooper for *Drosophila melanogaster* (*22*), and more recently by Berendes for *Drosophila hydei* (*10*). On the other hand, cytophotometric techniques for estimating DNA content of individual whole nuclei, while subject to certain methological variables, can more precisely describe such mosaics of polysomatic nuclei within a given tissue, allowing temporal and topologic correlation with other aspects of changing cellular function.

More than a decade ago, quantitative studies using methyl green staining for DNA (*53*) or the Feulgen reaction (*1*, *111*, *119*) revealed the presence of at least four clearly different DNA classes of polytene nuclei within the salivary gland of third-instar larvae of *Drosophila melanogaster*. These findings led to the suggestion that polytene chromosomes undergo periodic, synchronized DNA replication (*1*, *107*). Direct confirmation of these surmises on the polyfibrillar or polynemic nature of dipteran giant chromosomes was later provided by the studies of Bier (*12*) on ovarian nurse cells of another dipteran, *Calliphora*, which nuclei show during the course of their development small polytene chromosomes, that disband into component fibrils, and at a later time, reaggregate to form " secondary " polytene chromosomes that are indistinguishable in general appearance from the giant chromosomes present in other tissues of this and similar organisms at appropriate stages of development. [See also Beermann and Pelling (*8*) and Henderson (*45*).]

More recent studies, using the two-wavelength method of cytophotometry according to Patau (*69*) and Ornstein (*67*), on Feulgen-stained whole nuclei of *Drosophila melanogaster* (*92–94*), *Drosophila virilis* (*78*), and *Drosophila hydei* (*10*) have generally confirmed the earlier studies, as have those using scanning methods (*11a*) or photographic colorimetry (*66a*). Amounts of DNA-Feulgen per salivary gland nucleus were found to distribute into several classes, the means of which closely approximated values projected as geometric multiples, 2^n, of DNA levels found for other, presumably diploid somatic tissues such as neutral ganglion cells (*11a, 92, 99b, 119*), gland anlagen (*109, 111*), or cells of the larval hemolymph (*78, 82, 84a*). From 7 to 11 replication cycles have been variously reported for different species or mutant strains of *Drosophila* (*1, 53, 78, 93, 94, 111, 119*). Relevant studies on another type of dipteran larva, the midge *Chironomus tentans*, by Daneholt and Edström (*26, 34*), using their highly specialized analysis by microelectrophoretic separation of nucleotide bases in conjunction with the Zeiss USMP, also involve measurements from individually selected, whole nuclei or specific chromosomal segments. These authors report the presence of several DNA classes within chironomid salivary gland cells, the very largest nuclei of which contain chromosomes reflecting 13 cycles of DNA replication, for the DNA level found was some 16,000 times that estimated for individual, diploid, hemocyte nuclei of similar animals. Aside from the preliminary report by Himes and Crouse (46) on DNA levels of nurse cells in the ovary of *Sciara*, those of Freed and Schultz (35, 102) on nurse cells of *Drosophila*, Welch's interesting commentary (119) on DNA classes in the ring gland and other tissues of normal and lethal-mutant strains of *Drosophila melanogaster*, and the recent report by Roberts (91) on DNA changes in the giant nuclei of footpad cells of adult *Sarcophaga*, there are surprisingly few cytophotometric studies of DNA in the many different tissue systems of so biologically significant a group of organisms as the *Diptera*.

In addition to a characteristic association with larval salivary gland cells (1, 6, 109), polynemic chromosomes are known to occur in other dipteran tissues such as ovarian nurse cells (35, 47, 49, 68, 102, 105). They have also been reported in cytologic descriptions of nuclei from the esophagus, midgut, hindgut, rectum, Malpighian tubules, the fat body, muscles, tracheal cells, the hypodermis, and in certain cells of the brain (6, 9, 11a, 22, 36, 41, 60, 66a, 109). Since some of these cell systems from the larva persist with only slight remodelling to form organ systems of the adult (20), the occurrence of polytene chromosomes clearly is not uniquely a larval characteristic, although it is very often associated with secretory activity of glandular tissues at particular stages of larval development. A case in point is that of the extensive polytenization of chromosomes in footpad nuclei of developing pupae of *Sarcophaga bullata* (120).

The potential for multiple transcription sites in cells containing many, many copies of coded, genetic information, whatever the developmental stage of its polynemic chromosome system, is highly significant in terms of our current understanding of mechanisms of protein synthesis, and brings into new focus the relevancy of a currently held presumption: the informational equivalency of all nuclei within an organism. As noted several years ago by Zalokar (122), in systems such as the silk glands of *Bombyx mori*, the silkworm, where cells are synthesizing proteins at a rate surpassing all other known protein-producing cells, it is as if genes had to multiply in order to provide enough sites for the high synthetic activity required. Genes, as DNA, carry information about the amino acid sequence in proteins, but also determine the rate of protein formation through control of RNA production and the stage in cell life when the formation should occur. Although in many systems it is the rate of RNA synthesis which seems the main determining factor for the amount of enzyme or protein to be produced in a cell, in certain systems, the amount of DNA, or rather the number of specific templates per se, may be a major determining factor, for this could limit the rate of specific RNA synthesis, which would in turn limit the rate of production of specific enzymes or proteins within the cell.

The characteristic incidence of polytene chromosomes in dipteran tissues (for reviews, see *6, 22, 41, 42, 75a*) is coupled with the almost unique opportunity they afford to study localized changes in biosynthetic capacity, whether replicative or transcriptive, of highly amplified, specific chromosome regions (*2, 5, 7, 9, 11, 25, 34, 73, 76, 79, 80, 110*). In the present studies, we have used two-wavelength cytophotometry to measure amounts of DNA-Feulgen per nucleus in various types of cells from several, representative dipteran genera. In some cases, particular tissue systems were studied at different stages of development. General conformance with the hypothesis of DNA constancy (*18, 66, 106, 116*) was found for the polytene chromosome systems of *Drosophila* and *Chironomus*, and during early larval life of the mycetophylid dipteran, *Sciara coprophila*. Salivary gland nuclei of the latter species, however, during the week prior to pupation show DNA levels which are significantly higher than those expected solely from regular, periodic replication of the diploid, somatic chromosome complement. These deviations by *Sciara* polytene chromosomes from a simple, geometric series of DNA classes coincide in time with the appearance of "DNA puffs" in certain regions of the chromosomes, suggesting differential and selective DNA replication at these sites. Preliminary reports of these findings have appeared elsewhere (*82, 83, 84a*).

II. Materials and Methods

A. The Organisms and Staging

For *Drosophila melanogaster*, a single-pair mating of a wild-type Oregon R stock was obtained through the courtesy of Dr. H. J. Barr. Late-third-instar larvae, white and tan prepupae, were sampled at 110 to 120 hr after egg deposition and culture incubation at 18°C. Larval or prepupal sex was determined by the presence of ovary or testis in the posterior fat body (*17*).

Stocks now inbred for more than 5 years by Dr. R. W. Rasch were used for *Drosophila virilis*. Late-third-instar larvae from a single-pair mating were staged by the first appearance of pigment in the anterior spiracles (*78*). Prepupal stage was then determined by the elapsed time from operculum closure, an event which follows spiracle pigmentation in the particular strain used, by approximately 6 hr. Pupation proper was judged by eversion of the imaginal discs approximately 10 to 12 hr after spiracle eversion. At this latter stage, although salivary glands show vacuolate cytoplasm, they are still grossly whole in appearance and can be dissected out carefully to make whole-mount tissue preparations.

Late-fourth-instar female larvae and prepupae of the midge, *Chironomus thummi*, were sampled according to criteria outlined by Kroeger (*51*) using stocks originally obtained through the courtesy of Dr. B. J. Stevens and maintained in our laboratory for more than 3 years by mass cultures incubated at 22°–24°C.

All stages of development of both female and male larvae of *Sciara coprophila*, a mycetophylid gnat, were sampled from within 1 hr of egg hatching on through fourth instar and up to 36 hr following pupal molt. Unlike digeneic species in which single families consist of both sons and daughters, females in monogeneic lines of *Sciara* are either female producers or male producers (*23, 63*). In our stocks, originally obtained in 1962 through the kindness of Dr. H. V. Crouse, "wavy," a dominant sex-linked, mutant wing character is carried by the X′ chromosomes (*64*) and is used to identify female-producing females (X′X), as distinguished from wild-type, normal winged females (XX) who give rise only to male offspring. Thus sex of offspring, even at very early stages of development, can be realiably predicted from the phenotype of the mother used in single-pair matings. [For additional details on the production of unisexual progenies and mechanisms of sex determination in sciarid species, see reviews by Metz (*63*) and Crouse (*23, 24*).] A detailed time schedule of developmental stages for *Sciara coprophila* reared at 17°C and representative stages for the female larvae are shown elsewhere in this volume (*84*). Designations used in the present report for various regions of the long salivary gland of *Sciara* are given in Fig. 1. In confirmation of studies on salivary glands of other dipteran species (*17, 60, 71, 81*), counts of the number of nuclei in each of the designated regions from more than 60 pairs of salivary glands from male or female *Sciara* larvae showed no statistically significant difference in the average number of cells per pair or per gland member associated either with sex or with particular stage of development from the time of molt to the second instar until 72 hr after pupation, about which time gland histolysis occurs.

B. Tissue Preparation

Various dipteran organ systems were dissected from decapitated larvae in a drop of chilled insect Ringer's solution (*5*). Whole mounts were prepared from salivary glands, gastric caeca, the midgut, fat body, Malpighian tubules, and neural ganglia. Developing limb and antennal anlage were also sampled from prepupae and pupae of *Sciara*. Tissue fixation on gelatin-coated slides for 5 min in ethanol–acetic acid (3:1) was followed by transfer through several changes of absolute ethanol, hydration, and then postfixation for 30 min in 10% neutral formalin, followed by washing for 30 min in running tap water to remove excess formaldehyde. Slides were then stored at 4°C

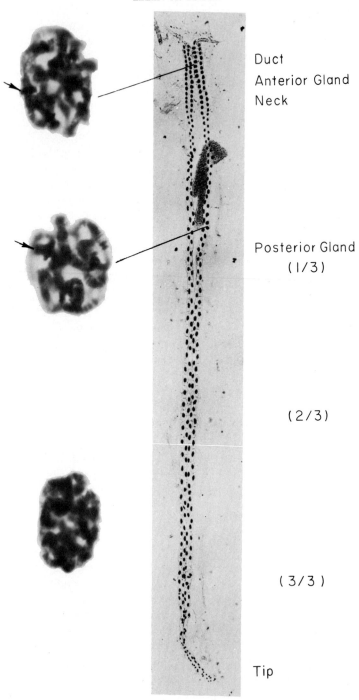

Duct
Anterior Gland
Neck

Posterior Gland
(1/3)

(2/3)

(3/3)

Tip

in 70% ethanol for up to 10 days, whenever necessary to accumulate an adequate number of samples from sibling animals at various stages of development. Each slide of tissues from a given animal contained a drop of hemolymph which had been drained from that larva after removal of jaws or head capsule and before dissection of organ systems. Hemocytes were allowed to air-dry on each of two or three slides; the other tissues or organ systems were added as quickly as possible; and the preparation was then processed through the fixation schedule outlined above. In addition, two types of standard tissues—10 μ sections of rat liver and air-dried films of chicken blood—were processed each day with each group of slides of dipteran tissues, both to serve as a check on possible variations in tissue handling and to provide "standards" of known DNA content (89) so that DNA-Feulgen levels found for various dipteran tissues might also be expressed in terms of pg DNA per cell. Although there obviously are potential sources of error implicit in presuming direct equivalency for levels of DNA-Feulgen staining by invertebrate and vertebrate nuclei, this procedure has been helpful in other cell systems where direct biochemical determinations, for one reason or another, were not available (85, 87, 88).

All slides of each series were hydrolyzed simultaneously for 12 min in 1 N HCl at 60°C (108) or for 60 min in 5 N HCl at room temperature (29, 31), rinsed briefly in two changes of 0.1 N HCl at room temperature, and stained for 2 hr in freshly prepared Schiff's reagent made according to Lillie (57), but using at least 60 hr for decolorization of the reagent (32, 33). Stained preparations were rinsed in three, 10-min changes of sulfite water, washed for 10 min in cool, running tap water, and then dehydrated through a graded series of alcohols before clearing in xylene, and mounting in

---

FIG. 1. Pair of salivary glands from a female larva at very late fourth instar of *Sciara coprophila* (Lintner). Nuclei magnified at left taken from zones indicated. Arrows point to DNA puffs. As with other primitive dipteran glands of similar, complex structure, each gland is divided, for convenience of description, into two regions: the basal reservoir (here often called "anterior" gland) and the gland proper (here often called "posterior" gland). These two main regions are themselves composed of more than one cell type, exhibiting marked differences in cytological appearance and presumably in physiologic function during the course of larval development. At the proximal or anterior end of the gland, close to the duct which joins the two glands, is an elongate cone of small nuclei, behind which are one or more pairs of polytene nuclei, clearly smaller than those in the following zone which contains 12 pairs of large nuclei with thick, polytene chromosomes that show a distinctive pattern of puff formation in late larval life. A "transitional" zone or "neck" region of 12–14 cells, the nuclei of which do not form DNA puffs, connects the basal reservoir region to the 96 or so cells of the gland proper. The large polytene chromosomes of many of the latter cells have a characteristic pattern of puff formation, different from that of the anterior gland region, during late larval life. The distal or "tip" region is composed of 16–18 cells with small nuclei also containing polytene chromosomes.

matching refractive index liquids at n_D 1.550 to 1.564 (R. P. Cargille Laboratories, Inc., Cedar Grove, New Jersey).

C. CYTOPHOTOMETRY

Amounts of DNA for individual, Feulgen-stained nuclei were determined by the two-wavelength method of Ornstein (67) and Patau (69), using the instrumentation described in a previous chapter (86) and the methods outlined there for computing M, the total amount of Feulgen chromophore per photometric field, corrected for distributional error at λ 560 mμ.

Because the majority of hemocyte or neural ganglia nuclei of dipteran larvae are both small and pale, all two-wavelength measurements were done near the absorption maximum of the Feulgen chromophore, using 560 mμ for λ_2. For each hydrolysis series an appropriate wavelength determination for λ_1 was obtained by analysis of absorption curves from at least 12 samples through homogeneous areas of rat liver nuclei from the same series of slides containing dipteran tissues. Consistently, λ_1 was between 498 to 502 mμ in all of the 15 hydrolyses series analyzed. Pairs of wavelengths so determined, where $E\lambda_2 = 2\ E\lambda_1$, were then checked on insect material identically treated for fixation and staining. No significant differences were found in absorption curve characteristics.

The extreme stain density and large size of nuclei from salivary glands or Malpighian tubules of older larvae, compared with the low dye concentration and small size of hemocyte nuclei, necessitated use of several alternative pairs of wavelengths and objective–ocular combinations. Pairs of wavelengths were selected in the range of 475 to 560 mμ, using the procedure just outlined, to obtain values of Q (cf. 69) within the 1.4 to 1.7 range recommended by Swift and Rasch (112). Also, excessive values for Q were avoided by minimizing the rim of background illumination surrounding measured objects, in which case, areas of the photometric field were determined with the aid of a standard, fixed reference area, as suggested by Garcia and Iorio (40). Since a variety of optic conditions and wavelength pairs were required to encompass the diversity in object sizes and staining densities encountered in these preparations, all raw photometric values computed for M were reduced by the use of empiric curves to yield a common base photometric unit, which in the present study is equivalent to measuring amounts of DNA-Feulgen with a 90× n.a. 1.32, oil-immersion objective, a 25× ocular, at a wavelength of 560 mμ. Under the staining and measuring conditions described, diploid (2C) nuclei from rat hepatocytes or diploid (2C) nuclei from chicken erythrocytes averaged 7.3 units and 2.7 units, respectively. Assuming that a rat 2C nucleus contains about 6 pg of DNA (56, 114) or that a chick 2C nucleus contains about 2.3 pg of DNA (66, 115), each two-wavelength, arbitrary photometric unit here is then equivalent to approximately 0.82 to 0.85 pg of DNA.

Finally, several preparations of diploid tissue systems of *Drosophila melanogaster* were measured with a Barr and Stroud Integrating Micro-densitometer (Model GN 2), according to the scanning technique of Deeley (*30*), using a Beck 100× oil-immersion objective, the high projector lens system, and 560 mμ. Data from the latter instrument are expressed in arbitrary units of relative absorbance (RA), each unit of which is equivalent to about 0.35 pg of DNA when calibrated against standard chick blood preparations under the conditions of staining and measurement described.

D. PROBIT ANALYSIS OF PHOTOMETRIC DATA

Extent of chromonemal replication in whole nuclei or of the probable strandedness in giant chromosomes, was assessed by comparing observed levels of DNA-Feulgen in polytene nuclei with projected means and limits for DNA classes obtained from measuring nuclei in presumably diploid tissues such as hemocytes and neural ganglia, or from developing limb anlagen of prepupae. In these actively dividing cell systems (*3, 97, 121*), the frequency distributions of DNA-Feulgen values for populations of nuclei characteristically are bimodal (1:2 ratio of modes) or trimodal (1:2:4 ratio of modes), with a significant, but variable number of nuclei containing inter-mediate (or interclass) amounts of DNA, which might be expected among nuclei preparing for mitosis or undergoing endomitosis (*107, 116*).

Minima in plots of frequency distributions may be taken as indices of DNA class limits for dipteran tissues, as was discussed by Schultz (*102*) in his consideration of chromatin replication by nurse cell nuclei of *Drosophila melanogaster*. In the present study, however, the more sensitive method of probit analysis according to Bliss (*15*) was used to identify coherent popula-tions of diploid, tetraploid, and polytene nuclei; to set statistic limits for the DNA classes so discerned; and to estimate the average amount of DNA associated with either the male or female diploid, somatic genome.

Two criteria were used to segregate distributions of DNA-Feulgen measure-ments into classes: (1) the division between classes must be a minimum in the frequency distribution, and (2) the subdivided class must provide a "best-fit" to a log-normal distribution of DNA amounts. The latter criterion was established by accumulating the frequencies of log-DNA amounts and displaying them on a probit scale versus the log-DNA amount (for examples of this type of plotting, see Figs. 11 and 12). Deviations from a linear relationship were used to segregate the measurements into classes that provided the "best-fit" on a trial and error basis. After the data had been segregated into classes, the mean and variance were computed by

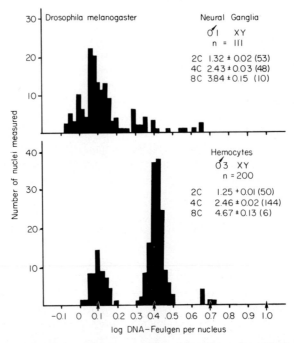

FIG. 2. Amounts of DNA-Feulgen per nucleus, plotted on a logarithmic scale, for cells of the subesophageal ganglion and hemolymph of late-third-instar larvae of male *Drosophila melanogaster*. Arrows on abscissa indicate position of means for 2C, 4C, and 8C DNA classes, respectively. Nuclei measured by Deeley scanning method (*30*), using Barr and Stroud Integrating Microdensitometer.

conventional formulas, although values so derived did not differ significantly from those which could be estimated directly from the probit graphs.

DNA-Feulgen values ascertained as diploid or 2C by probit analysis, for example, as in the hemocyte population of an individual prepupa, can be divided by 2 to get a "C" value, here defined as the average amount of DNA in the haploid somatic chromosome complement (*106*). A projected C value can then be computed as $C = 2^n$, where n is 1, 2, 3,..., i.e., some number of replication cycles, such as $n = 12$ for 8192C, presuming that all potential replication sites of the somatic chromosome complement duplicate their DNA proportionately in each ensuing replication cycle. Deviations from expectations can then be evaluated statistically by direct comparison of experimentally observed DNA-Feulgen values with projected polytene DNA class limits.

Several cytologic peculiarities of *Sciara coprophila*, such as differential retention of a variable number of so-called "limited chromosomes" in germ line cells (*24, 46, 63*) and the atypical, monopolar divisions which occur during meiosis in the male (*62*) preclude in this species the use of nuclei of

germ cell lineage, such as young spermatids, as a simple way to estimate the
haploid or 1C level of DNA applicable to somatic cells. For this reason, the
amount of DNA found for small, diploid hemocyte nuclei was adopted as
the basis for comparisons of DNA levels observed in all other diploid or
polytene somatic cell types.

III. Results and Observations

A. *Drosophila melanogaster*

Distributions of DNA-Feulgen values for nuclei of neural ganglion cells
clearly show two predominating modes, representing the 2C and 4C DNA
levels expected in an actively dividing, diploid tissue system (Figs. 2 and 3).
In all of the animals examined, several nuclei with intermediate DNA values
were found, as would be anticipated for cells in the DNA-synthetic or S
period of the mitotic cycle (*11a, 66a, 97, 107*). Whereas in a late-third-instar
larva (animal 1, Fig. 2), several nuclei with DNA values between the 4C and
8C DNA class means were found, in a sibling prepupa such interclass values
were absent, and there was instead a discrete population of ganglion nuclei
at the 8C DNA class level (animal 4, Fig. 3).

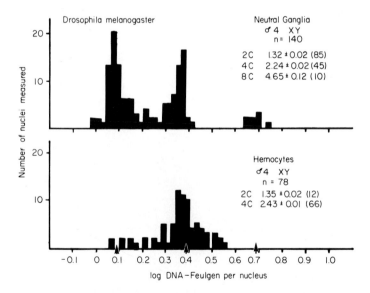

FIG. 3. Amounts of DNA-Feulgen per nucleus, plotted on a logarithmic scale, for cells
of the subesophageal ganglion and hemolymph from a tan prepupa, male *Drosophila
melanogaster*. Arrows on abscissa indicate position of means for 2C, 4C, and 8C DNA
classes, respectively. Nuclei measured by Deeley scanning method (*30*), using Barr and
Stroud Integrating Microdensitometer.

Distributions of DNA-Feulgen amounts for nuclei of hemocytes also clearly show the presence of three DNA classes, the means and modes of which coincide with those found for ganglion cells (Figs. 2 and 3). The predominating population of hemocyte nuclei from both third-instar larvae and prepupae, however, are at the 4C DNA class level. The absence of interclass values for hemocyte nuclei of animal 3, a third-instar larva (Fig. 2) is in sharp contrast to the presence of many nuclei with DNA amounts intermediate between the 2C and 4C or 4C and 8C DNA class levels shown by animal 4, a tan prepupa (Fig. 3). Generally similar patterns of characteristic frequencies of nuclei at the 2C or 4C DNA levels have also been found for ganglion cells and hemocytes of normal females (XX) or sterile males (XO). Fluctuations in the relative frequency of interclass values for these two tissues seem closely correlated with the pending onset of metamorphosis,

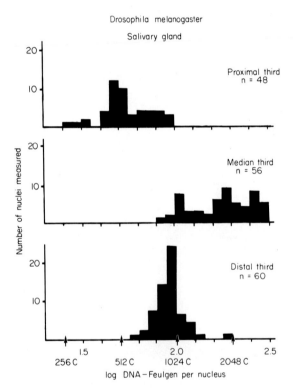

FIG. 4. Amounts of DNA-Feulgen per nucleus, plotted on a logarithmic scale, for cells in various regions from a soft squash preparation of the salivary glands from a white prepupa, female *Drosophila melanogaster*. Arrows on abscissa indicate position of means for DNA classes projected from 2C DNA levels found for the hemocyte nuclei of this animal. Measurements by two-wavelength cytophotometry.

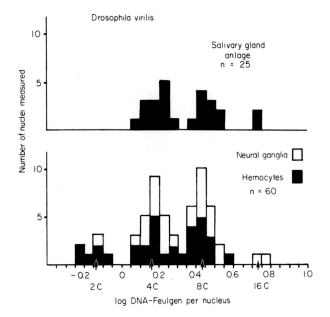

FIG. 5. Amounts of DNA-Feulgen per nucleus, plotted on a logarithmic scale, for cells of the imaginal gland anlage, posterior ganglion, and hemocytes from a late-third-instar, female larva of *Drosophila virilis*. Pigmentation of anterior spiracles had been observed in this animal almost 5 hr prior to its sacrifice for tissue sampling.

and presumably reflect tissue-specific, differential responses to hormonal mediation of mechanisms regulating DNA synthesis, in these and similar systems (*25, 50, 54, 101, 122*).

The distribution of DNA-Feulgen values for salivary gland nuclei from a pair of glands of a white prepupa female is shown in Fig. 4. Four modes of nuclear DNA amounts are evident, in general confirmation of the polytene class segregation of nuclear values described earlier for this species by Swift and Rasch (*111*). The class levels indicated are those obtained from projections of 2C DNA amounts measured for ganglion or hemocyte nuclei of the same animal. A preponderance of polytene nuclei at the 1024C DNA class level seems characteristic of cells from the distal one-third of the gland, but many nuclei from the median or transitional region of the gland show DNA values between the 1024C and 2048C levels, and a significant number of nuclei clearly are at the 2048C DNA class level. Many nuclei from the proximal region of the gland belong to the 512C DNA class, although some of the smaller nuclei of the gland proper are at the 256C DNA class level. At this stage of development, just before histolysis of the salivary gland, nuclei of the imaginal gland anlage show DNA values two, four, or eight times those found for diploid hemocyte or neural ganglion nuclei. It seems likely that the various types of cells used to estimate a diploid or 2C basis for comparison with polytene nuclei may account for the differences in previous and present estimates of nine to ten replication cycles for the largest polytene nuclei of salivary gland cells of *Drosophila melanogaster*.

Comparative estimates were made of DNA-Feulgen content for hemocyte nuclei of *Drosophila melanogaster*, using chick erythrocyte nuclei as a known "standard" of 2.3 pg DNA per cell (*66, 115*). A value of approximately 0.34–0.36 pg DNA per 2C or diploid *Drosophila* nucleus was obtained, which agrees well with previous estimates for this species by Rudkin and co-workers (*97, 104*) using ultraviolet microspectrophotometry. Further, when

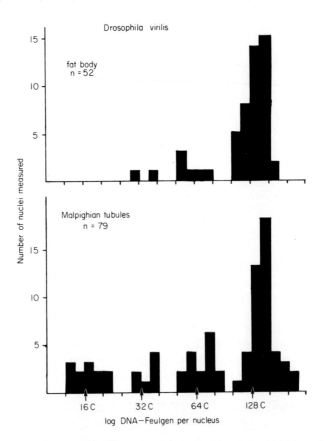

FIG. 6. Amounts of DNA-Feulgen per nucleus, plotted on a logarithmic scale, for cells of the larval fat body and Malpighian tubules from a late-third-instar, female larva of *Drosophila virilis*. Arrows on abscissa indicate position of means for DNA classes projected from the 2C DNA levels found for hemocyte nuclei of this animal (see Fig. 5).

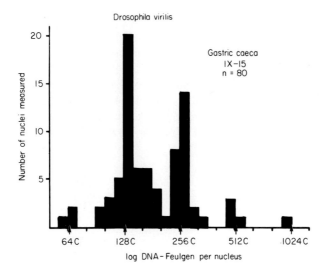

FIG. 7. Amounts of DNA-Feulgen per nucleus, plotted on a logarithmic scale, for cells of the gastric caeca of a late-third-instar, female larva of *Drosophila virilis*. Arrows on abscissa indicate position of means for DNA classes projected from 2C DNA levels found for hemocyte nuclei of this animal (see Fig. 5). Note the relative frequency of interclass DNA values between 128C and 256C, but absence of intermediate DNA values between other nuclear classes.

one considers relative frequencies of the different polytene DNA classes of nuclei from salivary gland cells during third instar, and computes an expected average DNA content in picograms per cell, using the data displayed in Fig. 4, an estimate of some 230 pg of DNA per salivary gland nucleus is obtained. This value again is in reasonable agreement with previous reports of an average of 284 pg per polytene nucleus (70) or 221 ± 14 pg DNA per cell for salivary glands of the male (XY) of this same species analyzed by microchemical techniques (71). These agreements have the value of indicating that DNA content estimated from microspectrophotometric measurement of net Feulgen content cannot be far off and does provide meaningful data.

B. *Drosophila virilis*

Distributions of DNA-Feulgen values for nuclei of several different tissue systems from late-third-instar larvae of *Drosophila virilis* are shown in Figs. 5, 6, 7, and 8.

Representatives of the 2C, 4C, and 8C DNA classes were found among nuclei of both hemocytes and neural ganglion cells (Fig. 5). There also were several nuclei from the 16C DNA class in the latter populations in general

confirmation of the earlier reports by Makino (*60*) of polytene nuclei among
posterior ganglion cells in this species (see also *11a, 66a*).

DNA values from nuclei of the larval fat body or Malpighian tubules distribute into DNA classes at the 16C, 32C, 64C, and 128C levels (Fig. 6),
whereas nuclei from the gastric caeca also show Feulgen values which fit
projected means for the 256C, 512C, and 1024C DNA class levels (Fig. 7).
In all three of these tissue systems, which are mosaics of polytene nuclei,
the most commonly encountered DNA class is 128C, although each individual

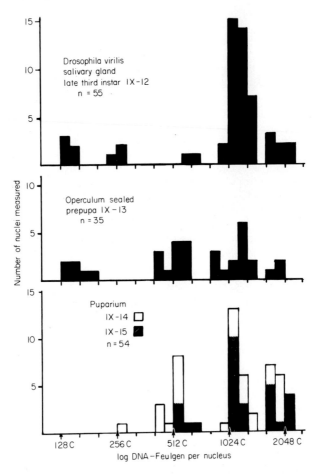

FIG. 8. Amounts of DNA-Feulgen per nucleus, plotted on a logarithmic scale, for
salivary gland cells of *Drosophila virilis* from animals sacrificed at various stages after
appearance of pigmentation in the anterior spiracles of late-third-instar female larvae.
Arrows on abscissa indicate means of polytene DNA classes projected from 2C or 4C
DNA values found for hemocyte nuclei of individual animals.

tissue has its own characteristic pattern of frequencies for the various DNA classes between 16C and 1024C. In the gastric caeca, the occurrence of many interclass DNA values between 128C and 256C presumably reflects a cycle of endonuclear chromonemal replication and is consistent with the gradual shift to a preponderance of nuclei at the higher DNA classes found for this and other tissue systems during larval development.

The DNA-Feulgen values shown in Fig. 8 are from salivary gland nuclei sampled at three different times at or after initial pigmentation of the anterior spiracles in late-third-instar female larvae. In addition to showing a range of nuclei in clearly segregated, polytene DNA classes from 2^6 to 2^{10}, these several distributions of Feulgen values also demonstrate a shift in frequencies of nuclei from the lower to the higher DNA classes. As shown in Table I, nuclei from glands just prior to or at pupation proper are for

TABLE I

COMPARISON OF DNA CLASS FREQUENCIES WITH STAGE OF DEVELOPMENT IN *Drosophila virilis*[a]

Developmental stage	Polytene DNA classes					Number of nuclei
	128C	256C	512C	1024C	2048C	
Late third instar	9.1	5.4	3.6	69.2	12.7	55
Early puparium	—	1.8	26.0	40.7	31.5	54

[a] Frequencies expressed as percent of total number of nuclei measured.

the most part in either the 1024C or the 2048C DNA classes. Finally, nuclei from the imaginal gland anlage of a late-third-instar larva just prior to closure of the operculum, 5–6 hr after spiracle pigmentation, show DNA-Feulgen values which fit 4C, 8C, and 16C class means projected from the diploid or 2C level determined from hemocyte or neural ganglion cells of the same animal (Fig. 5). This tissue clearly is not a predominantly 2C system, at least not at the stage of larval development sampled here.

In summary, as shown previously for *D. virilis* by Pettit *et al.* (78), using a combined statistic analysis of histograms and probits to determine mean amounts of DNA-Feulgen, values observed for salivary gland nuclei of larvae and prepupae do not differ significantly from polytene class limits set as projected, geometric multiples of the 2C or 4C DNA amounts found for hemocyte nuclei of the same animals. The present data would extend this conclusion to include several other major larval tissue systems as well.

C. *Chironomus thummi*

A bimodal distribution of DNA-Feulgen values was found for the small nuclei of hemocytes and cells of the anterior ganglion of *Chironomus thummi* (Fig. 9). Interclass values between the assigned 2C and 4C class levels were found for hemocyte nuclei, but were not a common feature of the limited population of neural ganglia sampled from this particular fourth-instar animal. Our observation may reflect a short duration for the S period of the cell cycle in this tissue and perhaps accounts for the consistent estimate of 0.5 pg DNA per cell found by Daneholt and Edström in their microchemical estimates of small somatic cells of a closely related species, *Chironomus tentans* (26).

The DNA-Feulgen values displayed in Fig. 10 for Malpighian tubules from a fourth-instar larva and a pupa show a preponderance of nuclei at the 1024C DNA class level in the larva, but a marked increase of relative frequency of nuclei from the next higher DNA class (2048C) in the pupa. Daneholt and Edström also report DNA values for large Malpighian tubule nuclei of *Chironomus* which indicate completion of nine cycles of chromonemal replication (26).

DNA-Feulgen values are also shown in Fig. 10 for the giant nuclei of salivary gland cells from a fourth-instar larva and a prepupa. Representatives of four DNA classes were found in the former, but essentially only two classes of very large nuclei, at the 8192C and 16,384C DNA class levels,

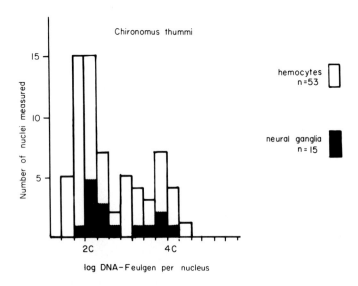

FIG. 9. DNA-Feulgen values, plotted on a logarithmic scale, for hemocytes and neural ganglion cells of a late-fourth-instar larva of *Chironomus thummi*.

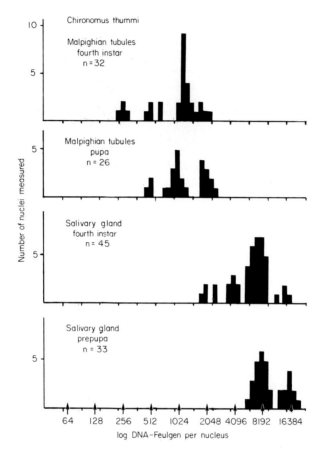

Fig. 10. Distribution of DNA-Feulgen values, plotted on a logarithmic scale, for nuclei of Malpighian tubules and salivary gland cells of a fourth-instar larva, a prepupa, and a pupa of *Chironomus thummi*. Arrows on abscissa indicate polytene, DNA class levels projected from the 2C DNA value found for nuclei of the fourth-instar animal (Fig. 9).

were found for prepupal salivary glands (see Table II). Again, the present data on *Chironomus thummi* are in excellent agreement with those of Daneholt and Edström (26) for *Chironomus tentans*, in which the largest nuclei contained 3360 pg of DNA, or approximately 16,000 times that found for hemocytes which averaged 0.50 pg/nucleus. The maximal degree of polytenization found in these two species of *Chironomus* corresponds then to 13 replications of the diploid or 2C value.

The data from *Chironomus*, as well as those from *Drosophila*, show that the average values for DNA content of nuclei from any one polytene class

TABLE II

COMPARISON OF DNA CLASS FREQUENCIES WITH STAGE OF DEVELOPMENT IN *Chironomus thummi*[a]

Developmental stage	Polytene DNA classes				Number of nuclei
	2048C	4096C	8192C	16384C	
Fourth instar	11.1	15.5	64.5	8.9	45
Prepupa	—	—	65.6	34.4	32

[a] Frequencies expressed as percent of total number of nuclei measured.

are approximately twice, four times, or some even geometric multiple of other and lower DNA classes. This indicates that most or all of the chromatids of a polytene chromosome take part in periodic cycles of replication.

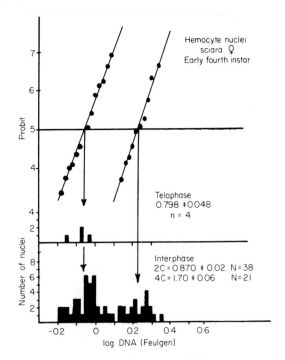

FIG. 11. Frequency histogram and probit plot of DNA-Feulgen values obtained by two-wavelength cytophotometry from hemocyte nuclei of a young fourth-instar larva of *Sciara coprophila*. Each arbitrary photometric unit from slides of this particular hydrolysis series (XIII) was equivalent to 0.5 pg of DNA, as determined by measurement of 2C rat hepatocytes from the "standard" slide. See text for additional discussion.

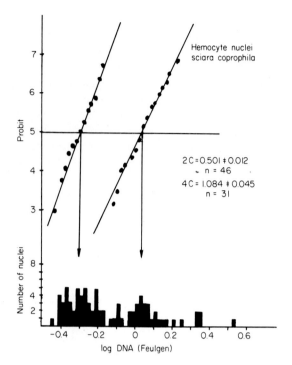

FIG. 12. Frequency histogram and probit plot of DNA-Feulgen values obtained by two-wavelength cytophotometry from hemocyte nuclei of a late-fourth-instar larva of *Sciara coprophila*. Each photometric unit from slides of this hydrolysis series (IX) was equivalent to 0.82 pg of DNA, as determined from measurements on 2C rat hepatocytes from a "standard" slide. See text for additional discussion.

Further, if all chromatid strands of polynemic chromosomes do not replicate simultaneously, or if certain portions along the axis of the chromosome. the so-called "replicons" (*79, 80*), synthesize DNA asynchronously, some degree of dependence of DNA content or frequency of nuclei with interclass values with stage or age of animals would be expected. That the latter is indeed the case is clear from the present study. Pettit *et al.* (*78*) and Rodman (*92, 95*) have reported a similar interdependence of polytenic replication cycles with stage of larval development in *Drosophila*.

D. *Sciara coprophila*

Frequency histograms and probit plots of DNA-Feulgen values for hemocyte nuclei from two different stages of development of female larvae of *Sciara coprophila* are shown in Figs. 11 and 12. As would be anticipated for populations of nuclei from a dividing tissue system, many interclass values were found between modes identified as 2C and 4C by the probit analysis. Clear confirmation of the presumption for assigning a 2C level to the lower mode was provided by several pairs of daughter telophase nuclei (Fig. 11), the DNA-Feulgen values of which closely approximated the mean estimated graphically for the interphase population designated as 2C for this animal. A third group of Feulgen values, reflecting the presence of some hemocyte nuclei at the 8C DNA class level is shown in Fig. 12. Similar analyses for populations of nuclei from neural ganglion cells or the developing limb anlage of prepupae (Fig. 13) also show nuclei representing the 8C and 16C DNA classes, in addition to the expected arrays of nuclei at or between modes for the 2C and 4C DNA levels.

As shown in Table III, the methodology illustrated above can be used to derive " C " values of both the male and female somatic genomes, and when calibrated against a "standard" tissue of known DNA level, to estimate average DNA content in picograms per diploid *Sciara* nucleus. In the several series of measurement shown, the male genome was consistently found to

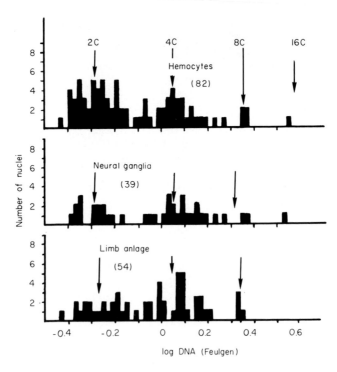

FIG. 13. Frequency histograms of hemocytes, neural ganglion cells, and developing limb anlage of an early female prepupa of *Sciara coprophila*. DNA class means obtained by probit analysis are indicated by arrows.

TABLE III

Estimated DNA Content of Diploid (2C) Nuclei of *Sciara coprophila* and Computed "C" Values for Somatic Male and Female Genomes

Series	Stage of animals	Sex	Cell type	Amount of DNA-Feulgen per nucleus			"C" value	DNA per diploid cell (pg)
				Mean	± S.E.	n		
IX	Prepupa	Female	Hemocytes	0.501	0.012	46	0.250	0.411
			Neural ganglia	0.493	0.018	15	0.246	0.404
			Limb anlage	0.529	0.025	18	0.264	0.434
XII	Prepupa[a]	Female	Hemocytes	0.522	0.010	112	0.261	0.429
XII	Prepupa[a]	Male	Hemocytes	0.478	0.012	37	0.239	0.392[b]
XV	Prepupa[a]	Female	Hemocytes	0.485	0.026	74	0.242	0.412
XV	Prepupa[a]	Male	Hemocytes	0.432	0.017	34	0.216	0.368[c]

[a] Values shown are averages derived from pooling for probit analysis all hemocytes measured from 2–5 animals within a particular hydrolysis series.
[b] Male value 8.4% less than female value from same hydrolysis series; $T = 2.8$.
[c] Male value 10.9% less than female value from same hydrolysis series; $T = 1.9$.

contain some 8% to 10% less DNA than the female, as might be predicted from cytologic studies (63) which describe two X homologs in the female somatic genome (XX), but only a single X in the male soma (XO). This cytophotometric confirmation of cytogenetic expectations is encouraging, for the amounts of DNA involved (10^{-14} gm) are well below those commonly encountered in more familiar systems from higher organisms. Also worth noting is the consistent estimate of approximately 0.4 pg of DNA per

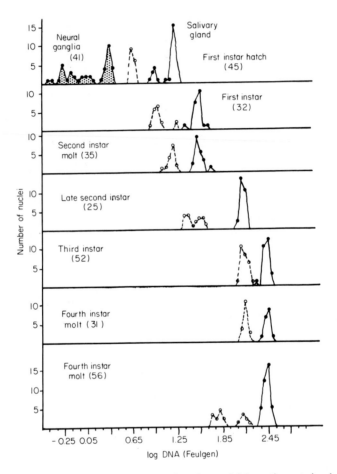

FIG. 14. Distributions of DNA-Feulgen values for nuclei from the anterior (solid lines) or posterior (dashed lines) regions of salivary glands from female larvae of *Sciara coprophila* at various developmental stages from hatch to first instar through molt to fourth instar. Representative values for neural ganglion cells are shown by the shaded distribution at the top of the figure. Number of nuclei measured for each animal shown in parentheses. A total of 317 values are displayed.

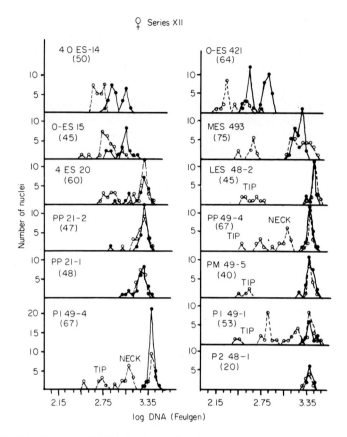

FIG. 15. Distributions of DNA-Feulgen values for nuclei from the anterior (solid lines) or posterior (dashed lines) regions of salivary glands during fourth-larval instar and pupation in *Sciara coprophila*. A total of 681 values are shown from 13 different female animals.

diploid somatic genome from several different tissue systems and from three separate hydrolysis series.

In the salivary gland of *Sciara*, populations of nuclei from the anterior or proximal and posterior or distal portions of the gland can be readily mapped and their progression through successive cycles of polytenization easily followed. The patterns of change in DNA class distributions found from such monitoring of salivary gland nuclei at various stages of female larval development are shown in Figs. 14 and 15. Animals were sampled from within 1 hr of egg hatching through 36 hr following pupal molt. Nuclei from both the anterior and posterior regions progress with time from lower to higher DNA class levels, and for each region of the gland there is a characteristic pattern for this progression. As shown in Fig. 14 (top), nuclei from salivary glands of a larva just after hatching fall into three DNA classes. Those from the posterior gland region are in the 16C class, while the larger nuclei in the proximal or basal reservoir portion of the gland are at the 64C level, with a few nuclei at the 32C level. Clearly, four to five cycles of endonuclear chromonemal replication must have occurred during the latter stages of embryogenesis to account for these elevations in DNA content above the 2C and 4C levels found for diploid neural ganglion cells of the same animal. During the next 24 hr of first-instar growth, another complete replication cycle occurs in nuclei of both gland regions, for nuclei

TABLE IV

COMPARISON OF DNA-FEULGEN AMOUNTS IN WHOLE SALIVARY GLAND NUCLEI FROM
MALE AND FEMALE PUPAE OF *Sciara coprophila* AT 1 HR AFTER MOLT

Animal	Sex	Gland region	Amount of DNA-Feulgen per nucleus		
			Mean	± S.E.	*n*
XV-12	Female	Anterior	2348.8	40.7	20
		Posterior	2067.8	57.0	25
		Tip	518.2	39.4	13
XV-07	Female	Anterior	2226.0	53.5	25
		Posterior	2360.8	77.8	30
XV-03	Male	Anterior	1249.7	22.1	25
		Posterior	1256.7	29.7	25
XV-02	Male	Anterior	1317.3	35.6	15
		Posterior	1305.4	60.2	20
		Tip	505.3	23.8	12

from the proximal zone now fall in the 128C DNA class, while DNA values for nuclei from the posterior gland region coincide with projected means for the 32C DNA class. During each ensuing larval growth stage, one or more additional replication cycles occur so that by the time of the molt to fourth instar (Fig. 14, bottom), nuclei from the basal reservoir region of the gland are predominantly at the 1024C DNA class level, while nuclei from the proximal one-third of the gland proper are at the 256C and 512C DNA class levels. Although specific data are not depicted here, a similar progression from lower to higher DNA class levels also occurs for nuclei of the transitional or "neck" region and for the small nuclei in the distal or tip portions of the gland.

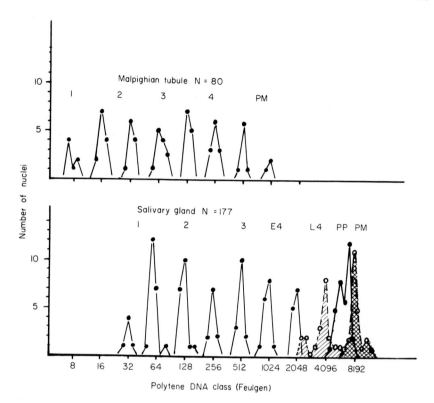

FIG. 16. Frequency distributions of relative amounts of DNA-Feulgen in nuclei from salivary glands and Malpighian tubules of female larvae of *Sciara coprophila* sampled at various stages of development. All salivary gland nuclei shown were sampled only from the proximal or anterior region of the gland. Values expressed in log arbitrary photometric units. Scale of abscissa given is a geometric progression of the amount of DNA-Feulgen determined for diploid (2C) hemocyte nuclei from individual preparations. Larval stages are designated 1 (first instar), 2 (second instar), 3 (third instar), 4 (within 20 min of molt to fourth instar), L4 (fourth instar after appearance of imaginal eye spots), PP (prepupa), and PM (within 30 min after pupal molt). Shaded distributions indicate DNA classes present during initiation of DNA puffs (L4) or after puff condensation (PM).

Shortly before the appearance of pigment in the imaginal eye anlagen, here called the "eyespots" of the larva, and concomitant color changes in the gastric caeca, a marked shift occurs in characteristic DNA class frequencies of the anterior and posterior gland regions (Fig. 15). Nuclei from the posterior region, or the gland proper, evidence many interclass Feulgen values, indicative of DNA synthesis. These chromosomes apparently undergo two successive cycles of replication (cf. animals 14, 15, and 421, in Fig. 15) so that by the time larval eyespots are evident a majority of the nuclei from the posterior gland region are in or approaching DNA class 2048C. Two additional cycles of chromosome replication occur prior to pupation, for nuclei from the same location in glands from prepupae are in DNA class 8192C (cf. animals 21-1, 21-2, and 49-4 in Fig. 15). Meanwhile, nuclei from the anterior or proximal gland region undergo two cycles of chromosome replication, from 1024C to 2048C to 4096C, prior to and at the early eyespot stage (animals 15 and 20 in Fig. 15), but only evidence one additional replication cycle prior to pupation. From the late eyespot stage of larval development on, nuclei of both the posterior and anterior gland regions show coincident distributions of DNA-Feulgen values, which for female larvae approximate the 8192C DNA class level. As shown by the measurements from another hydrolysis series in Table IV, nuclei from comparable gland regions in male larvae approximate a 4096C DNA class level, reflecting only 11 cycles of chromosome replication instead of the 12 cycles shown by females of the same species.

A summary of the changes in DNA class levels for cells of the anterior region of the salivary gland is presented in Fig. 16, in which data for nuclei from Malpighian tubules of the same animals at particular stages of development are also displayed. The latter measurements are from the 6 to 10 largest nuclei near the base of the tubules at their point of attachment to the hindgut. DNA values for tubule nuclei distribute into well-demarcated classes, the modes and means of which closely approximate even geometric multiples up to 1024C of the DNA-Feulgen values found for presumptive diploid nuclei from hemocytes of these animals. Consistently, throughout larval development, nuclei of the Malpighian tubules lag by three or more

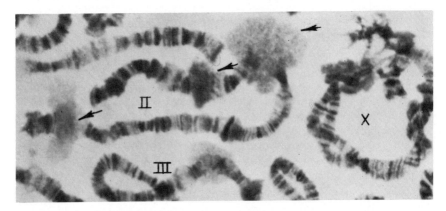

FIG. 17. Salivary gland chromosomes of a late-fourth-instar female larva of *Sciara coprophila* from a squash preparation stained with Feulgen. Prominent DNA-puff sites of Chromosome II are at stage of maximal expansion (arrows). Note the two holomogs apparent in asynapsed regions of the X chromosome at the left of the figure.

DNA classes behind the high degree of polytenization shown by nuclei from salivary glands of the same animals.

As suggested from the shaded histograms of Fig. 16, the patterns of changes in DNA levels for polytene chromosomes of salivary gland nuclei of *Sciara* are even more complex than thus far indicated, for coincident to the final replication cycles which occur during the last week of larval life, certain regions of these chromosomes expand and accumulate Feulgen-positive material, forming so-called "DNA puffs," shown in Fig. 17 and discussed in detail elsewhere in this volume (*84*). The process of puff formation in a related genus, *Rhynchosciara*, has been described by Breuer and Pavan (*19, 72*) and more recently by Mattingly and Parker (*61*). For the present, we are concerned with DNA levels of whole nuclei at the time of

TABLE V

RELATIVE AMOUNTS OF DNA-FEULGEN IN *Sciara* SALIVARY GLANDS NUCLEI WITH OR WITHOUT DNA PUFFS[a]

Animal	Sex	Nuclei without puffs			Nuclei with puffs			Percent increase
		DNA (Feulgen)	S.E.	n	DNA (Feulgen)	S.E.	n	
19–6	Female	1840	±50	10	1972	±190	5	7.2
20–5	Female	2001	±68	22	2347	±41	8	17.3
17–1	Male	1483	±62	15	1710	±53	6	15.3
17–5	Male	1277	±49	20	1426	±54	14	11.7
17–8	Male	1583	±53	30	1766	±25	13	11.6

[a] All measurements shown are of nuclei from the proximal one-third of the posterior region of glands taken from very early prepupae. Differences expressed as percentage increase above values for nuclei without puffs.

puff formation and during the subsequent period of puff regression. As shown in Table V, when amounts of DNA-Feulgen from whole salivary gland nuclei in which obvious puffs can be readily discerned (see Fig. 1) are compared with values obtained from other nuclei in the same region of the gland but in which puffs are not yet visible or do not form (*25, 37*), there are significant elevations in the average DNA content of the nuclei with puffs. These differences comprise some 7% to 17% of the total DNA content of nuclei without puffs. In another series of 314 measurements from 7 female larvae just prior to pupal molt, an average increase of 16.1% was found for nuclei with puffs in the posterior region of the gland, and an increase of 8.6% for nuclei with puffs from the proximal region of the gland. Similar measurements of 183 nuclei from 5 male prepupae showed an average increase of 15.7% in DNA levels associated with nuclei containing obvious puffs.

Analyses by the probit technique for populations of nuclei from a pair of salivary glands from a female are shown in Fig. 18. (These studies were done in collaboration with Dr. R. W. Rasch of the Department of Physiology,

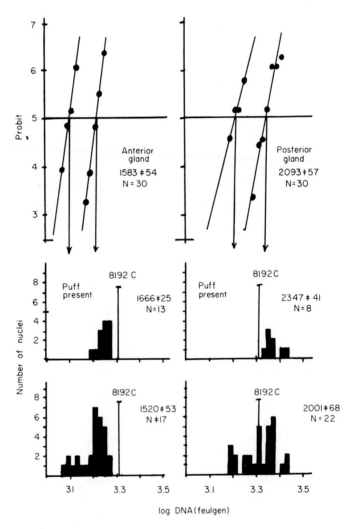

FIG. 18. Frequency histograms and probit plots of DNA-Feulgen values from nuclei of salivary glands from a very late fourth-instar female larva of *Sciara coprophila*. See text for additional discussion.

Marquette Medical School, Inc.) Cells from the anterior portion of the gland all contain significantly less than the projected 8192C DNA class level estimated from measurements on diploid hemocyte nuclei of the same animal. As indicated in a previous section of this report, a lag in rate of replication of chromosomes from nuclei of the proximal gland region seems characteristic of this stage of larval development. Nuclei from cells of the proximal one-third of the posterior region of the gland, as a whole, cannot be distinguished from the 8192C DNA class. The nuclei-containing puffs, considered as a separate subpopulation, contain significantly more DNA ($\approx 17\%$) than those of the nonpuffed population, and there is a good probability that the nuclei with puffs contain more than the 8192C amount of DNA.

Several other points of interest emerge from this type of analysis. Since nuclei containing puffs in the anterior gland cells have a significantly different modal DNA content than nuclei containing puffs in the posterior region of the gland, it seems clear that puffing per se is probably not related solely to late replicating sections of the chromosomes, for the puffed nuclei in the anterior cells will have to synthesize $2608 \pm 338C$ amount of DNA to get to the state of $8991 \pm 300C$ found in posterior cells. This is almost 30% of the total amount of DNA in the latter nuclei, which is too much to be the last bit of asynchronous synthesis, as was suggested some years ago by Taylor (*113*) to explain the dense, heterochromatic DNA puffs of *Rhyncosciara*. Further, since nuclei containing puffs from the posterior part of the gland are indeed *above* the 8192C DNA level, this "overage" may provide an estimate of total puff DNA, unless these nuclei are shooting for the next DNA class. The latter possibility seems unlikely, for DNA synthesis in the system as a whole is shutting down, as shown by other studies using tritiated thymidine (*37, 83*).

The amount of excessive DNA estimated by the above analysis is some $809C \pm 300C$, or roughly $9\% \pm 3.5\%$ of the total nuclear DNA content. As shown earlier, the difference between nuclei with or without puffs from either the anterior or posterior gland regions averages about 10–15%. These two estimates of the contribution of puff DNA to total nuclear DNA content agree reasonably well with separate determinations made on individual puff sites of similar chromosomes. In the latter case, measurements on six major DNA puff sites can account for some 10–12% of the DNA-Feulgen content measured for the whole polytene chromosome complement (*84*).

The amount of DNA we are talking about is an appreciable quantity. Estimates of actual DNA levels, based on comparative measurements against "standard" rat tissues, indicate that a polytene *Sciara* nucleus at the 8192C DNA class level from a female larva just prior to pupal molt contains about 2000 pg of DNA, of which some 1640 pg is attributed to 12 replication cycles of the diploid, somatic chromosome complement ($0.4\,\text{pg} \times 2^{12}$), but of which some 360 pg seems associated with the formation of DNA puffs. If one considers further that one picogram ($= 1\,\mu\mu\text{g}$ or 10^{-12} gm) is equivalent to about 31 cm of double helical DNA (123), then the quantity of DNA involved in the sciarid puffing phenomenon could be approximately 100 meters of "extra" or "selectively amplified" DNA filament per polytene nucleus.

IV. Discussion

We have analyzed distributions of DNA-Feulgen values obtained by two-wavelength cytophotometry for nuclei of a number of different dipteran larval tissue systems. In both *Drosophila* and *Chironomus*, tissues such as salivary glands, gastric caeca, and Malpighian tubules show amounts of DNA per nucleus which are those to be expected from periodic replication of the whole chromosome complement, because the values observed fit a 2^n geometric series when DNA-Feulgen estimates for hemocyte nuclei of the same animal are used as the base for computing such a series (see Table VI). In the case of *Sciara coprophila*, similar arrays of DNA classes were found for nuclei containing polytene chromosomes in various tissue systems during early larval life. Prior to pupation in *Sciara*, however, the giant chromosomes of many nuclei of the salivary gland show localized accumulations of DNA at certain specific sites of particular chromosomes. Appearance of these DNA puffs coincides temporally with the final replication cycle of many nuclei of the salivary gland during the late-fourth-instar period. The formation of DNA puffs, however, constitutes a separate phenomenon, for after puffing, amounts of DNA per nucleus are found which are appreciably in excess of those expected solely from geometric projections of the 2C DNA level found for diploid hemocyte nuclei from these same animals.

Expression of potentials for chromonemal replication far beyond levels normally found, even in dipteran giant chromosomes, is not unprecedented. The truly spectacular polytene chromosomes recently described from *Drosophila* salivary glands cultured in adult female abdomens are a case in point (44). In another context, Plaut and co-workers (78, 80) have suggested from radioautographic evidence that the polynemic salivary chromosomes of *Drosophila* are composed of longitudinal arrays of individual DNA replicative units, or "replicons," so disposed along the axis that chromosome duplication need not involve DNA synthesis at all sites simultaneously, nor even in linear sequence. Models and mechanisms have also been proposed for localized, stepwise doubling of DNA replication units in the polytene chromosomes of *Chironomus* from different species or in hybrids between species [see Pelling (77) for review]. Capacities for selective and autonomous replication by only a portion of the genome are suggested also for other types of cell systems by recent studies of the DNA body in oocytes of Tipulid flies (4, 58, 59) or the selective gene amplification shown by the DNA cores of amphibian oocyte nucleoli (20, 39, 65). These and other diverse lines of evidence for "atypical" DNA behavior (13, 14, 38, 45, 48, 58, 96) suggest that functionally discrete replicating units of chromosomes may at times operate independently of one another when acted upon by appropriate control mechanisms. The selective replication of genetic material at DNA puff sites in sciarid chromosomes, for example, may be related to changes of hormonal levels in the cell environment (25) and to altered requirement within the cell for information in the replicons at these regions. As originally stated several years ago by Pavan: "we are dealing with cases of gene reproduction without a corresponding chromosomal reproduction, or in other words, we have an increase in number of a particular gene among the many others which are present in the chromosomes" (73). Essentially, this same idea has been restated by others, notably Waddington (118), von Borstel (117), and Roels (96). Use of the term "metabolic" DNA, however, to denote cases of "excessive" DNA accumulations in nuclei, while often applied (55, 75, 96), seems unfortunate because of its usual connotations of lability, turnover, or utilization, rather than the kind of stability commonly associated with genetic template material.

The consistent appearance and characteristic behavior of DNA puffs in *Sciara* coincident only with specific stages of larval development, despite marked alterations in developmental schedules induced by X-irradiation, cortisone, gibberellic acid, or starvation (83), would seem to require a complex interplay of intracellular target sites and levels of hormonal stimuli from the extracellular milieu to regulate the kind of selective and differential DNA synthesis shown by these salivary gland chromosomes. Further careful studies like those of Crouse (25) on puff evocation by ecdysone will certainly provide important insights for our understanding of mechanisms which govern intranuclear differentiation and selective genic replication.

Finally, nucleic acid determinations from polytene nuclei and their diploid, somatic counterparts in various dipteran systems provide data useful for estimating the average amount of DNA associated with functional chromosomal units, the chromomeres, as this term is used by Pelling (77) and others (6, 34). To estimate the amount of DNA per chromomere, one must consider the total amount of DNA per nucleus, the strandedness of its chromosomes, and finally the number of chromomeres, or bands, for the chromosome complement as a whole. From microchemical determination on nuclei of *Chironomus tentans*, Daneholt and Edström (26) computed approximately 6×10^{-5} pg of DNA per chromomere, which they consider equivalent to about 60,000 nucleotide base pairs. Comparable estimates of

3×10^{-5} to 5×10^{-5} pg of DNA per chromomere of *Drosophila melanogaster* were reported earlier by Rudkin and co-workers using ultraviolet microspectrophotometry (*97, 99a, 104*), and recently confirmed for *Drosophila hydei* by Mulder *et al.* using photographic colorimetry (*66a*). These estimates of the mean haploid DNA content per band are equivalent to roughly 9μ of DNA filament in each chromomere. Since a mean length of 72μ has been found for the DNA molecules from salivary gland chromosomes of *Drosophila melanogaster* (*121a*), it is possible that a single DNA molecule may be involved in as many as eight or more separate bands.

Estimates of actual DNA contents for both diploid and polytene nuclei of the three dipteran genera studied here were obtained from Feulgen measurements calibrated against determinations on "standard" rat or chick tissues (see Table VI). Using presumptions like those of Edström (*34*) that

TABLE VI

COMPARISON OF NUMBER OF REPLICATION CYCLES ESTIMATED FOR POLYTENE CHROMOSOMES OF SALIVARY GLAND NUCLEI FROM THREE DIPTERAN GENERA

Species	Approximate DNA content of diploid (2C) hemocyte nuclei[a] (pg)	Polytene level of largest salivary gland nucleus	Number of replication cycles
Chironomus thummi	0.30–0.32	16384 C	13
Drosophila melanogaster	0.34–0.36	2048 C	10
Sciara coprophila ♂	0.37–0.39	4096 C (+ DNA puffs)	11
Sciara coprophila ♀	0.40–0.42	8192 C (+ DNA puffs)	12

[a] These estimates based on comparative measurements of DNA-Feulgen content in dipteran hemocytes and chicken erythrocytes, presuming the nucleus of the latter contains about 2.3 pg of DNA (*56, 66, 115*).

the probable number of bands lies in the range of 2000 to 5000 for the chromosome complements in question, estimated amounts of DNA per chromomere were obtained ranging from 3.4×10^{-5} pg of DNA for *Drosophila melanogaster* to 6.7×10^{-5} pg of DNA for *Sciara coprophila*. A value of about 5×10^{-5} pg of DNA per chromomere was computed for *Chironomus thummi*. In comparison with viral "chromosomes" the amounts of DNA in single chromomeres would seem to hold coding potentials for amino acid synthesis several orders of magnitude in excess of projected requirements even for complex proteins of high molecular weight. As discussed by Edström, since 1 pg of DNA content per homozygous band in *Chironomus tentans* contains about 60,000 nucleotide pairs, there evidently may be enough DNA in a band of average size to permit localization of several cistrons coding for ribosomal components (*34*). When the amounts of deoxyribonucleoprotein accumulated after puffing at particular sites of *Sciara* chromosomes are likewise viewed in terms of the possibility of selective amplification of genetic template material for RNA transcription, the magnitude of coding potential verges on the astronomical! For the present, however, any suggestions of a definite physiologic role for DNA puffs or of a possible genetic redundancy seem premature in light of our incomplete understanding of the structure, much less the function, of these complex chromosome systems.

V. Concluding Remarks

More than a decade ago, in describing findings on the DNA content of nurse cells of the ovary in *Drosophila*, Schultz wrote: "The Y chromosome is synthesizing more DNA than we should expect from its cytological appearance. Is this related to the DNA-carrying body that Bayreuther has discovered in the oogenesis of Tipulids? Are we facing a change in our concepts of the relation of DNA to nucleoli? Or is this simply a dispersal of the body of the Y chromosome, and it will all straighten out into a conventional pattern of DNA constancy?" (*102*).

These notions seem remarkably contemporary in view of many recent discussions on the synthesis of "excessive" amounts of DNA associated with selective gene amplification for the production of ribosomal RNA during oogenesis in amphibians and other forms (*4, 13, 14, 20, 39, 58, 75a, 90*).

Whether all other cases of selective and excessive DNA accumulation will prove to be associated with one specific aspect of nucleolar function, namely, ribosome biogenesis, awaits further study. Certainly also, our working hypothesis of DNA constancy as applied to cells of higher organisms (*18*) is undergoing some significant and dynamic revision to encompass those particular cell systems which heretofore have been exceptions to the general rule of direct proportionality between the number of sets of chromosomes in a nucleus and its DNA content.

REFERENCES

1. Alfert, M., Composition and structure of giant chromosomes. *Intern. Rev. Cytol.* **3** 131–176 (1954).
2. Ashburner, M., Patterns of puffing activity in the salivary gland chromosomes of Drosophila. I. Automosal puffing patterns in a laboratory stock of *Drosophila melanogaster. Chromosoma* **21**, 398–428 (1967).

3. Bairati, A., L'ultrastruttura dell'organo dell'emolinfa nella larva di *Drosophila melanogaster*. *Z. Zellforsch Mikroskop. Anat.* **61**, 769–802 (1964).
4. Bayreuther, K., Die oogenese der *Tipuliden. Chromosoma* **7**, 508–557 (1956).
5. Becker, H. J., Die Puffs der Speicheldrüsenchromosomen von *Drosophila melanogaster*. I. Mitteilung. Beobachtungen zum Verhalten des Puffmusters im Normalstamm und bei zwei Mutanten, *giant* und *giant-lethal-larvae. Chromosoma* **10**, 654–678 (1959).
6. Beermann, W., Riesenchromosomen. *Protoplasmatologia* **6D**, 1–161 (1962).
7. Beermann, W., Cytological aspects of information transfer in cellular differentiation. *Am. Zoologist* **3**, 23–32 (1963).
8. Beermann, W., and Pelling, C., H³-Thymidin-markierung einzelner Chromatiden in Riesenchromosomen. *Chromosoma* **16**, 1–21 (1965).
9. Berendes, H. D., The induction of changes in chromosomal activity in different polytene types of cells in *Drosophila hydei. Develop. Biol.* **11**, 371–384 (1965).
10. Berendes, H. D., Salivary gland function and chromosomal puffing patterns in *Drosophila hydei. Chromosoma* **17**, 35–77 (1965).
11. Berendes, H. D., Factors involved in the expression of gene activity in polytene chromosomes. *Chromosoma* **24**, 418–437 (1968).
11a. Berendes, H. D., and Keyl, H.-G., Distribution of DNA in heterochromatin and euchromatin of polytene nuclei of *Drosophila hydei. Genetics* **57**, 1–13 (1967).
12. Bier, K., Der Karyotyp von *Calliphora erythrocephala* (Meigen) unter besonder Berücksichtigung der Nährzellkernchromosomen im gebündelten und gepaarten Zustand. *Chromosoma* **11**, 335–364 (1960).
13. Bier, K., Kunz, W., and Ribbert, D., Struktur und Funktion der Oocytenchromosomen und Nukleolen sowie der Extra-DNS während der Oogenese panoistischer und meriostischer Insekten. *Chromosoma* **23**, 214–254 (1967).
14. Bier, K., and Ribbert, D., Struktur und genetische Aktivität des DNS-Keimbahnkörpers von *Dytiscus. Naturwissenschaften* **53**, 115–116 (1966).
15. Bliss, C. I., The method of probits. *Science* **79**, 38–39 (1934).
16. Bodenstein, D., Factors influencing growth and metamorphosis of the salivary gland in *Drosophila. Biol. Bull.* **84**, 13–33 (1943).
17. Bodenstein, D., The postembryonic development of *Drosophila. In* "Biology of Drosophila" (M. Demerec, ed.) pp. 275–367. Wiley, New York, 1950.
18. Boivin, A., Vendrely, R., and Vendrely, C., L'acide desoxyribonucléique du noyau cellulaire dépositaire des caractères héréditaires; arguments d'ordre analytique. *Compt. Rend.* **226**, 1061–1063 (1948).
19. Breuer, M. E., and Pavan, C., Behavior of polytene chromosomes of *Rhynchosciara angelae* at different stages of larval development. *Chromosoma* **7**, 371–386 (1955),
20. Brown, D. D., and Dawid, I. B., Specific gene amplification in oocytes. *Science* **160**, 272–280 (1968).
21. Chen, P. S., Farinella-Ferruzza, N., and Oelhafen-Gandolla, M., Contents of DNA and RNA in the salivary glands of normal and lethal larvae of the mutant "lethal-meander" (1me) of *Drosophila melanogaster. Exptl. Cell Res.* **31**, 538–548 (1963).
22. Cooper, K. W., Concerning the origin of the polytene chromosomes of *Diptera. Proc. Natl. Acad. Sci. U.S.* **24**, 452–458 (1938).
23. Crouse, H. V., Translocations in *Sciara*; their bearing on chromosome behavior and sex determination. *Missouri Univ., Agr. Expt. Sta., Res. Bull.* **379**, 1–74 (1943).
24. Crouse, H. V., The controlling element in sex chromosome behavior in *Sciara. Genetics* **45**, 1429–1443 (1960).

25. Crouse, H. V., The role of ecdysone in DNA-puff formation and DNA synthesis in the polytene chromosomes of *Sciara coprophila. Proc. Natl. Acad. Sci. U.S.* **61**, 971–978 (1968).

25a. Crouse, H. W., and Keyl, H.-G., Extra replications in the "DNA-puffs" of *Sciara coprophila. Chromosoma* **25**, 357–364 (1968).

26. Daneholt, B., and Edström, J.-E., The content of deoxyribonucleic acid in individual polytene chromosomes of *Chironomus tentans. Cytogenetics (Basel)* **6**, 350–356 (1967).

27. Danieli, G. A., and Rodino, E., Biochemical estimate of DNA content in *D. hydei* glands. *Drosophila Inform. Service* **42**, 63 (1967).

28. Danieli, G. A., and Rodino, E., Larval moulting cycle and DNA synthesis in *Drosophila hydei* salivary glands. *Nature* **213**, 424–425 (1967).

29. De Cosse, J. J., and Aiello, N., Feulgen hydrolysis; effect of acid and temperature. *J. Histochem. Cytochem.* **14**, 601–604 (1966).

30. Deeley, E. M., An integrating microdensitometer for biological cells. *J. Sci. Instr.* **32**, 263–267 (1955).

31. Deitch, A. D., Wagner, D., and Richart, R. M., Conditions influencing the intensity of the Feulgen reaction. *J. Histochem. Cytochem.* **16**, 371–379 (1968).

32. De la Torre, L., Mrtek, R. G., and Velat, G. F., Investigative studies in quantitations of Feulgen-DNA. *J. Cell Biol.* **19**, 20A (1963).

33. De la Torre, L., Velat, G. F., and Dysart, M. P., The role of hydrogen ion concentration and metabisulfite in the mechanism of the Feulgen reaction. *J. Cell Biol.* **23**, 24A (1964).

34. Edström, J.-E., Chromosomal RNA and other nuclear RNA fractions. *In* "The Role of Chromosomes in Development" (M. Locke, ed.), pp. 137–152. Academic Press, New York, 1965.

35. Freed, J. J., and Schultz, J., Effect of the Y chromosome on the DNA content of ovarian nuclei in *Drosophila melanogaster* females. *J. Histochem. Cytochem.* **4**, 441 (1956).

36. Frolova, S. L., Development of the giant salivary gland nuclei of *Drosophila. Nature* **141**, 1014–1015 (1938).

37. Gabrusewycz-Garcia, N., Cytological and autoradiographic studies in *Sciara coprophila* salivary gland chromosomes. *Chromosoma* **15**, 312–344 (1964).

38. Gabrusewycz-Garcia, N., and Kleinfeld, R. G., A study of the nucleolar material in *Sciara coprophila. J. Cell Biol.* **29**, 347–359 (1966).

39. Gall, J. G., Differential synthesis of the genes for ribosomal RNA during amphibian oogenesis. *Proc. Natl. Acad. Sci. U.S.* **60**, 553–560 (1968).

40. Garcia, A. M., and Iorio, R., Potential sources of error in two-wavelength cytophotometry. *In* "Introduction to Quantitative Cytochemistry" (G. L. Weid, ed.), pp. 215–237. Academic Press, New York, 1966.

41. Geitler, L., Über den Bau des Ruhekerns mit besonderer Berücksichtigung der Heteropteren und Dipteren. *Biol. Zentr.* **58**, 152–179 (1938).

42. Geitler, L., Endomitose und endomitotische Polyploidisierung. *Protoplasmatologia* **6C**, 1–86 (1953).

43. Goodman, R. M., Goidl, J-A., and Richart, R. M., Larval development in *Sciara coprophila* without the formation of chromosomal puffs. *Proc. Natl. Acad. Sci. U.S.* **58**, 553–559 (1967).

44. Hadorn, E., Gehring, W., and Staub, M., Extensives Grössenwachstum larvaler Speicheldrüsenchromosomen von *Drosophila melanogaster* im Adultmillieu. *Experientia* **19**, 530–531 (1963).

45. Henderson, S. A., The salivary gland chromosomes of *Dasyneura crataegi* (Diptera: Cecidomyiidae). *Chromosoma* **23**, 38–58 (1967).
46. Himes, M. H., and Crouse, H. V., The contribution of the limited chromosomes to the DNA of the giant nurse cells of *Sciara*. *1st Ann. Meeting Am. Soc. Cell Biol. Chicago, Illinois*, 1961, Abstr., p. 88.
47. Jacob, J., and Sirlin, J. L., Cell function in the ovary of *Drosophila melanogaster*. I. DNA classes in the nurse cell as determined by autoradiography. *Chromosoma* **10**, 210–228 (1959).
48. Keyl, H.-G., and Hagele, K., Heterochromatinproliferation an den Speicheldrüsenchromosomen von *Chironomus melanotus*. *Chromosoma* **19**, 223–230 (1966).
49. King, R. C., Rubinson, A. C., and Smith, R. F., Oogenesis in adult *Drosophila melanogaster*. *Growth* **20**, 121–157 (1956).
50. Krishnakumaran, A., Berry, S. J., Oberlander, H., and Schneiderman, H. A., Nucleic acid synthesis during insect development. II. Control of DNA synthesis in the *Cecropia* silkworm and other *Saturniid* moths. *J. Insect Physiol.* **13**, 1–57 (1967).
51. Kroeger, H., Experiments on the extranuclear control of gene activity in dipteran polytene chromosomes. *J. Cellular Comp. Physiol.* **62**, Suppl. 1, 45–59 (1963).
52. Kroeger, H., and Lezzi, M., Regulation of gene action in insect development. *Ann. Rev. Entomol.* **11**, 1–22 (1966).
53. Kurnick, N. B., and Herskowitz, I. H., The estimation of polyteny in *Drosophila* salivary gland nuclei based on determination of desoxyribonucleic acid content. *J. Cellular Comp. Physiol.* **39**, 281–300 (1952).
54. Laufer, H., Developmental interactions in the dipteran salivary gland. *Am. Zoologist* **8**, 257–271 (1968).
55. Laufer, H., Rao, B., and Nakase, Y., Developmental studies of the dipteran salivary gland. IV. Changes in DNA content. *J. Exptl. Zool.* **166**, 71–76 (1967).
56. Leslie, I., The nucleic acid content of tissues and cells. *In* "The Nucleic Acids" (E. Chargaff and J. N. Davidson, eds.), Vol. 2, pp. 1–50. Academic Press, New York, 1955.
57. Lillie, R. D., Simplification of the manufacture of Schiff reagent for use in histochemical procedures. *Stain Technol.* **25**, 163–165 (1951).
58. Lima-de-Faria, A., Metabolic DNA in *Tipula oleracea*. *Chromosoma* **13**, 47–59 (1962).
59. Lima-de-Faria, A., and Moses, M. J., Ultrastructure and cytochemistry of metabolic DNA in *Tipula*. *J. Cell Biol.* **30**, 177–192 (1966).
60. Makino, S., A morphological study of the nucleus in various kinds of somatic cells of *Drosophila virilis*. *Cytologia (Tokyo)* **9**, 272–282 (1938).
61. Mattingley, E., and Parker, C., Sequence of puff formation in *Rhynchosciara* polytene chromosomes. *Chromosoma* **23**, 255–270 (1968).
62. Metz, C. W., Monocentric mitosis with segregation of chromosomes in *Sciara* and its bearing on the mechanism of mitosis. I. The normal monocentric mitosis. II. Experimental modification of the monocentric mitosis. *Biol. Bull.* **64**, 333–347 (1933).
63. Metz, C. W., Chromosome behavior, inheritance, and sex determination in *Sciara*. *Am. Naturalist* **72**, 485–520 (1938).
64. Metz, C. W., and Smith, H. B., Further observations on the nature of the X-prime (X′) chromosome in *Sciara*. *Proc. Natl. Acad. Sci. U.S.* **17**, 195–189 (1931).
65. Miller, O. L., Jr., Extrachromosomal nucleolar DNA in amphibian oocytes. *J. Cell Biol.* **23**, 60A (1964).
66. Mirsky, A. E., and Ris, H., The desoxyribonucleic acid content of animal cells and its evolutionary significance. *J. Gen. Physiol.* **34**, 451–462 (1951).

66a.Mulder, M. P., van Duijn, P., and Gloor, H. J., The replicative organization of DNA in polytene chromosomes of *Drosophila hydei*. *Genetica* **39**, 385–428 (1968).

67. Ornstein, L., The distributional error in microspectrophotometry. *Lab. Invest.* **1**, 250–262 (1952).

68. Painter, T. S., and Reindorp, E. C., Endomitosis in the nurse cells of the ovary of *Drosophila melanogaster*. *Chromosoma* **1**, 276–283 (1939).

69. Patau, K., Absorption microphotometry of irregular-shaped objects. *Chromosoma* **5**, 341–362 (1952).

70. Patterson, E. K., and Dackermann, M. E., Nucleic acid content in relation to cell size in the mature larval salivary gland of *Drosophila melanogaster*. *Arch. Biochem. Biophys.* **36**, 97–113 (1952).

71. Patterson, E. K., Lang, H. M., Dackerman, M. E., and Schultz, J., Chemical determinations of the effect of the X and Y chromosomes on the nucleic acid content of the larval salivary glands of *Drospohila melanogaster*. *Exptl. Cell Res.* **6**, 181–194 (1954).

72. Pavan, C., and Breuer, M. E., Polytene chromosomes in different tissues of *Rhynchosciara*. *J. Heredity* **43**, 150–157 (1952).

73. Pavan, C., and Breuer, M. E., Differences in nucleic acids content of the loci in polytene chromosomes of "*Rhynchosciara angelae*" according to tissues and larval stages. *In* "Symposium on Cell Secretion" (G. Schreiber, ed.), pp. 90–99. Belo Horizonte, Brazil, 1955.

74. Pavan, C., Nucleic acid metabolism in polytene chromosomes and the problem of differentiation. *Brookhaven Symp. Biol.* **18**, 222–241 (1966).

75. Pavan, C., Synthesis. *In* "Genetics Today" (S. J. Geerts, ed.), Vol. 2, pp. 335–342. Pergamon Press, Oxford, 1965.

75a.Pavan, C., and da Cunha, A. B., Chromosomal activities in *Rhynchosciara* and other Sciaridae. *Ann. Rev. Genet.* **3**, 425–450 (1969).

76. Pelling, C., Ribonukleinsäure-synthese der Riesenchromosomen. Autoradiographische Untersuchungen an *Chironomus tentans*. *Chromosoma* **15**, 71–122 (1964).

77. Pelling, C., A replicative and synthetic chromosomal unit—the modern concept of the chromomere. *Proc Roy. Soc.* **B164**, 279–289 (1966).

78. Pettit, B. J., Rasch, R. W., and Rasch, E. M., DNA synthesis in the giant salivary chromosomes of *Drosophila virilis* prior to pupation. *J. Cellular Physiol.* **69**, 273–280 (1967).

79. Plaut, W., On the replicative organization of DNA in the polytene chromosome of *Drosophila melanogaster*. *J. Mol. Biol.* **7**, 632–635 (1963).

80. Plaut, W., and Nash, D., Localized DNA synthesis in polytene chromosomes and its implications. *In* "The Role of Chromosomes in Development" (M. Locke, ed.), pp. 113–135. Academic Press, New York, 1965.

81. Poulson, D. F., Histogenesis, organogenesis, and differentiation in the embryo of *Drosophila melanogaster* (Meigen). *In* "Biology of Drosophila" (M. Demerec, ed.), pp. 168–274. Wiley, New York, 1950.

82. Rasch, E. M., Developmental changes in patterns of DNA synthesis by polytene chromosomes of *Sciara coprophila*. *J. Cell Biol.* **31**, 91A (1966).

83. Rasch, E. M., Nucleoprotein metabolism during larval development in *Sciara*. *Proc. 3rd Intern. Congr. Histochem. Cytochem.*, *New York*, 1968, Summary Reports, pp. 215–216. Springer, Berlin, 1968.

84. Rasch, E. M., Two-wavelength cytophotometry of *Sciara* salivary gland chromosomes. *In* "Introduction to Quantitative Cytochemistry-II" (G. L. Wied and G. F. Bahr, eds.), p. 335. Academic Press, New York, 1970.

84a.Rasch, E. M., and Barr, H. J., DNA-Feulgen cytophotometry of *Drosophila hemocyte* nuclei. *J. Histochem. Cytochem.* **17**, 187 (1969).

85. Rasch, E. M., Darnell, R. M., Kallman, K. D., and Abramoff, P., Cytophotometric evidence for triploidy in hybrids of the gynogenetic fish, *Poecilla formosa. J. Exptl. Zool.* **160**, 155–170 (1965).

86. Rasch, E. M., and Rasch, R. W., Special applications of two-wavelength cytophotometry in biologic systems. *In* "Introduction to Quantitative Cytochemistry-II" (G. L. Wied and G. F. Bahr, eds.), p. 297. Academic Press, New York, 1970.

87. Rasch, E. M., Swift, H., and Klein, R. M., Nucleoprotein changes in plant tumor growth. *J. Biophys. Biochem. Cytol.* **6**, 11–34 (1959).

88. Rasch, E. M., and Woodard, J. W., Basic proteins of plant nuclei during normal and pathological cell growth. *J. Biophys. Biochem. Cytol.* **6**, 263–276 (1959).

89. Ris, H., and Mirsky, A. E., Quantitative cytochemical determination of desoxyribonucleic acid with the Feulgen nucleal reaction. *J. Gen. Physiol.* **32**, 125–146 (1949).

90. Ritossa, F. M., Atwood, K. C., Lindsley, D. L., and Spiegelman, S. On the chromosomal distribution of DNA complementary to ribosomal and soluble RNA. *Natl. Cancer Inst. Monograph* **23**, 449–472 (1966).

91. Roberts, B., DNA granule synthesis in the giant food pad nuclei of *Sarcophaga bullata. J. Cell Biol.* **39**, 112A (1968).

92. Rodman, T. C., DNA replication in salivary gland nuclei of *Drosophila melanogaster* at successive larval and prepupal stages. *Genetics* **55**, 375–386 (1967).

93. Rodman, T. C., Control of polytenic replication in dipteran larvae. I. Increase number of cycles in a mutant strain of *Drosophila melanogaster. J. Cellular Physiol.* **70**, 179–186 (1967).

94. Rodman, T. C., Control of polytenic replication in dipteran larvae. II. Effect of growth temperature. *J. Cellular Physiol.* **70**, 187–190 (1967).

95. Rodman, T. C., Relationship of developmental stage to initiation of replication in polytene nuclei. *Chromosoma* **23**, 271–287 (1968).

96. Roels, H., "Metabolic" DNA: A cytochemical study. *Intern. Rev. Cytol.* **19**, 1–34 (1966).

97. Rudkin, G. T., Cytochemistry in the ultraviolet. *Microchem. J., Symp. Ser.* **1**, 261–276 (1961).

98. Rudkin, G. T., The structure and function of heterochromatin. *In* "Genetics Today" (S. J. Geerts, ed.), pp. 359–374. Pergamon Press, Oxford, 1965.

99. Rudkin, G. T., The proteins of polytene chromosomes. *In* "The Nucleohistones" (J. Bonner and P. O. Ts'o, eds.), pp. 184–192. Holden-Day, San Francisco, California, 1964.

99a.Rudkin, G. T., The relative mutabilities of DNA in regions of the X chromosome of *Drosophila melanogaster. Genetics* **52**, 665–681 (1965).

99b.Rudkin, G. T., Non replicating DNA in *Drosophila. Genetics* **61**, Suppl. 1, 227–238 (1969).

100. Rudkin, G. T., and Corlette, S. L., Disproportionate synthesis of DNA in a polytene chromosome region. *Proc. Natl. Acad. Sci. U.S.* **43**, 964–968 (1957).

101. Schneiderman, H., and Gilbert, L., Control of growth and development in insects. *Science* **143**, 325–333 (1964).

102. Schultz, J., The relation of the heterochromatic chromosome regions to the nucleic acids of the cell. *Cold Spring Harbor. Symp. Quant. Biol.* **21**, 307–328 (1956).

103. Schultz, J., Genes, differentiation, and animal development. *Brookhaven Symp. Biol.* **18**, 116–147 (1965).

104. Schultz, J., and Rudkin, G. T., Direct measurement of deoxyribonucleic acid content of genetic loci in *Drosophila*. *Science* **132**, 1499–1500 (1960).
105. Stalker, H. D., Banded polytene chromosomes in the ovarian nurse cells of adult *Diptera*. *J. Heredity* **45**, 259–264 (1954).
106. Swift, H. H., The desoxyribose nucleic acid content of animal nuclei. *Physiol. Zool.* **23**, 169–198 (1950).
107. Swift, H., Quantitative aspects of nuclear nucleoproteins. *Intern. Rev. Cytol.* **2**, 1–76 (1953).
108. Swift, H., Cytochemical techniques for nucleic acids. In "The Nucleic Acids" (E. Chargaff and J. N. Davidson, eds.), Vol. 2, pp. 51–92. Academic Press, New York, 1955.
109. Swift, H., Nucleic acids and cell morphology in dipteran salivary glands. In "The Molecular Control of Cellular Activity" (J. Allen, ed.), pp. 73–125. McGraw-Hill, New York, 1962.
110. Swift, H., Molecular morphology of the chromosome. *In Vitro* **1**, 26–49 (1965).
111. Swift, H., and Rasch, E. M., Nucleoproteins in *Drosophila* polytene chromosomes. *J. Histochem. Cytochem.* **2**, 456–457 (1954).
112. Swift, H., and Rasch, E., Microphotometry with visible light. *Phys. Tech. Biol. Res.* **3**, 353–400 (1956).
113. Taylor, J. H., Asynchronous duplication of chromosomes in cultured cells of Chinese hamster. *J. Biophys. Biochem. Cytol.* **7**, 455–463 (1959).
114. Thompson, R. Y., Heagy, F. D., Hutchison, W. C., and Davidson, J. N., The desoxyribonucleic acid content of the rat cell nucleus and its use in expressing the results of tissue analysis, in particular reference to the composition of liver tissue. *Biochem. J.* **53**, 460–474 (1953).
115. Vendrely, R., The desoxyribonucleic acid content of the nucleus. In "The Nucleic Acids" (E. Chargaff and J. N. Davidson, eds.), Vol. 2, pp. 155–188. Academic Press, New York, 1955.
116. Vendrely, R., and Vendrely, C., The results of cytophotometry in the study of the deoxyribonucleic acid (DNA) content of the nucleus. *Intern. Rev. Cytol.* **5**, 171–197 (1956).
117. von Borstel, R. C., Cytogenetics and developmental genetics course. A postcript. *Am. Zoologist* **3**, 87–95 (1963).
118. Waddington, C. H., "New Patterns in Genetics and Development." Columbia Univ. Press, New York, 1962.
119. Welch, R. M., A developmental analysis of the lethal mutant L(2) GL of *Drosophila melanogaster* based on cytophotometric determination of nuclear desoxyribonucleic acid (DNA) content. *Genetics* **42**, 544–559 (1957).
120. Whitten, J. M., Giant polytene chromosomes in hypodermal cells of developing footpads of dipteran pupae. *Science* **143**, 1437–1438 (1964).
121. Whitten, J. M., Hemocytes and the metamorphosing tissues in *Sarcophaga bullata*, *Drosophila melanogaster*, and other cyclorrhaphous Diptera. *J. Insect. Physiol.* **10**, 447–469 (1964).
121a. Wolstenholme, D. R., Dawid, I. B., and Ristow, H., An electron microscope study of DNA molecules from *Chironomus tentans* and *Chironomus thummi*. *Genetics* **60**, 759–770 (1968).
122. Zalokar, M., Ribonucleic acid and the control of cellular processes. In "Control Mechanisms in Cellular Processes" (D. M. Bonner, ed.), pp. 87–140. Ronald Press, New York, 1961.
123. Zubay, G., and Doty, P., Isolation and properties of deoxyribonucleoprotein particles containing single nucleic acid molecules. *J. Mol. Biol.* **1**, 1–20 (1959).

FUNDAMENTAL CONCEPTS OF FLUORESCENCE AND PHOSPHORESCENCE SPECTROMETRY

W. J. McCarthy* and E. S. Moyer†

DEPARTMENT OF CHEMISTRY, WEST VIRGINIA UNIVERSITY, MORGANTOWN, WEST VIRGINIA

Fluorescence and phosphorescence spectrometry have achieved positions of importance in the analysis of wide varieties of organic and inorganic compounds. Applications of luminescence analytic techniques have in recent years also achieved a degree of acceptance in the area of cytochemistry. It is the function of this chapter to present the fundamental aspects of luminescence theory and instrument design considerations. This chapter deals with the molecular origin of fluorescence and phosphorescence; the general nature of the transitions between molecular electronic states is examined. The concepts of quantum efficiency and decay time are presented through a kinetic description of transitions between molecular energy levels. The chapter is concluded with a discussion of the chemical properties of excited molecular electronic states and the nomenclature of electronic transitions.

I. General Introduction to the Properties of Molecular Electronic States

Most organic molecules at room temperature or below and in the absence of external electromagnetic influences can be described in the following manner. With few exceptions all filled molecular orbitals contain two spin-paired electrons. This means that for each filled orbital there exist two electrons with spin orientations in opposite directions; therefore each molecular orbital then contributes no *net* spin to the entire molecular system. When the above condition is fulfilled, the molecule is said to exist in its electronic ground state. Thus, for most organic molecules the ground

---

* Deceased.
† Present address: Louisina State University, New Orleans, Louisiana.

electronic state may be described in terms of total electronic spin. The spin, S, corresponding to any molecular state is simply calculated by

$$S = \sum_i s_i \qquad (1)$$

where s_i is the spin of the ith molecular electron and the summation is taken over all electrons in the molecule. Electrons may have either $+\frac{1}{2}$ or $-\frac{1}{2}$ spin quantum numbers; spin-paired electrons have opposite values of the spin quantum number resulting in a net axial electronic spin of 0 for each filled molecular orbital. Finally when Eq. (1) is summed over all molecular electrons, the net resultant spin will be $S = 0$ for most organic molecules in the *ground state*.

An additional method of classifying molecular electronic states is through their energy (the ground state energy is a convenient reference point and is arbitrarily taken to be zero). It can be shown that organic molecules have, in general, many unoccupied molecular electronic orbitals (by convention the lowest energy molecular orbitals are occupied first when filling molecular orbitals according to the *aufbau* principle). These unoccupied molecular orbitals lie at higher absolute energies than the filled molecular orbitals. There are numerous physical means of promoting an electron from a lower energy state to one of higher energy. The process of the transition itself will be described below; however, the nature of the resultant state will now be examined.

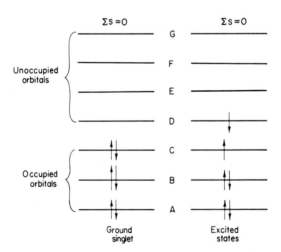

Fig. 1. Energy level diagrams for a representative typical organic molecule in the ground electronic state (left) and the lowest energy excited singlet electronic state (right). For both states all electron spins are paired and therefore the total spin is zero and the multiplicity is one, thus producing singlet states.

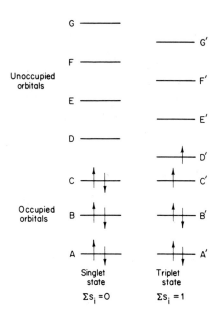

FIG. 2. Energy level diagrams for a representative typical organic molecule in the ground singlet electronic state (left) and the lowest energy excited triplet electronic state (right). The sum of the paired and unpaired electron spins for the triplet state is 1 and the multiplicity is 3, thus producing a triplet state.

Consider an organic molecule whose molecular orbitals are pictorially represented in Fig. 1. The ground state is shown with three filled molecular orbitals (A, B, and C) and four unoccupied (unfilled) orbitals (D, E, F, and G). If by some physical process energy equivalent to the difference in energy between levels C and D is supplied to the ground state molecule then an electron in level C *may* be promoted to level D resulting in the production of an *excited state molecule*. This resultant state differs from the ground state in two important aspects: the excited state molecule possesses more energy than the ground state molecule; and the molecular orbitals of the excited state molecule are filled in a manner quite different from the manner in which the orbitals of the ground state molecule were filled. These differences cause the following conclusions to be drawn. The ground state molecule and the excited state molecule are chemically different species; i.e., one may reasonably expect differences in chemical reactivity of the two states. The energy-rich nature of the excited state leads to additional physical differences between the ground and excited states; these differences can be observed upon application of several physicochemical tests, many of which are spectroscopic in nature and are described below.

While there are important differences between the ground and excited molecular electronic states, there is one significant similarity. The net spin of the excited electronic state (see Fig. 1) will be the same as the net spin of the ground electronic state. This observation simply means that the promoted electron has maintained its spin orientation. It is convenient, therefore, to note this similarity by designating the *multiplicity* of the ground and excited electronic states. The multiplicity, M, is simply calculated by

$$M = 2S + 1 \qquad (2)$$

where S is the net electronic spin. Because the net spin of each state is zero, then the multiplicity for each state is 1. States with multiplicities of 1 are traditionally referred to as *singlet* states. Therefore, the above discussion has dealt with the *ground singlet state* and an *excited singlet state*.

There is, of course, the possibility that the promoted electron may have acquired sufficient energy to reside in a molecular orbital higher in energy than the lowest energy unoccupied orbital; such a situation is still referred to as a singlet state but higher excited singlet states are often assigned ordinal numbers to denote their relative energy. Thus, if the promoted electron were to reside in orbital E (see Fig. 1) the resultant state would be called the *second excited singlet state*.

For every *excited* singlet state it is easily seen that there will be a state in which the promoted electron has reversed (or flipped) its spin (see Fig. 2). Following the procedure used for describing the excited singlet state(s) this new state will have the following characteristics. The net spin is no longer zero since the electron spins are now aligned, resulting in a net spin of 1 (or -1 depending upon convention). The multiplicity of a state with a net spin of 1 is given by Eq. (2) to be $2(1) + 1 = 3$. Such a state is conveniently named a *triplet state*. Again several important observations can be made about the differences between the ground singlet state and the *triplet state* illustrated in Fig. 2. The triplet state molecule possesses more energy than the ground state molecule; and the molecular orbitals are filled differently. These differences result again in chemically and physically different species. An extremely important physical difference is the fact that molecules with net electronic spins greater than zero are paramagnetic. The most important differences between the triplet state and the singlet states are, however, peculiar to the nature of the triplet state. The most obvious difference between the *triplet manifold* (i.e., the orbital diagram in Fig. 2) and the corresponding singlet manifold is in the absolute energies of the orbitals designated D′, E′, F′, and G′. It will be noted that the energy of a triplet state is always lower than the energy of its corresponding singlet state. The source of this energy change can be seen from a simple consideration of electron spins and repulsive forces. Other major differences between singlet

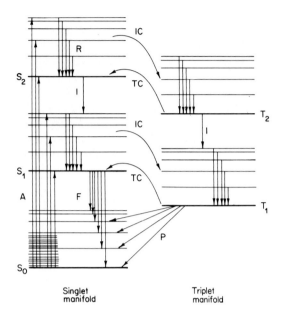

Singlet
manifold

Triplet
manifold

FIG. 3. A detailed energy level diagram for a typical organic molecule. States S_0, S_1, and S_2 are the ground state, first excited, and second excited singlet states, respectively. States T_1 and T_2 are the first and second excited triplet states, respectively. Vibrational levels are shown for each state. Some rotational levels are indicated for S_0. Transitions are labeled as follows: A, absorption; R, radiationless vibronic deactivation; F, fluorescence; I, internal conversion, IC, singlet to triplet intersystem crossing; TC, triplet to singlet intersystem crossing; and P, phosphorescence.

and triplet states arise from a consideration of transitions between these states; these differences will be described below. Obviously there may be several triplet states for any given molecule; as in the case of singlet states, ordinal numbers are usually assigned to each triplet state in order of increasing energy. The lowest energy triplet state corresponds to the first excited singlet state and not to the ground singlet state, although it is fairly common practice to refer to this state as the *ground triplet state*.

Electronic states for molecules are not adequately described by discrete energy levels as shown in Figs. 1 and 2. The schematic representation of these states is complicated by the presence of vibrational and rotational levels superimposed upon each electronic level. Because molecules are polyatomic there is no convenient method for representing these superimposed rotational and vibrational states for even triatomic molecules since this representation requires a third orthogonal axis. Therefore, only an idealized diagram of the molecule energy level pattern may be conveniently presented.

For each electronic state there will in general be several fundamental vibrational modes. With each vibrational mode one may associate a discrete energy. For each vibrational mode there will, in general, be a number of discrete rotational energy levels. Thus, the total representation of a molecular electronic diagram becomes quite complicated. There are some simplifying alterations which may be conveniently made to this picture in order to obtain a useful model for discussion. Consider for the moment that the electronic state diagram for the hypothetical molecule in question is shown in Fig. 3. There are three singlet states, S_0, S_1, and S_2 and two triplet states T_1 and T_2. For each electronic state some vibrational levels are shown. In the S_0 ground state several closely spaced rotational levels are indicated (rotational levels are omitted from the other electronic states for simplicity).

At room temperature or below the molecule will exist in the lowest vibrational level of the ground, S_0, electronic state. The molecule further may exist in any number of rotational levels, their populations being determined by a Boltzmann distribution. The concept of transition between electronic states is most easily understood by reference to Fig. 3. The energy necessary for promotion of an electron from the ground S_0 state to an excited singlet state, S_1 or S_2, is easily seen to lie within a range of energies rather than being a discrete energy; such discrete energy, line transitions, are typical of atomic spectra. In solution the rotational energy levels become "smeared-out" and are no longer discrete, due to rapid intermolecular interaction. Therefore, the range of energies necessary for transition between two electronic states becomes essentially a continuum so that any energy supplied to the molecule between two boundaries will be capable of promoting an electron. In spectroscopic terms this means that photons with energy hv, where h is Planck's constant and v is the frequency in sec^{-1}, between two boundaries will be capable of causing promotion of a molecular electron. Thus, the absorption spectra of organic compounds in solution will in general be broad continua occurring between two limits of wavelength. The shape of the absorption spectrum is determined by factors such as the Franck-Condon Principle which are not necessary for the subsequent discussion and therefore these factors will not be considered here.

Electronic absorption spectra of polyatomic organic molecules generally occur in the ultraviolet–visible region of the electromagnetic spectrum. The absorption spectra are in general broad bands with widths at half-intensity of the order of several hundred angstroms; some absorption bands may have widths as large as 1000 Å or more.

The problems of concern in luminescence spectrometry are usually not the details of the absorption process; in particular one is generally more concerned with the *fate* of this absorbed energy. Excitation processes may produce a number of chemically and physically different species; these differences result from the fact that there are numerous vibrational, rotational, and electronic states (see Fig. 3) which may be produced upon excitation. The initial deactivation processes involve a rapid loss to the environment of small amounts of the excitation energy. The net result of such loss is the production of a large population of molecules in the lowest vibrational level of the first excited singlet state, S_1. All of these excited molecules are chemically identical and their existence in this state is independent of their individual previous histories; these facts will aid in the development of quantum efficiency expressions later in this chapter. The rapid deactivation from high energy vibrational, rotational, and electronic levels is quantitative and in general it will be assumed that all activated molecules eventually reach the lowest vibrational level of the first excited electronic state. It will further be assumed that the deactivation process to the lowest vibrational level of the S_1 state occurs much more rapidly than any of the processes which will be discussed below. The initial deactivation processes are in general radiationless, i.e., they occur without emission of a photon during the deactivation process.

The phenomena generally known as luminescence can be thought of as originating from the excited state, S_1 in the lowest vibrational state. The luminescence processes of interest simply are radiational modes of deactivation of radiationally excited molecules. If a molecule undergoes radiational deactivation from S_1 to S_0 the process is called *fluorescence*. The deactivation from S_1 to S_0 also results in a band spectrum; any number of vibrational levels in S_0 may be populated by the S_1 to S_0 transition. The argument presented above for broad bands in absorption spectra is the basis for the previous statement. Other modes of deactivation from S_1 are available. If the excited S_1 molecule interacts with its environment, then there may be a radiationless deactivation of S_1 resulting in a production of S_0; the nature of this environment interaction (quenching) will be examined below in detail. It is possible that high-lying vibrational levels of S_0 will overlap the S_1 state; spontaneous conversion from the S_1 to the S_0 state can readily occur with no net loss in energy. All of the above radiational and radiationless deactivation steps may produce an S_0 molecule in excited vibrational levels. Rapid radiationless deactivation of these vibrational levels produce the "ground-state" species.

In addition to the simple sequences of events outlined in the previous paragraph there are a number of other extremely important modes of excited state deactivation. Consider the possibility that a molecule in state S_1 could cross from the singlet manifold to the triplet manifold; the process is shown schematically in Fig. 3. The conditions under which such an *intersystem crossing* may occur will be described below; it shall be sufficient at this point to examine the consequences of such a process. Rapid radiationless deactivation of the excited vibrational levels in T_1 yield a large population of chemically identical molecules in the lowest vibrational level of the T_1 state. There are a number of deactivation routes now open to these T_1 molecules; they are in most respects identical in qualitative nature to the deactivation routes described for the S_1 to S_0 transition. Radiational deactivation of T_1 to S_0 is the phenomenon known as *phosphorescence*.

The process of intersystem crossing involves the flip of an electron spin. According to atomic selection rules for electronic transitions, such intersystem crossing transitions are quantum mechanically forbidden. For many atoms and for large numbers of organic molecules; however, there is a finite probability that such intersystem crossing transitions will occur. The quantum mechanical view of this situation may be summarized as follows. The quantized nature of electron spin is reduced, resulting in a lessening of the identity of distinct electronic spin states. The *spin–orbit* coupling which occurs may be a result of the presence of a heavy atom in the environment of the molecule; external or internal inhomogeneous magnetic fields also tend to cause mixing of the spin states. The effect of the heavy atom upon phosphorescence is severe and will be examined below.

There are further modes of radiational deactivation available to the excited molecule. Consider the consequences of a T_1 to S_1 transition. Once the S_1 state is produced, all of the deactivation routes described above are available. In particular the radiational deactivation of S_1 to S_0 (after a T_1 to S_1 transition) is called slow (or delayed) fluorescence. The circumstances under which this type of luminescence may occur are rare and as a consequence it need not be considered further in this discussion.

The discussion in this introductory portion of the chapter has dealt in particular with the energy relationships of absorption, fluorescence, and phosphorescence. The details of this section can be conveniently summarized

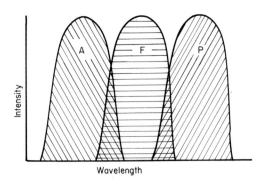

Fig. 4. Predicted spectroscopic wavelength relationship between absorption (A), fluorescence (F), and phosphorescence (P) spectra for a typical organic molecule.

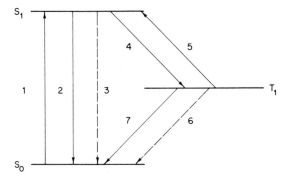

FIG. 5. Simplified Jablonsky diagram for a typical organic molecule. Transitions depicted are defined as follows: 1, absorption; 2, fluorescence; 3, radiationless deactivation; 4, intersystem crossing from S_1 to T_1; 5, intersystem crossing from T_1 to S_1; 6, phosphorescence; 7, radiationless deactivation.

in Fig. 4. The spectral relationships between absorption, fluorescence, and phosphorescence are a direct result of the molecular basis of these luminescence phenomena. Luminescence excitation spectra will occur at shorter wavelengths (higher energies) than either fluorescence or phosphorescence. Fluorescence will, in turn, be observed at shorter wavelengths than phosphorescence.

II. Kinetics of Transitions between Molecular Electronic States

All of the transitions described in the previous section can be summarized and represented schematically as shown in Fig. 5. For convenience all vibrational levels have been omitted; in addition no excited states higher than T_1 and S_1 are indicated. There are consequently only seven basic transition types of interest. These transition types are as follows: (1) absorption (excitation); (2) fluorescence; (3) radiationless deactivation from S_1; (4) intersystem crossing from S_1 to T_1; (5) radiationless intersystem crossing from T_1 to S_1; (6) phosphorescence; and (7) radiationless deactivation from T_1 to S_0. All other processes occur quantitatively and much more rapidly than any of the deactivation processes listed above. Each of the processes depicted in Fig. 5 may be assigned a rate constant; as stated in the previous section of this chapter the existence of a molecule in S_1 or T_1 is necessarily independent of its past history. The independence of the fate of a particular molecule on its past history allows one to assign rate constants to these processes. The nature of the rate constant so assigned depends upon the particular process which is being described. Presented in Table I are the assigned rate constants for a variety of deactivation processes. The processes which are presented in Table I are a great oversimplification of the real situation in the condensed phase. There are numerous types of quenching processes which may occur from S_1 and T_1. There may be direct internal conversion of excitation energy due to a S_1 to S_0 (or T_1 to S_0) transition. Collision with energy-sink species in the environment, indicated by Q, may cause the S_1 or T_1 state to lose its excitation energy. Alternatively, the concentration of the S_0 species in the environment may be so high that so-called

TABLE I

RATES OF ACTIVATION AND DEACTIVATION[a]

Process	Type[b]	Description	Rate
$S_0 + h\nu \rightarrow S_1$	1	Excitation	$k_A P_{ABS}[S_0]$
$S_1 \rightarrow S_0 + h\nu$	2	Fluorescence	$k_F[S_1]$
$S_1 + Q \rightarrow S_0 + Q$	3	Quenching	$k_{QS}[Q][S_1]$
$S_1 \rightarrow T_1$	4	Intersystem crossing	$k_{IC}[S_1]$
$T_1 \rightarrow S_0 + h\nu$	7	Phosphorescence	$k_P[T_1]$
$T_1 + Q \rightarrow S_0 + Q$	6	Quenching	$k_{QT}[Q][T_1]$
$T_1 \rightarrow S_1$	5	Intersystem crossing	$k_{TC}[T_1]$

[a] All terms are defined in the text.
[b] Refer to Fig. 5.

concentration quenching or self-quenching may occur. Photochemical decomposition from S_1 (or T_1) may occur, yielding completely unrelated chemical species. Finally, S_1 or T_1 may transfer excitation energy to another species in the environment; the energy transfer process may in itself give rise to the luminescence of the energy acceptor species. Such energy transfer processes have recently been investigated for possible analytic significance; work is currently being carried forth in the authors' laboratories with the purpose of investigating these phenomena in detail. Finally, at small intermolecular distances, mutual interaction of two excited species may occur. Well-documented studies of the process $T_1 + T_1 \rightarrow S_1 + S_0$ have been presented by McGlynn and co-workers. This process will presumably be important only in situations of relatively high analytic concentrations; for this reason this triplet–triplet annihilation will be neglected in the following discussion.

The rate of change of the population of each of the excited states depicted in Fig. 5 and listed in Table I will be given by the values of the following quantities: $d[S_1]/dt$ and $d[T_1]/dt$, where t is the time in seconds. One obtains $d[S_1]/dt$ and $d[T_1]/dt$ considering all of the possible excitation and deactivation routes available to the molecule for the transition of interest. For instance, the total rate of change of the population of state S_1 with time will be given by

$$d[S_1]/dt = \sum (\text{rates for reaching } S_1) - \sum (\text{rates for leaving } S_1)$$

Expressions for each of the "rates" may be found by reference to Table I. Thus,

$$d[S_1]/dt = k_A P_{ABS}[S_0] - (k_F[S_1] + k_{QS}[S_1][Q] + k_{IC}[S_1]) \tag{3}$$

where $k_{TC}[T_1]$ has been assumed to be much less than $k_A P_{ABS}[S_0]$ and is thus neglected. Similarly one may write for the total rate of change of the population of state T_1 with time,

$$d[T_1]/dt = k_{IC}[S_1] - (k_P[T_1] + k_{QT}[T_1][Q]) \tag{4}$$

It is assumed in Eq. (4) that $k_{TC}[T_1] \ll (k_P[T_1] + k_{QT}[T_1][Q])$ and therefore may be neglected. Now each of these equations [(3) and (4)] may be simplified by making the following substitutions:

$$K_S = k_F + k_{QS}[Q] + k_{IC} \tag{5}$$

and

$$K_T = k_P + k_{QT}[Q] \tag{6}$$

Equations (3) and (4) may then be simplified

$$d[S_1]/dt = k_A P_{ABS}[S_0] - K_S[S_1] \tag{7}$$

$$d[T_1]/dt = k_{IC}[S_1] - K_T[T_1] \tag{8}$$

It is usual to assume that a luminescence experiment will be performed after the entire molecular system has the opportunity to reach a condition of equilibrium with the external perturbing radiation source; this condition is called the steady state. The use of the term "steady state" in reference to a dynamic kinetic system, such as the one under consideration, requires that there be constant illumination of the sample by the source [this requirement is obvious since P_{ABS} (incident power absorbed) must remain constant for $d[S_1]/dt$ and consequently $d[T_1]/dt$ to remain constant]. In the steady-state condition the rate of change of S_1 and T_1 with time will be equal to 0. Therefore, Eqs. (7) and (8) become

$$0 = k_A P_{ABS}[S_0] - K_S[S_1] \tag{7'}$$

and

$$0 = k_{IC}[S_1] - K_T[T_1] \tag{8'}$$

respectively.

III. The Quantum Efficiencies of Fluorescence and Phosphorescence

The intensity and therefore the analytic sensitivity of a luminescence process will, of course, be intimately related to the efficiency of conversion of absorbed photons to luminescence photons. The number which gives this conversion efficiency is called the quantum efficiency.* The quantum efficiency for fluorescence is defined as the ratio of the number of fluorescence emission transitions occurring per second to the number of absorption transitions per second. The quantum efficiency for fluorescence is therefore defined by

$$\phi_F = k_F[S_1]/k_A P_{ABS}[S_0] \tag{9}$$

In a similar manner the quantum efficiency for phosphorescence, ϕ_P, is defined by

$$\phi_P = k_P[T_1]/k_A P_{ABS}[S_0] \tag{10}$$

These equations are not in their present form very useful since one in general does not know the value of either $[S_1]$ or $[T_1]$. However, using the results of the kinetics discussion in Section II one may discuss the value of ϕ_F and ϕ_P under a steady-state assumption. It is necessary to solve simultaneously Eqs. (7) and (8) for the value of $[S_1]$ and $[T_1]$. Doing this one obtains

$$[S_1] = k_A P_{ABS}[S_0]/K_S$$

and

$$[T_1] = k_{IC} k_A P_{ABS}[S_0]/K_S K_T$$

Substituting these values in Eqs. (9) and (10), respectively, one obtains the following expressions for the quantum efficiencies:

$$\phi_F = k_F/K_S \tag{11}$$

and

$$\phi_P = k_{IC} k_P/K_S K_T \tag{12}$$

Several very important conclusions may be drawn regarding ϕ_F and ϕ_P from a purely qualitative viewpoint when one considers typical values for the first order (or pseudo first order) rate constant. The value of k_F for

---

* The quantum efficiency is often erroneously called the quantum yield. Quantum yields are proper when reference is made to some spectrochemical process in which a distinct chemical product is obtained. For instance, quantum yields are useful in photochemistry when discussing the efficiency of a photochemical reaction.

many typical organic molecules will be of the order of 10^{+8} sec^{-1}; k_P on the other hand may be of the order of 10^{-1} to 10^{+1} (values outside these ranges are not at all uncommon). The values of k_{IC} may be of the order of k_F if the mixing of the singlet and triplet states is large; on the other hand k_{IC} may be very small for some molecules. The effect of the value of k_{IC} will be seen below. The quenching rate constants, k_{QS} and k_{QT}, are difficult to define, considering the variety of molecular processes which one is attempting to describe with k_{QS} and k_{QT}. If it is assumed that all quenching processes in the molecular environment are diffusion-controlled then the system may be described in the following manner. It can be shown that if colliding species (molecules, ions, etc.) in solution are considered to be hard spheres with the activation energy for a collision much less than kT (the thermal energy per molecule available through normal equilibrium considerations), then the rate constant, k_Q, for collision will be given by

$$k_Q = P \frac{kT}{3\eta} \frac{(r_a + r_b)^2}{r_a r_b} \tag{13}$$

where P is the fraction of the total number of collisions per second that are effective in deactivating the excited state, k is the Boltzmann constant, T is the absolute temperature, η is the viscosity, r_a is the radius of the species of interest, and r_b is the radius of the quenching species. This equation has been shown to describe adequately the observed behavior in a number of luminescence systems. For the purposes of calculations to be performed below it will be assumed that $r_a = r_b$ and thus, k_Q is given by

$$k_Q = \frac{4kT}{3\eta} \tag{14}$$

In order to calculate k_Q, it is necessary to know the manner in which the viscosity, η, of a system will vary with temperature. The best known equation relating viscosity to temperature is in the following form

$$\eta = A \exp (B/T) \tag{15}$$

where A and B are constants. In liquids where intermolecular association is significant then Eq. (15) is modified to allow for significant logarithmic nonlinearity

$$\log_{10} \eta = \alpha + \beta/T + \gamma/T^2 \tag{16}$$

where α, β, and γ are empirical constants. For ethanol α, β, and γ have been found to be -0.42873, 734.48806, and 72.3301, respectively (due to exponent limitations in the computer used for these calculations, if $\log_{10}\eta \geq 35$ then $\log_{10} \eta$ was set to be equal to 35).

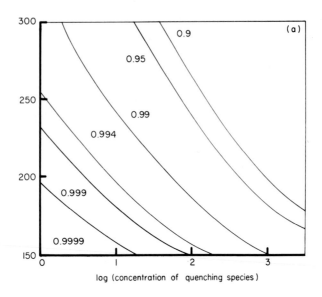

log (concentration of quenching species)

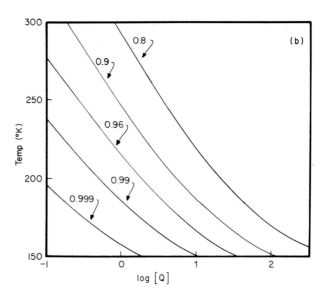

Fig. 6

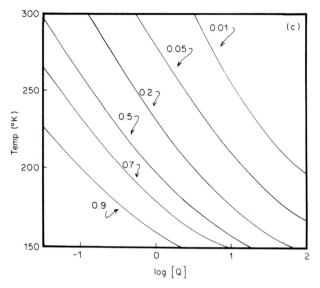

FIG. 6. (a) Fluorescence quantum efficiency surface as a function of temperature and the concentration of the quenching species $[k_F = 10^{10}$ sec^{-1} and k_{QS} is calculated from Eq. (13)]. (b) Same as (a) except $k_F = 10^8$ sec^{-1}. (c) Same as (a) except $k_F = 10^6$ sec^{-1}.

If it is assumed, as stated above, that all radiationless processes except quenching by extraneous or foreign species are unimportant compared to radiational and intersystem crossing processes, that is, $K_S = k_F + k_{QS}[Q] + k_{IC}$ and $K_T = k_P + k_{QT}[Q]$ [refer to Eqs. (5) and (6)], then in general ϕ_F will be given by

$$\phi_F = \frac{k_F}{k_F + k_{QS}[Q] + k_{IC}} \qquad (17)$$

Evaluating k_{QS} as described above one obtains for the quantum efficiency surface (as a function of temperature and concentration of the quenching species) the plots shown in Figs. 6a, 6b, and 6c for values of $k_F = 10^{10}$, 10^8, and 10^6, respectively.

For large values of k_F (Fig. 6a), the quantum efficiency is affected only to a very small extent; for an increase in concentration of 10^5 and a 1.5×10^2 °K increase in temperature the value of ϕ_F changes by only about 0.5%. For $k_F = 10^8$ and for a 10^5 and 1.5×10^2 increase in concentration and temperature, respectively, the quantum efficiency changes by 25%. Finally for $k_F = 10^6$ (Fig. 6c) and for the same increases in concentration and temperature noted above, the value of k_F drops by 92%. From these observations certain conclusions about the efficacy of fluorescence analytic methods may be made. If the species Q were in fact S_0 then quenching due to S_0 would

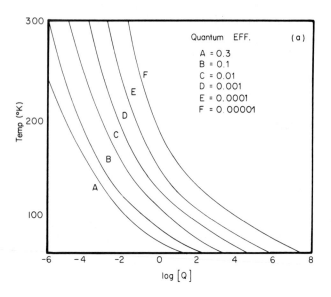

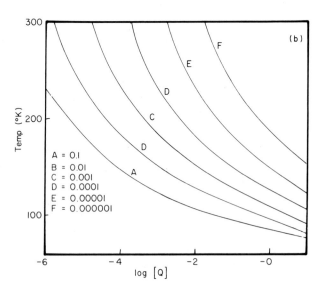

FIG. 7

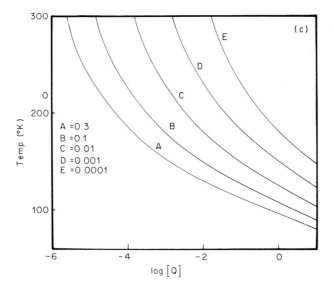

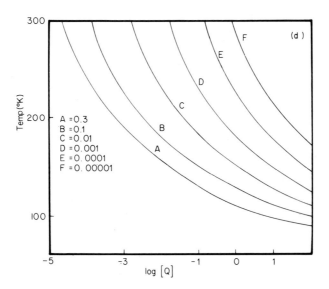

FIG. 7. (a) Phosphorescence quantum efficiency surface as a function of temperature and the concentration of the quenching species. $[k_F = 10^7 \text{ sec}^{-1}; k_{IC} = 10^7 \text{ sec}^{-1}; k_P = 10 \text{ sec}^{-1};$ and k_{QS} and k_{QT} are calculated from Eq. (13)]. (b) Same as (a) except $k_P = 1 \text{ sec}^{-1}$. (c) Same as (a) except $k_P = 0.1 \text{ sec}^{-1}$. (d) Same as (a) except $k_P = 0.01 \text{ sec}^{-1}$.

not usually be a problem in analytic studies because concentrations used are nearly always much less than 10^{-2} M. It should be emphasized, however, that if Q is an extraneous species, and if this extraneous species is the solvent, then severe reduction in fluorescence quantum efficiencies is to be expected in the vicinity of room temperature. Because the concentration of the solvent may be quite large, that is, greater than 10 M, even a relatively inefficient quencher ($P < 0.5$) will severely affect the value of ϕ_F. Severe dependence of ϕ_F upon temperature will necessitate closely controlled solvent temperature in order to obtain analytic reproducibility. In general, the analyst using fluorescence will greatly improve analytic sensitivities, and consequently, precision, if low sample temperatures are used. This control of conditions can easily be accomplished since there are commercially available low temperature thermostated baths for many spectrofluorometers.

In the total absence of collisional and other forms of the process represented by k_{QS} then the value of ϕ_F will be given by

$$\phi_F = \frac{k_F}{k_F + k_{IC}} \tag{18}$$

This situation, although probably never experimentally realized, points out the significant influence of k_{IC} upon ϕ_F. The maximum value of ϕ_F is in all practicality limited by the value of k_{IC}. If $k_{IC} \ll k_F$ then ϕ_F approaches unity; moreover if $k_{IC} \gg k_F$ then ϕ_F can be so small as to make fluorescence assay impracticable.

These same procedures may be followed now for the value of the phosphorescence quantum efficiency, ϕ_P. If it is assumed that all radiationless processes except quenching by dissolved species are unimportant compared to radiational and intersystem crossing processes, that is, $K_S = k_F + k_{QS}[Q] + k_{IC}$ and $K_T = k_P + k_{QT}[Q]$, then in general ϕ_P will be given by

$$\phi_P = \frac{k_{IC}\, k_P}{(k_F + k_{QS}[Q] + k_{IC})\,(k_P + k_{QT}[Q])} \tag{19}$$

Evaluating k_{QS} and k_{QT} as described above with $P = 1.0$, one obtains for the quantum efficiency surface (temperature versus [Q] versus ϕ_P) the plots shown in Figs. 7a, b, c, and d for values of $k_P = 10, 1, 0.1$, and 0.01 sec^{-1}. The important points of interest in Figs. 7a–d are best summarized as follows. One should first of all note the severe temperature dependence of ϕ_P regardless of the value of k_P; ϕ_P reaches maximal values only at very low temperatures (i.e., very high viscosities). At such high viscosities, k_{QT} and k_{QS} will be much less than k_P and k_F, respectively, even for highly efficient quenchers ($P = 1$) at high concentrations. Thus ϕ_P could be approximately given by

$$\phi_P = \frac{k_{IC} \, k_P}{(k_F + k_{IC}) \, (k_P)} = \frac{k_{IC}}{k_F + k_{IC}} \qquad (20)$$

Comparing Eqs. (18) and (20) it is seen that under conditions prescribed above that

$$\phi_F + \phi_P = 1 \qquad (21)$$

and moreover,

$$\phi_P / \phi_F = k_{IC} / k_F \qquad (22)$$

This most important statement, Eq. (22), can presumably be used to alter analytic sensitivities of fluorescence and phosphorescence by altering the value of k_{IC}; k_F is usually invariant under normal experimental circumstances.

Some further observations on Figs. 7a–d seem in order at this point. At very low temperatures, ϕ_P is nearly independent of temperature and quencher concentration; in fact, the value of ϕ_P is nearly independent of k_P [see Eq. (20)]. One notes that the value of T decreases as k_P increases. Furthermore at moderate to high values of T, the value of ϕ_P significantly decreases with a decrease in k_P. Of course, in most analytic studies the concentration of the phosphorescing species will be much less than 10^{-2} or 10^{-3} M therefore quenching by S_1 may be considered negligible at low T (i.e., $k_{QT}[Q] \ll k_P$). Conversely, as with ϕ_F above, severe quenching by the solvent may be a significant factor in the choice of a solvent system; it should be observed, however, that such quenching by the solvent is much less severe with the simple C,H,O-containing systems generally used, such as ethanol.

Perhaps the most important point to be obtained from Fig. 7a–d is that large values of ϕ_P will never be obtained at even moderate solution temperature unless solvents of very high viscosity are available. Therefore, it would be of considerable analytic importance if solvents of high viscosity at temperatures near 273°K were readily available. Such a solvent would extend the "flat" portion of the surfaces in Fig. 7a–d to much higher values of T and thus allow analytic phosphorimetry at more convenient temperatures. The possibilities for such solvents would seem to be solutions of polymers or colloidal solgel systems which lack significant micelle structure. Historically phosphorescence of organic molecules was observed at room temperature in boric acid melts. Polyethylene and other plastic mounts for organic molecules have also allowed high temperature observation of phosphorescence.

Thus far in the discussion, the effect of the intersystem crossing rate from the triplet state to the excited singlet state has been ignored. The mechanism for this transition is poorly understood; however, a description of the control of this process through a Boltzmann thermal energy distribution appears to be most promising. In particular this means that k_{TC} will be approximated by

$$k_{TC} = k_{IC} \exp \left(- \Delta E / kT \right) \qquad (23)$$

where k_{IC} and k_{TC} have been previously defined, ΔE is the difference in energy between the lowest vibrational state in T, and the vibrational level in S, which is to receive the crossing electron, k is the Boltzmann constant, and T is the absolute temperature. It can be easily shown, using the same arguments as above, that ϕ_F and ϕ_P, incorporating k_{TC}, are given by

$$\phi_F' = k_F / K_S - \frac{k_{IC}^2 \exp \left(-\Delta E / kT \right)}{K_T'} \qquad (24)$$

and

$$\phi_P' = \frac{k_P k_{IC}}{K_T' K_S - k_{IC}^2 \exp \left(-\Delta E / kT \right)} \qquad (25)$$

where K_T' is given by

$$K_T' = k_P + k_{QT}[Q] + k_{IC} \exp \left(-\Delta E / kT \right) \qquad (26)$$

Now it is important to observe that at very low values of T, ϕ_F' and ϕ_P' approach ϕ_F and ϕ_P while at higher temperatures, i.e., when $\Delta E / kT \approx 0$, ϕ_F' and ϕ_P' may differ radically from ϕ_F and ϕ_P. It must be noted that the definition of the quantum efficiencies given in Eqs. (11) and (12) are exact; it is only the values of $[S_1]$ and $[T_1]$ which may be in error. If ΔE is sufficiently large then ϕ_F and ϕ_P adequately describe the quantum efficiencies. However, when ΔE is of the order of magnitude of the thermal energy available in the system, kT, then ϕ_F' and ϕ_P' adequately describe the efficiencies given in Eqs. (11) and (12). The problem lies in the fact that the process of fluorescence involves simple deactivation after absorption. Incorporation of k_{TC} into the argument allows for population of state S_1 via the route $S_0 \to S_1 \to T_1 \to S_1$ and subsequently fluorescence emission may occur. A photon which is emitted after the sequence $S_0 \to S_1 \to T_1 \to S_1$ will have the same energy distribution probability (spectrum) as the simple fluorescence photon (i.e., $S_0 \to S_1 \to S_0 + h\nu$). In fact, an experimenter performing "fluorescence" measurements will in general not be able to distinguish the simple fluorescence photon from the extraordinary fluorescence photon, unless special techniques discussed below are employed. Therefore the analytic fluorescence quantum efficiency for depopulation of the S_1 state (regardless of the intermediate molecular transitions) is given by ϕ_F'; this is especially true since fluorescence is generally performed at moderate to high temperatures. Restating the above in more concise language, one has, "the $S_1 \rightarrow S_0$ transition described by the rate term, $k_F[S_1]$, is independent of the mode of population of S_1 after initial excitation of S_0 by the rate term, $k_A P_{ABS}[S_0]$."

Some important revisions to previous discussions must necessarily be examined in the light of the presence of the k_{TC} rate constant in a generalized molecular system. The total quantum efficiency for the S_1 to S_0 transition is given by Eq. (24). For purposes of demonstrating certain facts, it shall be assumed that all radiationless processes are unimportant compared to the radiational and intersystem crossing processes, that is $K_S = k_F + k_{IC}$ and $K_T = k_P + k_{IC} \exp(-\Delta E/kT)$. Calculations for the value of ϕ_F as a function of k_F and temperature for $\Delta E = 0.1$, 0.3, and 0.5 electron volts are presented in Figs. 8a, b, and c, respectively. At very low temperatures the value of ϕ_F is independent of small changes in temperatures. As the temperature increases the quantum efficiency also begins to increase (due to repopulation of S_1 from T_1 via thermal reexcitation) until a constant value of ϕ_F is attained. As ΔE increases, the range of temperatures over which the value of ϕ_F is essentially constant also increases. Similarly as ΔE increases, there is an increase in the temperature at which ϕ_F begins to show significant change with temperature because $\exp(-\Delta E/kT)$ becomes progressively smaller.

Thus in nonquenched fluorescing systems ϕ_F' will *increase* with temperature; the amount of variation is, however, determined by the values of k_F, ΔE, T, k_{IC}, and k_{TC}.

Most analytic systems, in which fluorescence is measured, are operated at about room temperature or slightly below. Thus, the observed fluorescence emission is due not only to simple fluorescence ($S_0 \rightarrow S_1 \rightarrow S_0 + h\nu$), but also to a small contribution from E-type delayed fluorescence ($S_0 \rightarrow S_1 \rightarrow T_1 \rightarrow S_1 \rightarrow S_0 + h\nu$). The amount of the contribution from E-type delayed fluorescence is also determined by the value of k_F, ΔE, T, k_{IC}, and k_{TC}. Equation (9), therefore, is the number of $S_1 \rightarrow S_0 + h\nu$ transitions per second per number of $S_0 \rightarrow S_1$ transitions per second *regardless of the source of the S_1 molecules*. Usually this fact is ignored in kinetic approaches to fluorescence due to the importance of quenching at high temperatures, but it is evident that if quenching becomes important at temperatures above the point at which ϕ_F begins to vary with temperature then a maximum will occur for ϕ_F at a given temperature. Thus, for an analytic study of fluorescence in a molecule, where $\Delta E < 0.4$ eV and where quenching is minor, an approach to the optimum temperature will yield greater fluorescence sensitivity.

A reexamination of ϕ_P in light of the inclusion of the k_{TC} rate constant must also be made. Calculations of ϕ_P as a function of temperature and k_P for $\Delta E = 0.1$, 0.3, and 9.5 are presented in Figs. 9a, b, and c. It is assumed that all quenching processes are negligible; that is, $k_{QS}[Q] \approx k_{QT}[Q] \approx 0$. At very low temperatures the value of ϕ_P is relatively independent of temperature.

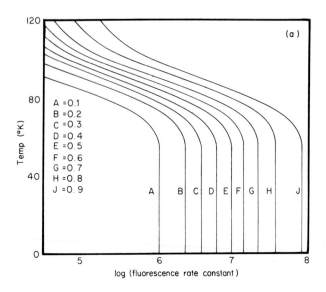

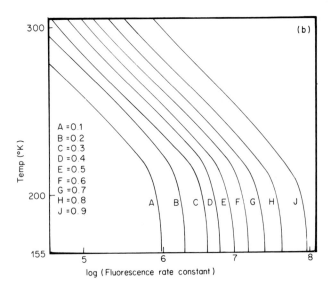

Fig. 8

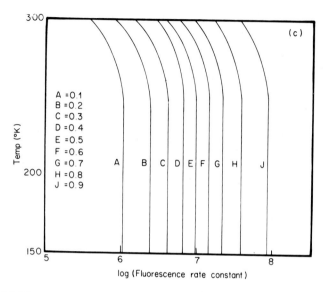

FIG. 8. (a) Fluorescence quantum efficiency surface as a function of temperature and the fluorescence radiational rate constant ($k_{IC} = 10^7$ sec^{-1}; $k_{QS} = 0$). Note k_{TC} is calculated as indicated in the text with $\Delta E = 0.1$ eV. (b) Same as (a) except $\Delta E = 0.3$ eV. (c) Same as (a) except $\Delta E = 0.5$ eV.

Moreover, at large values of k_P, ϕ_P is also relatively independent of temperature. There is, however, a severe dependence of ϕ_P both in surface contour and magnitude on the value of ΔE. As ΔE decreases, the magnitude of ϕ_P decreases and ϕ_P becomes dependent upon temperature at lower values of temperature. This effect is simply due to the presence of exp $(-\Delta E/kT)$ in the K_T expression. It is obvious that all of the surfaces presented in Figs. 9a, b, and c would display less dependence upon temperature as k_{TC} decreases. Therefore, in the absence of effective quenching species, a reduction in temperature will result in an increase in ϕ_P.

IV. The Decay Times of Fluorescence and Phosphorescence

In general the lifetime or decay time of a luminescence process is of considerable experimental interest due to the convenient manner in which it may be measured. For the moment consider the system of electronic states discussed above; further let it be assumed that a steady state of activation and deactivation has been achieved under constant illumination by a source, that is, $d[S_1]/dt = d[T_1]/dt = 0$. In Section I of this chapter it is pointed out that once a molecule reaches any particular energy state its subsequent

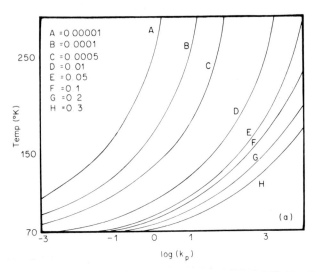

FIG. 9. (a) Phosphorescence quantum efficiency surface as a function of temperature and the phosphorescence radiational rate constant. ($k_F = k_{IC} = 10^7$ sec^{-1}, $k_{QS} = k_{QT} = 0$.) Note k_{TC} is calculated as indicated in the text with $\Delta E = 0.1$ eV. (b) Same as (a) except $\Delta E = 0.3$ eV. (c) Same as (a) except $\Delta E = 0.5$ eV.

actions are independent of its past history. Therefore radiational and radiationless deactivation occur without regard to previous molecular history. Let us examine the effect upon the steady state system of terminating the excitation light. Immediately $k_A P_{ABS}[S_0]$ becomes zero (since P_{ABS} is now zero) and one has

$$d[S_1]/dt = -K_S[S_1] \tag{27}$$

and

$$d[T_1]/dt = k_{IC}[S_1] - K_T[T_1] \tag{28}$$

Now Eq. (27) may be integrated to yield

$$\frac{[S_1]}{[S_1]_0} = \exp(-K_S t) \tag{29}$$

where $[S_1]_0$ is the population of state S_1 at $t = 0$ or just at the moment at which the excitation source was extinguished. Thus the population of S_1 or any variate with S_1, (intensity, signal, etc.) will decay exponentially with time. The decay constant is simply K_S. The units of K_S are sec^{-1}; it is frequently more convenient to discuss the reciprocal of K_S, the lifetime or decay

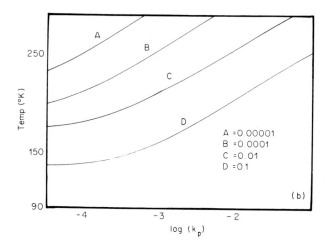

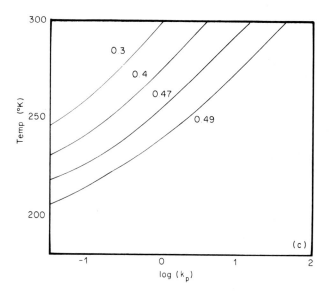

Fig. 9

time of state S_1. The reciprocal of K_S, $1/K_S$, will, of course, have units of seconds. The decay time, τ, is thus given for fluorescence by

$$\tau_F = (K_S)^{-1} \tag{30}$$

Experimentally the meaning of τ is obtained by observing that when $t = \tau$,

$$\frac{[S_1]}{[S_1]_0} = \exp(-1)$$

or

$$[S_1] = [S_1]_0 \exp(-1)$$

or

$$[S_1] = 0.37\,[S_1]_0 \tag{31}$$

Thus τ_F is obtained experimentally by plotting intensity versus time; measure the time (τ_F) at which I_F is 37% of I_F^0. In practice, for fluorescence, this procedure is nearly impossible since τ_F's are of the order of nanoseconds or tens of nanoseconds. On the other hand phosphorescence lifetimes may often be conveniently measured. Consider Eq. (28) where $[S_1]$ is nearly 0 within a few nanoseconds after termination of the excitation radiation; $d[T_1]/dt$ will be approximately given by $-K_T[T_1]$. Thus, the decay time of phosphorescence may be given by

$$\tau_P = (K_T)^{-1} \tag{32}$$

ACKNOWLEDGMENTS

The authors wish to thank the West Virginia University Computer Center for providing all computer time for calculations relating to Figs. 6 through 9. The research leading to this chapter was supported in part by a grant from the Research Corporation. Acknowledgment is made to the donors of the Petroleum Research Fund administered by the American Chemical Society for partial support of this research.

BIBLIOGRAPHY

This section is divided into several topic areas. Each of the pertinent references has been rather arbitrarily assigned to a general area of interest.

A. General References Regarding the Theory and Application of Fluorescence and Phosphorescence Spectrometry

Cetorelli, J. J., McCarthy, W. J., and Winefordner, J. D., The selection of optimum conditions for spectrochemical methods. IV. Sensitivity of absorption fluorescence, and phosphorescence spectrometry in the condensed phase. *J. Chem. Educ.* **45**, 98 (1968).

Förster, T., "Fluoreszenz Organischer Verbindungen." Vandenhoeck & Ruprecht, Göttingen, 1951.

Guilbault, G. G., "Fluorescence." Marcel Dekker, New York, 1967.

Hercules, D. M., "Fluorescence and Phosphorescence Analysis." Wiley (Interscience), New York, 1966.

Jaffé, H. H., and Orchin, M., "Theory and Applications of Ultraviolet Spectroscopy." Wiley, New York, 1962.

Kasha, M., "Manchester Lectures." Manchester, England, 1951.

McCarthy, W. J., and Winefordner, J. D., Phosphorimetry as a means of chemical analysis. In "Fluorescence" (G. Guilbault, ed.), pp. 371–442. Marcel Dekker, New York, 1967.

McCarthy, W. J., and Winefordner, J. D., The selection of optimum conditions for spectrochemical methods. II. Quantum efficiency and decay time of luminescent molecules. J. Chem. Educ. 44, 136 (1967).

Mellon, M. G., ed., "Analytical Absorption Spectroscopy." Wiley, New York, 1950.

Minkoff, G. J., "Frozen Free Radicals." Wiley (Interscience), New York, 1960.

Nebbia, G. Proc. 10th Colloqu. Spectros. Intern. Univ. Maryland, 1962 p. 605. Spartan Books, Washington, D.C., 1963.

Pringsheim, P., "Fluorescence and Phosphorescence." Wiley (Interscience), New York, 1949.

Udenfriend, S., "Fluorescence Assay in Biology and Medicine." Academic Press, New York, 1962.

Weller, A., In "Progress in Reaction Kinetics" (G. Porter, ed.). Macmillan, New York, 1961.

West, W., In "Fluorescence and Phosphorescence in Chemical Applications of Spectroscopy." Wiley (Interscience), New York, 1956.

Winefordner, J. D., McCarthy, W. J., and St. John, P. A., Phosphorimetry as an analytical approach in biochemistry. Methods Biochem. Analys. 15, 369–483 (1967).

B. Citations Dealing with the Problem of the Activation and Deactivation of Organic Molecules in Condensed and Gaseous Phases

Bowen, E. J., Fluorescence quenching in solution and in the vapor state. Trans. Faraday Soc. 50, 97–102 (1954).

Bowen, E. J., and Brocklehurst, B., Energy transfer in hydrocarbon solutions. Trans. Faraday Soc. 49, 1131–1133 (1953).

Bowen, E. J., and Livingston, R., An experimental study of the transfer of energy of excitation between unlike molecules in liquid solutions. J. Am. Chem. Soc. 76, 6300–6304 (1954).

De Groot, M. S., and van der Waals, J. H., Paramagnetic resonance in phosphorescent aromatic hydrocarbons. III. Conformational isomerism in benzene and triptycene. Mol. Phys. 6, 545–562 (1963).

De Groot, M. S., Hesselmann, D. A. M., and van der Waals, J. H., Paramagnetic resonance in phosphorescent aromatic compounds. IV. Ions in orbitally degenerate states. Mol. Phys. 10, 241–251 (1966).

Förster, T., Transfer mechanisms of electronic excitation. (10th Spiers Memorial Lecture.) Discussions Faraday Soc. 27, 7–17 (1959).

Furst, M., and Kallmann, H., Cross quenching of fluorescence in organic solutions. Phys. Rev. 109, 646–651 (1950).

Graham-Bryce, D. F., and Corkill, J. M., Use of solvents containing ethyl iodide in the investigation of phosphorescence spectra of organic compounds. Nature 186, 965–966 (1960).

Kellog, R. E., and Schwenker, R. P., Temperature effect on triplet state lifetimes in solid solutions. *J. Chem. Phys.* **41**, 2860–2863 (1964).

Lewis, G. N., and Lipkin, D., Reversible photochemical processes in rigid media: The dissociation of organic molecules into radicals and ions. *J. Am. Chem. Soc.* **64**, 2801–2808 (1942).

Linschitz, H., and Pekkarin, L., The quenching of triplet states of anthracene and por-phyrins by heavy metal ions. *J. Am. Chem. Soc.* **82**, 2411–2416 (1960).

McGlynn, S. P., Daigre, J., and Smith, F. J., External heavy-atom spin-orbital coupling effect. IV. Intersystem crossing. *J. Chem. Phys.* **39**, 675–679 (1963).

Nieman, G. C., and Robinson, G. W., Direct determination of exciton interactions for triplet states of organic crystals. *J. Chem. Phys.* **39**, 1298–1307 (1963).

O'Dwyer, M. F., Ashraf El-Bagoumi, M., and Strickler, S. J., Observation of anomalous phosphorescence-fluorescence intensity ratio in excitation of upper electronic states of certain aromatic hydrocarbons. *J. Chem. Phys.* **36**, 1395–1396 (1962).

Porter, G., and Windsor, M., Studies of the triplet state in fluid solvents. *Discussions Faraday Soc.* **17**, 178–186 (1954).

Porter, G., and Wright, M. R., I. Modes of energy transfer from excited and unstable ionized states: Intramolecular and intermolecular energy conversion involving change of multiplicity. *Discussions Faraday Soc.* **27**, 18–27 (1959).

Sternlicht, H., Nieman, G. C., and Robinson, G. W., Triplet-triplet annihilation and delayed fluorescence in molecular aggregates. *J. Chem. Phys.* **38**, 1326–1335 (1963).

Stevens, B., and Dubois, J. T., Simultaneous quenching of molecular fluorescence by oxygen and biacetyl in solution. *Trans. Faraday Soc.* **59**, 2813–2819 (1963).

Tsubomura, H., and Mulliken, R. S., Molecular complexes and their spectra. XII. Ultra-violet absorption spectra caused by the interaction of oxygen with organic molecules. *J. Am. Chem. Soc.* **82**, 5966–5974 (1960).

Ware, W. R., An experimental study of energy transfer between unlike molecules in solution. *J. Am. Chem. Soc.* **83**, 4374–4377 (1961).

Ware, W. R., Oxygen quenching of fluorescence in solution: An experimental study of the diffusion process. *J. Phys. Chem.* **66**, 455–458 (1962).

C. Papers Dealing with the Practical and Instrumental Problems of Performing Fluores-cence and Phosphorescence Analysis

Argauer, R. J., and White, C. E., Fluorescent compounds for calibration of excitation and emission units of spectrofluorometer. *Anal. Chem.* **36**, 368–371 (1964).

Bhaumik, M. L., Clark, G. L., Snell, M., and Ferder, L., Stroboscopic time-resolved spectroscopy. *Rev. Sci. Instr.* **36**, 37–40 (1965).

Fletcher, M. H., Study of multicomponent mixtures in solution with a vertical-axis trans-mission-type filter-fluorometer. *Anal. Chem.* **35**, 278–288 (1963).

Fletcher, M. H., A vertical-axis transmission-type filter-fluorometer for solutions. *Anal. Chem.* **35**, 288–292 (1963).

Keirs, R. J., Britt, R. D., Jr., and Wentworth, W. E., Phosphorimetry: A new method of analysis. *Anal. Chem.* **29**, 202–209 (1957).

Lewis, G. N., and Kasha, M., Phosphorescence and the triplet state. *J. Am. Chem. Soc.* **66**, 2100–2116 (1944).

Melhuish, W. H., Calibration of spectrofluorimeters for measuring corrected emission spectra. *J. Opt. Soc. Am.* **52**, 1256–1258 (1962).

O'Haver, T. C., and Winefordner, J. D., The influence of phosphoroscope design on the measured phosphorescence intensity in phosphorimetry. *Anal. Chem.* **38**, 602–607 (1966).

O'Haver, T. C., and Winefordner, J. D., Derivation of expression for integrated lumines-cence intensity when using pulsing techniques in luminescence spectrometry. *Anal. Chem.* **38**, 1258–1260 (1966).

Parker, C. A., and Hatchard, C. G., The possibilities of phosphorescence measurement in chemical analysis: Tests with a new instrument. *Analyst* **87**, 664–676 (1962).

Parker, C. A., and Rees, W. T., Correction of fluorescence spectra and measurement of fluorescence quantum efficiency. *Analyst* **85**, 587–600 (1960).

Potts, W. J., Jr., Purification of hydrocarbons for use as solvents in far ultraviolet spectro-scopy. *J. Chem. Phys.* **20**, 809–810 (1952).

Scott, D. R., and Allison, J. B., Solvent glasses for low temperature spectroscopic studies. *J. Phys. Chem.* **66**, 561–566 (1962).

White, C. E., Ho, M., and Weimer, Q., Methods for obtaining correction factors for fluorescence spectra as determined with the aminco-Bowman spectrophotofluorometer. *Anal. Chem.* **32**, 438–440 (1960).

D. Papers Dealing Primarily with the Luminescence Properties of a Variety of Organic Molecules

Adam, F. C., and Simpson, W. T., Electronic spectrum of 4, 4'-bis-dimethylamino fuchsone and related triphenylmethane dyes. *J. Mol. Spectry.* **3**, 363–380 (1959).

Argauer, R. J., and White, C. E., Effect of substituent groups on fluorescence of metal chelates. *Anal. Chem.* **36**, 2141–2144 (1964).

Azumi, T., and McGlynn, S. P., Polarization of the luminescence of phenanthrene. *J. Chem. Phys.* **37**, 2413–2420 (1962).

Azumi, T., and McGlynn, S. P., Delayed fluorescence of solid solutions of polyacenes. *J. Chem. Phys.* **38**, 2773–2774 (1963).

Churchich, J. E., The phosphorescence properties of pyridoxal 5-phosphate. *Biochim. Biophys. Acta* **79**, 643–646 (1964).

Drushel, H. V., and Sommers, A. L., Direct determination of inhibitors in polymers by luminescence techniques. *Anal. Chem.* **36**, 836–840 (1964).

Drushel, H. V., and Sommers, A. L., Combination of gas chromatography with fluorescence and phosphorescence in analysis of petroleum fractions. *Ann. Chem.* **38**, 10–19 (1966).

Drushel, H. V., and Sommers, A. L., Isolation and identification of nitrogen compounds in petroleum. *Anal. Chem.* **38**, 19–28 (1966).

Ferguson, J., Iredale, T., and Taylor, J. A., The phosphorescence spectra of naphthalene and some simple derivatives. *J. Chem. Soc.* pp. 3160–3165 (1954).

Foster, R., Hammick, D. L., Hood, G. M., and Sanders, A. C. D., Phosphorescence and fluorescence of some aromatic nitro-amines. *J. Chem. Soc.* pp. 4865–4868 (1956).

Freed, S., and Salmre, W., Phosphorescence spectra and analyses of some indole derivatives *Science* **128**, 1341 (1958).

Hirshberg, Y., Full luminescence at liquid-air temperature of methyl-1,2-benzanthracenes and methylbenzo [C] phenanthrenes. *Anal. Chem.* **28**, 1954–1957 (1956).

Levinson, G. S., Simpson, M. T., and Curtis, M., Electronic spectra of pyridocyanine dyes with assignments of transitions. *J. Am. Chem. Soc.* **79**, 4314–4320 (1957).

McGlynn, S. P., Neely, B. T., and Neely, C., Total luminescence of organic molecules of petrochemical interest. Part I. Naphthalene, phenanthrene and 1,2,4,5-tetramethyl-benzene. *Anal. Chim. Acta* **28**, 472–479 (1963).

Melhuish, W. H., and Hardwich, R., Lifetime of the triplet state of anthracene in lucite. *Trans. Faraday Soc.* **58**, 1908–1911 (1962).

Parker, C. A., and Rees, W. T., Fluorescence spectrometry. *Analyst* **87**, 83–111 (1962).

Sawicki, E., and Pfaff, J. D., Analysis for aromatic compounds on paper and thin-layer chromatograms by spectrophotophosphorimetry: Application to air pollution. *Anal. Chim. Acta* **32**, 521–534 (1965).

Sawicki, E., Stanley, T. W., Pfaff, J. D., and Elbert, W. C., Thin-layer chromatographic separation and analysis of polynuclear aza heterocyclic compounds. *Anal. Chim. Acta* **31**, 359–375 (1965).

von Forster, G., Studies of the phosphorescence of organic substances in crystalline form or in frozen solutions. *J. Chem. Phys.* **40**, 2059–2061 (1964).

Winefordner, J. D., and Latz, H. W., Phosphosphorimetry as a means of chemical analysis: The analysis of aspirin in blood serum and plasma. *Anal. Chem.* **35**, 1517–1522 (1963).

Winefordner, J. D., and Moye, H. A. The application of thin-layer chromatography and phosphorimetry for the rapid determination of nicotine, nornicotine, and anabasine in tobacco. *Anal. Chim. Acta* **32**, 278–286 (1965).

E. References Dealing with the Problems of Luminescence from Bio-Logic Systems. A Number of Very Useful Applications of Luminescence for Naturally Occurring Biologic Molecules Are Covered in this Section

Bersohn, R., and Isenberg, I., On the phosphorescence of DNA. *Biochim. Biophys. Res. Commun.* **13**, 205–208 (1963).

Bersohn, R., and Isenberg, I., Phosphorescence in nucleotides and nucleic acids. *J. Chem. Phys.* **40**, 3175–3180 (1964).

Churchich, J. E., Luminescence properties of muramidose and reoxidized muramidase. *Biochim. Biophys. Acta* **92**, 194–197 (1964).

Debye, P., and Edwards, J. O., A note on the phosphorescence of proteins. *Science* **116**, 143–144 (1952).

Douzou, P., and Francq, J. C., Luminescence de lingue durée de la sérum-albumine et de ses principaux constituants. *J. Chim. Phys.* **59**, 578–583 (1962).

Eisinger, J., Shulman, R. G., and Szymanski, B. M., Transition metal binding in DNA solutions. *J. Chem. Phys.* **36**, 1721–1729 (1962).

Freed, S., and Vise, M. H., On phosphorimetry as quantitative microanalysis with application to some substances of biochemical interest. *Anal. Biochem.* **5**, 338–344 (1963).

Freed, S., Turnbull, J. H., and Salmre, W., Proteins in solutions at low temperatures. *Nature* **181**, 1731–1732 (1958).

Grossweiner, L. I., Metastable states of photoexcited ovalbumin and constituents. *J. Chem. Phys.* **24**, 1255–1256 (1956).

Helene, C., Douzou, P., and Michelson, A. M., The phosphorescence of dinucleotides and the problem of energy transfer between the bases of nucleic acids. *Biochim. Biophys. Acta* **109**, 261–267 (1965).

Helene, C., Douzou, P., and Michelson, A. M., Energy transfer in dinucleotides. *Proc. Natl. Acad. Sci. U.S.* **55**, 376–381 (1966).

Hercules, D. M., and Rogers, L. B., Luminescence spectra of naphthols and naphthalenediols: Low temperature phenomena. *J. Phys. Chem.* **64**, 397–400 (1960).

Hirs, C. H. W., Stein, W. H., and Moore, S., The amino acid composition of ribonuclease. *J. Biol. Chem.* **211**, 941–950 (1954).

Hirs, C. H. W., Moore, S., and Stein, W. H., The sequence of the amino acid residues in performic acid-oxidized ribonuclease. *J. Biol. Chem.* **235**, 633–647 (1960).

Hollifield, H. C., and Winefordner, J. D. A phosphorimetric investigation of several representative alkaloids of the isoquinoline, morphone and indole groups. *Talanta* **12**, 860–863 (1965).

Hollifield, H. C., and Winefordner, J. D. A phosphorimetric investigation of several sulfinamide drugs: A rapid direct procedure for the determination of drug levels in pooled human serum with specific application to sulfadiazine, sulfamethazine, sulfamerazine and sulfacetamide. *Anal. Chim. Acta* **36**, 352–359 (1966).

McCarthy, W. J., and Winefordner, J. D. Phosphorimetric background of the ether extracts of blood and urine at various pH values. *Anal. Chim. Acta* **35**, 120–123 (1966).

Rahn, R. O., Longworth, J. W., Eisinger, J., and Shulman, R. G., Electron spin resonance and luminescence studies of excited states of nucleic acids. *Proc. Natl. Acad. Sci. U.S.* **51**, 1299–1303 (1964).

Steele, R. H., and Szent-Györgyi, A., On excitation of biological substances. *Proc. Natl. Acad. Sci. U.S.* **43**, 477–491 (1957).

Vladimirov, I. A., and Burshtein, E. A., Luminescence spectra of aromatic amino acids and proteins. *Biophysics* (*USSR*) (*English Transl.*) **5**, 445–453 (1960), *Biofizika* **5**, 385 (1960).

Vladimirov, I. A., and Litvin, F. F., Long-lived phosphorescence of aromatic amino-acids and proteins at low temperatures. *Biophysics* (*USSR*) (*English Transl.*) **5**, 151–159 (1960), *Biofizika* **5**, 127 (1960).

Winefordner, J. D., and Tin, M., The use of rigid ethanolic solutions for the phosphorimetric investigation of organic compounds of pharmacological interest. *Anal. Chim. Acta* **31**, 239–245 (1964).

Winefordner, J. D., and Tin, M., Phosphorimetric determination of procaine, phenobarbital, cocaine, and chlorpromazine in blood serum, and atropine in urine. *Anal. Chim. Acta* **32**, 64–72 (1965).

Yearger, E., and Augenstein, L., Absorption and emission spectra of psoralen and 8-methoxy-psoralen in powders and in solutions. *J. Invest. Dermatol.* **44**, 181–187 (1965).

Zander, M., Phosphorescence spectroscopy as an analytical method for aromatic and heterocyclic compounds. *Angew. Chem. Intern. Ed. Engl.* **4**, 930–938 (1965).

PRINCIPLES AND SOME APPLICATIONS
OF CYTOFLUOROMETRY

Fritz Ruch

DEPARTMENT OF GENERAL BOTANY, SWISS FEDERAL INSTITUTE OF TECHNOLOGY, ZÜRICH,
SWITZERLAND

I. Introduction

Quantitative fluorescence microscopy has become a valuable technique in cytochemistry, especially because of its sensitivity and simplicity. In this chapter cytofluorometric methods for the determination of the total amount of a particular substance in cells or cell organelles will be considered.

The principle of fluorometry in solutions will be mentioned briefly and compared with the well known absorption technique (Fig. 1). The latter is based on the Beer-Lambert Law: measurements of light intensities I and I_0 allow calculation of the concentration. In the case of fluorometry the solution is excited with ultraviolet or visible light and emits fluorescent light of longer wavelength. There is no simple relation between the intensity of excitation (I_{exc}) and the intensity of fluorescent light (I_{fl}). Therefore, only I_{fl} is measured and hence calculation of concentration is not possible. Fluorometry thus is used for comparative determinations; it allows comparison of solutions of different concentrations by measuring their fluorescent intensities. If absolute values of concentrations are of interest we have to compare with a standard of known concentration.

For quantitative work I_{fl} must be proportional to the concentration c. This condition is only fulfilled in a restricted range of c, where the absorptions of exciting and of emitted light are low (15).

Often in cytofluorometry, not concentrations but amounts of substances are determined, i.e., concentration times volume.

The high sensitivity of fluorometry is of special interest. In absorption photometry the lower limit is given by the accuracy in measuring small differences in light intensities. In fluorometry this limit is determined mainly by the brilliance of the light source and the sensitivity of the sensor (photo-multiplier) to small intensities of light. Using comparable equipment, fluorometry is much more sensitive (up to 100 times) than absorption photometry.

Next we will consider the influence of the microscope and the object on the measurements. First, photometry of a small particle will be discussed (Fig. 2). The absorbing particle is examined in bright field. The particle not only absorbs light, it also scatters light. For that reason the light distribution around the particle is heterogeneous. The greater part of the scattered light enters the microscope objective; a lesser part is lost, and gives rise to so-called nonspecific absorption. This effect depends on particle size, difference in refractive indices of object and medium, wavelength of light used, and numeric aperture of the objective. It is, therefore, variable and hardly possible to determine. Errors from nonspecific absorption are, in general, more serious in UV than in visible light (5). The size of the smallest particle which permits accurate absorption measurements is, according to Caspersson (5), about three times the wavelength.

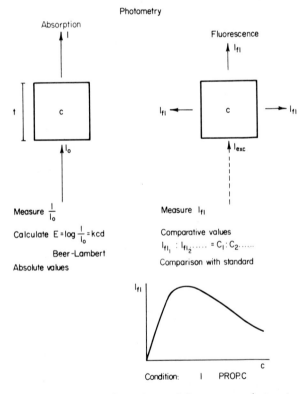

FIG. 1. Principle of absorption and fluorescence photometry.

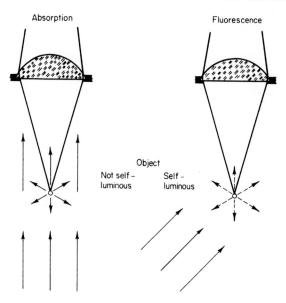

FIG. 2. Photometry of small particles in the microscope.

A small fluorescing particle behaves quite differently. For one thing it is observed in a dark field. The particle likewise scatters excitation light (not considered in Fig. 2), but this scattering is of no importance, since the excitation light is prevented from reaching the image plane by a barrier filter. The excited particle behaves like a self-luminous object; it emits fluorescent light of the same intensity in all directions [an anisotropic particle behaves differently, it produces difluorescence (27)]. A constant fraction of emitted light, independent of particle size, enters the microscope objective. The size of this fraction, given by the numeric aperture of the objective, has no influence on the relative fluorescent measurements (it must, of course, be the same for successive measurements). Theoretically there is no lower limit to the size of isolated particles for cytofluorometry. (This fact must not be confused with the resolving power of the microscope, which is similar in absorption and fluorescence.) In practice the limit is given by the quantum yield of the substance, the sensitivity of the light receiver, the intensity of the excitation light, stray light and fluorescence in the optic components, and insufficient spectral isolation of excitation and fluorescent light. For this reason it is not possible to give data of general validity.

In our experiments structures with dimensions of about 0.2 μ have been studied.

Finally we will discuss the determination of the total amount of a substance in a cell or a cell organelle (Fig. 3). These objects are usually inhomogeneous

In the case of absorption the scanning technique is most useful. Transmission is measured through a small scanning diaphragm; it is transformed to extinction and summed up over the cell area. For this procedure relatively complex equipment is necessary. An alternative is the two-wavelength method, for which the apparatus is simpler, but which is very time consuming. In some special cases the plug method is sufficient. [For extended discussion of these methods, see also the first volume of this treatise (26).]

In the case of fluorescence, determination of the total amount of substance is quite simple (28). If the intensity of the emitted light in each point of the object is proportional to the substance content, we can collect the light from all points and obtain the relative amount of the substance in the object. Such an integration of light is most easily accomplished by measuring the light intensity in the exit pupil of the objective of the microscope.

In the image plane of the microscope the measuring field is restricted by a diaphragm. Since the object lies in a dark field, the opening of the diaphragm can be slightly larger than the object size.

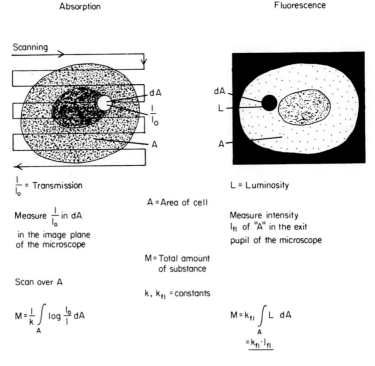

FIG. 3. Photometry of large inhomogeneous objects in the microscope. Determination of total amount of substance in cells.

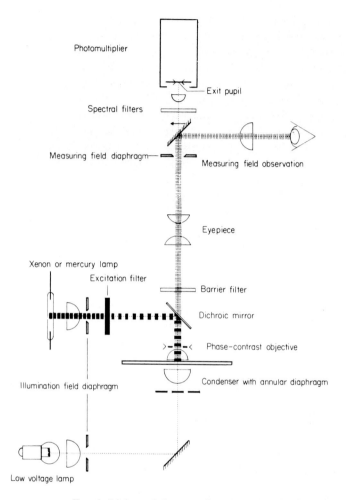

FIG. 4. Light path in a cytofluorometer.

II. Instrumentation

The apparatus for cytofluorometric determination of the amount of substance in cells consists of an ordinary fluorescence microscope combined with a photometer (*25, 26, 28, 33*). Only for some special purposes are more complex instruments necessary (*6, 24, 32, 36*). The light path in a simple routine cytofluorometer is represented in Fig. 4. Since very good time-stabilized light sources are now available, the one-beam system is preferred for such an instrument. Some high pressure mercury or xenon lamps (i.e.,

HBO 100 or XBO 150 from Osram) combined with a good magnetic stabilizer are suitable. To isolate the spectral range for excitation the glass filters as used in modern fluorescence microscopes are sufficient (a monochromator is only necessary for excitation in the UV range below 360 nm or for measurements of excitation spectra of narrow spectral bandwidth). It is important to use exact Köhler illumination for uniform light distribution in the area of the object measured. Furthermore, the illumination field diaphragm should restrict the illuminated area in the object plane to the size of the measuring area. This will protect objects in the neighborhood of the measured object from radiation; also, stray light in the preparation and in the objective is reduced (improvement of contrast).

Most suitable is an illuminating system for excitation in incident light using a vertical illuminator with a dichromatic reflector [with coatings of high reflection for the excitation light and high transmission for the longer wavelengths of fluorescent light (16)]. Such a vertical illuminator is most efficient with objectives with high numeric aperture. The main advantage of this illuminating system for cytofluorometry is that the focusing of a condenser is avoided. Furthermore, the contrast is greater than with a bright-field condenser for transmitted light, and it is easily possible to use any phase-contrast condenser for observations in visible light. This is very important since most of the fluorochromes are not stable in excitation light; searching and focusing in the microscope should not be carried out in fluorescent light.

Barrier filters above the objectives absorb excitation light scattered by the object and reflected by the lenses of the objective. A measuring field diaphragm is placed in the image plane of the objective. This diaphragm should be interchangeable or variable, i.e., depending on the size and shape of the object, iris diaphragms, variable rectangular diaphragms or disks with fixed openings are preferable.

A movable mirror or a fixed semireflecting mirror in the beam allows simultaneous observation of the image and the measuring field through the eyepiece of a telescope. Light measurement is usually carried out with a highly sensitive photomultiplier tube. The exit pupil of the microscope is projected on the cathode of this tube by means of a lens system. Spectral filters may be inserted in front of the multiplier. Such filters are required if the fluorescence in a restricted spectral range is of interest, or if a fluorescence spectrum is to be determined. An interference wedge filter is especially suitable for this purpose. The photocurrent of the multiplier can be measured by a galvanometer, a paper recorder, or a current integrator. Instruments which fulfill the described requirements have been commercially available for some time from Leitz, Reichert, and Zeiss.

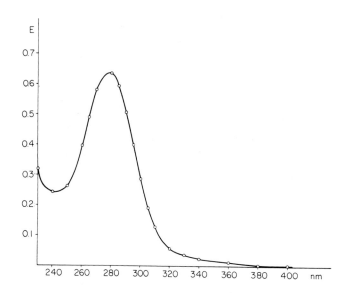

FIG. 5. Spectral extinction (*E*) of BAO in solution (0.003%, 10 mm). From Ruch (*26*).

III. Linearity

Cytofluorometry depends on a proportionality between the amount of substance and intensity of fluorescent light. This condition is only fulfilled if the absorption of the excitation and of the fluorescent light is negligible (*15*). Of course, a certain absorption of excitation light is necessary to produce fluorescence and therefore, deviations from linearity, especially in thick specimens, are always present, but should not exceed a few percent, depending on the accuracy demanded. Each fluorescing substance in the concentrations and thicknesses used must therefore be tested for such deviations. The spectral absorption curve of the substance gives preliminary evidence of its usefulness for quantitative work (Fig. 5). Series of measurements in the fluorescence microscope at different wavelengths of *excitation* are then necessary. The wavelength range with the highest fluorescence efficiency is first used for the test of linearity. If deviation from linearity is too high, the test has to be repeated with other wavelengths of excitation.

Fluorescent spectra produced by different excitation wavelengths should be recorded in order to collect information on the selection of suitable filters in the photometer and of multipliers with a corresponding spectral sensitivity. Fluorescence spectra often show a broad and flat intensity distribution, in

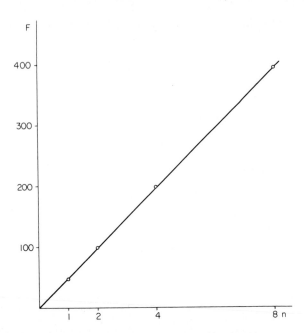

FIG. 6. Liver nuclei and sperms of the rat, stained with berberine sulfate after ribonuclease treatment. Dependence of fluorescence intensity, F, on polyploidy, n. Each point represents a mean value in arbitrary units for 20 nuclei. From Bosshard (4).

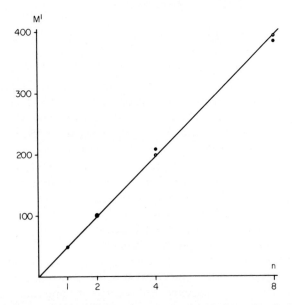

FIG. 7. Liver nuclei and sperms of the rat, stained with Auramine O. Dependence of fluorescence (○) and UV absorption (●) at 265 nm on the polyploidy n. Each point represents a mean value in arbitrary units for 20 nuclei. From Ruch and Bosshard (28).

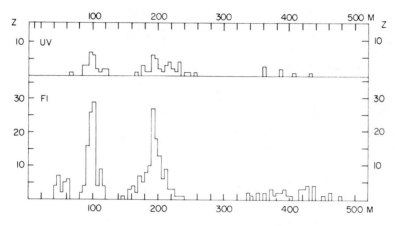

FIG. 8. Liver nuclei and sperms of the rat. Amounts of DNA (*M*, in arbitrary units), measured by UV absorption (UV) and microfluorometry with Auramine O (Fl). *Z* is number of nuclei measured. From Ruch and Bosshard (*28*).

which case no filter is inserted in the measuring beam, in order to achieve maximum sensitivity. If, on the other hand, the fluorescence spectrum is more complex, spectral filters may be necessary for the measurement of individual peaks. It is clear, that the recorded fluorescence spectra only demonstrate intensity distributions relative to the spectral characteristic of the instrument. Spectra determined with different types of instruments may therefore not be alike. For cytofluorometric work, as described in this paper, it is not necessary to know the actual energy distribution of the spectrum.

An important factor in the practical test for linearity is the use of an adequate object. In the first experiment, model substances should be used. Dried films of the stained substance in different thicknesses are most suitable for this purpose (*22*). The thickness of these models can be determined by interference microscopy. If biologic preparations are to be tested, microtome sections of different thickness may be used. For certain substances, like DNA and certain proteins, isolated nuclei of different degrees of polyploidy (*13*) are suitable (Figs. 6–8).

IV. Standard

A standard fixed in a magnetic disk, as used by Zeiss for reflection measurements, is very practical (Fig. 9). This disk can quickly be attached to the microscope objective. Since the fluorescing plate is placed in the object plane of the microscope it gives correct values for all types of objectives used, provided the diaphragms in the image plane and in the illuminating system used for the standard readings are always the same.

V. Thickness of the Object and Numerical Aperture of the Objective

In absorption photometry the thickness of small or inhomogeneous objects should not greatly exceed the depth of focus. In cytofluorometry, however, the thickness of the object can be a multiple of the depth of the focus. Here, the diaphragm at the image plane should be large enough to receive the light from blurred contours which are out of focus. The upper limit permissible for the thickness of the object is dependent on the numeric aperture of the objective used. Model experiments with the vertical illuminator

Fig. 9. Fluorescence standard (i.e., uranium glass plate) mounted in a magnetic holder attachable to the microscope objective.

showed that the error is about 5% for an object thickness of 10 μ with a numeric aperture of 1.0 and 6 μ with a numeric aperture of 1.3. For most cytochemical studies a numeric aperture of about 1.0 is suitable. Higher numeric apertures are only advantageous for the measuring of very small structures. Objectives with numeric apertures below 1.0 are sometimes necessary for measurements on thick preparations at low magnification. However, objectives with a numeric aperture below 0.5 may show an insufficient image brightness when used with a vertical illuminator. Excitation with transmitted light is preferable in such cases. In order to achieve equal light emission in all directions, the refractive index of the embedding medium should be similar to that of the object.

VI. Photodecomposition

Most fluorochromes show fading under excitation. This effect may be quite different in various types of dyes, and is moreover dependent on the kind of object, intensity of excitation, age of preparation, and type of embedding medium (*23, 25*).

In general, it is advisable to keep the time and intensity of irradiation to a minimum. For this reason searching and focusing should be carried out in green light using phase contrast. If highly sensitive multipliers and recording instruments are available, the intensity of the excitation light should be reduced as far as possible with neutral filters. Nevertheless, especially if it is very fast at the beginning of irradiation, fading can render measuring somewhat difficult. In this case maximum fluorescence intensity is only measured if the photocurrent is recorded within a fraction of a second after starting excitation. A simpler method commonly used in our laboratory, is the measurement after a fixed time of excitation (i.e., 5 sec). The fading of fluorescence is then less than at the beginning of excitation and can be recorded with an ordinary galvanometer. An alternative method, with the advantage of very high sensitivity, consists in an integration of the photocurrent within a fixed time of excitation.

VII. Nonspecific Fluorescence

When working with staining reactions specific for a compound in the cell, a certain amount of nonspecific fluorescence is usually present. This error may come from different sources. It can be produced by the staining reaction itself if the specificity is not 100%. This property has to be studied quantitatively for each dye to allow a correction.

Interference with other fluorescing substances present may occur. Such substances are, e.g., pigments, proteins, and nucleic acids. A certain primary fluorescence of proteins and nucleic acids is always present (*31*). Their primary fluorescence spectra are very broad with decreasing intensities toward the red. The intensity is highest with UV excitation and decreases with excitation in visible light of increasing wavelengths. The influence of this protein and nucleic acid fluorescence can, therefore, be reduced by using fluorochromes with red fluorescence and blue or green excitation. A correction for interfering primary fluorescence is usually possible if one measures unstained preparations and subtracts the intensity obtained from that of stained preparations.

When studying weakly fluorescing objects the fluorescence of the slide, the coverslip, the embedding medium, the immersion oil, and the objective lenses must be considered. This is shown by the fact that the background in the image plane exhibits an intensity which may constitute several percent of the intensity of the object. In such cases the background intensity must be subtracted from the object intensity. This is very simply done by adjusting the photometer reading for the field to zero. Since the background intensity decreases the sensibility and accuracy of cytofluorometry it is important to keep this effect as low as possible. For this purpose only fluorescence-free embedding media (i.e., Fluormount from Edward Gurr, London, or pure glycerine) and fluorescence-free immersion oil should be used. For critical determinations slides and coverslips of quartz are preferable to glass, which is seldom fluorescence-free. It is even important to choose an objective with low fluorescence of lenses and cements. Complex objectives with a high degree of correction are unfortunately often not suitable for cytofluorometry.

VIII. Some Applications of Cytofluorometry

Several applications have been published during recent years (*3, 7, 10, 12, 14, 17–20, 31, 35*). In this article, only a few techniques, developed in our laboratory, for studies of fixed cells will be considered (*4, 11, 21, 22, 25*).

A. NUCLEIC ACIDS

DNA and RNA can be stained with basic fluorochromes. Berberine sulfate, 0.1% in phosphate buffer pH 6.3 or in 50–95% ethanol (*4*), was found suitable for this purpose. The preparation is placed in the staining solution for 20 min, covered with a coverslip, and examined in the staining solution. The range 360–400 nm is used for excitation. Formalin (4%) or ethanol–acetic acid (3:1) is suitable for fixation. Berberine sulfate shows relatively little photodecomposition. If DNA is to be determined, RNA must be removed (i.e., 1 mg of ribonuclease in 2 ml of 30% ethanol, 1 hr at 50°C). A test for linearity between fluorescence intensity and DNA content is shown in Fig. 6. It is also possible to determine RNA: the total nucleic acid content is first determined and then the DNA content after RNA extraction measured and subtracted from the total nucleic acid content. This technique, however, is not very accurate.

In general, for DNA a modified Feulgen reaction is preferable, due to its specificity. The pararosaniline in the ordinary reagent must be replaced by a fluorochrome (4, 9, 17, 25). According to Bosshard (4) Auramine O is suitable for quantitative purposes (excitation 360–400 nm). Figures 7 and 8 show DNA determinations in sperm and isolated liver nuclei of rat. The mean values of the DNA contents of the four types of nuclei with increasing polyploidy values lie in a straight line (Fig. 7). Measurements with UV absorption are shown for comparison and are seen to correspond very well with the fluorescence values. Sufficient linearity (within 4%) exists up to a DNA content of a 16-ploid liver nucleus.

Auramine O is unfortunately not stable under UV irradiation and was, therefore, later replaced by 2,5-bis[4′-aminophenyl-(1′)]-1,3,4-oxadizole (abbreviation BAO) from Ciba (Basle, Switzerland),* which fluorochrome shows less fading (Fig. 10) (25). The range 360–400 nm is used for excitation. Figure 5 shows the spectral extinction and Fig. 11 the relative spectral emission of BAO. Since the UV absorption, even at 260 nm, is relatively low, comparative measurements of DNA in BAO stained preparations are possible. BAO increases the extinction readings for DNA by 10%.

Staining technique for BAO

(1) Hydrolyze in HCl (see below for acid concentration and time of hydrolysis).

(I)

(2) Rince in water (5 min).

(3) Stain in freshly prepared staining solution (2 hr): Mix 10 ml Ciba bis(aminophenyl)oxadizole, 0.01% in distilled water (or 10 ml Auramine O, 0.2%) with 1 ml of 1 N HCl and 0.5 ml of NaHSO$_3$, 10%; shake and filter.

(4) Wash in sulfite water (3 times 2 min), (sulfite water: 180 ml distilled water; 10 ml 1 N HCl; 10 ml NaHSO$_3$, 10%).

(5) Wash in water (10 min).

* Obtainable from Fluka A. G., Buchs, Switzerland.

(6) Dehydrate in a graded series of ethanol. Mount from xylene in Fluormount (or in another nonfluorescing medium).

Fixation is carried out in 4% formalin ($\frac{1}{4}$–6 hr) or ethanol–acetic acid (3 : 1,
$\frac{1}{4}$–3 hr), wash in water or 70% ethanol, respectively (10 min).

For quantitative determination, it is important that the optimum hydrolysis conditions be maintained. The use of 1 N HCl at 60°C is common,
but the use of 6 N HCl at 20°C is often more practical and, with somewhat
shorter hydrolysis time, gives the same result.

The optimum time of hydrolysis must be observed exactly. It is best if
this is determined for each type of object and fixation. A test of this kind
is shown in Fig. 12. The fluorescence intensities of a series of preparations
with increasing hydrolysis time (between 5 and 20 min) are measured, and
thus the time for maximum fluorescence intensity is found. Our experiments
showed that, for most material fixed with formalin or ethanol–acetic acid,
the optimum time is 8 min if 6 N HCl at 20°C is used and 12 min if 1 N
HCl at 60°C is used.

The time–intensity curves often show two peaks (Fig. 12). This phenomenon is probably due to DNA complexes with different hydrolysis characteristics. It is not caused by the fluorochrome, because it is also present if
normal Feulgen staining is carried out with fuchsin (1, 30). If objects with
similar types of curves are to be compared, the peak corresponding to the
shorter hydrolysis time can be chosen. Caution is required if objects with
different types of curves are to be compared. In this case, a supplementary
method, i.e., a basic fluorochrome such as berberine sulfate or UV absorption, should be used for checking, and possible correcting purposes.

Bull sperm provide a suitable object for absolute DNA determinations.
They must be stained simultaneously with the object under investigation.
Since the DNA content of the spermatozoa may vary somewhat from bull
to bull, only sperma from one bull should be used. If it is necessary to use

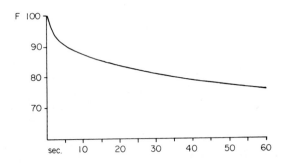

FIG. 10. Bull sperm stained with BAO. Decrease in fluorescence intensity F caused by
fading, due to time of UV excitation.

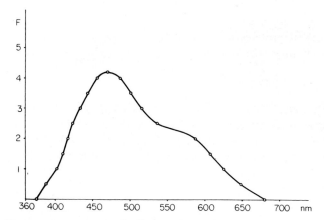

FIG. 11. Fluorescence spectrum of a liver nucleus stained with BAO. Excitation at 366 nm. Kodak Panatomic X Film. From Ruch (24).

further material from another bull, a comparative DNA measurement of the two preparations should be made. The DNA of bull sperm amounts from 2.9 to 3.4×10^{-12} gm (29, 34). If more accurate values are needed for the determination of absolute DNA amounts, the DNA content of the standard preparation must be measured using a biochemical method (34).

We use the following method for preparation of the bull sperm specimens:

(1) Cut the epididymis and shake in 1% citric acid.
(2) Filter through Kleenex.
(3) Centrifuge.
(4) Wash quickly in water.
(5) Centrifuge.
(6) Fix in 4% formalin, or ethanol–acetic acid (3:1) 1 hr.
(7) Centrifuge.
(8) Wash quickly in water, or ethanol 70%, respectively.
(9) Centrifuge.
(10) Remove water using ethanol, store in ethanol 70% or dry in desiccator.

Recently, Ploem (16) described the use of ordinary Feulgen-stained preparations in fluorescent microscopy. If such preparations are excited with green light, they emit a red fluorescence. The absorption of green light in Feulgen-stained cell nuclei, however, is relatively high, therefore, there is no proportionality between DNA content and fluorescent light. As Fig. 13 demonstrates, measurements are only possible with small amounts of DNA. We found the method very useful for cytofluorometric studies of DNA in cytoplasmic organelles, such as chloroplasts and mitochondria. With the BAO technique described earlier it is difficult to determine small amounts of DNA in these organelles, since they exhibit a relatively high primary fluorescence of proteins and pigments. As already discussed, this influence of primary fluorescence is lowest for red fluorescence and green excitation.

B. PROTEINS

Cytofluorometric determination of proteins in cells may be carried out according to their positive or negative charge or according to their amino acid composition. The following dyes have been tested in our laboratory for specificity and linearity with model films and nuclei of known protein composition.

1. Brilliant Sulfaflavine

Brilliant sulfaflavine is an acid fluorochrome and therefore stains proteins according to their net positive charge (free basic groups).

Sulfaflavine can be used in two different ways (*11*). At pH 2.8 it allows the determination of total protein, comparable to the naphthol yellow staining, published by Deitch (*8*). At pH 8.1 basic proteins (histones) are stained,

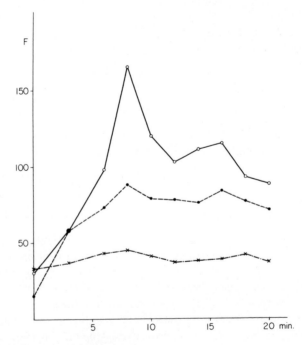

FIG. 12. Changes of fluorescence intensity (*F*) with hydrolysis time in BAO stained nuclei of ascites (○), lymphocytes (●), and sperm of hamsters (×).

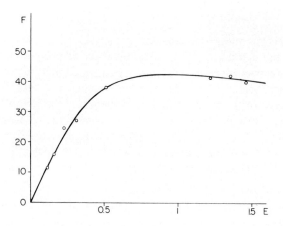

FIG. 13. Relation between fluorescence intensity (F) and extinction (E) of green light (550 nm) in nuclei of *Tradescantia virginica*. Feulgen reaction (pararosaniline).

similar to the fast green reaction according to Alfert and Geschwind (2). Removal of DNA is necessary in this case.

Procedure for total protein

(1) Rinse 2 min in citric acid–phosphate buffer pH 2.8 (McIlvaine buffer, diluted 10×).
(2) Stain 30 min in a 0.1% brilliant sulfaflavine solution, in buffer pH 2.8.
(3) Differentiate 5 min in buffer pH 2.8.
(4) Dehydrate in ethanol 95%, 2 × 100%, and 2 × xylene (2 min each).
(5) Mount from xylene in Fluormount.

Ethanol–acetic acid is used for fixation. Excitation at 360–400 nm.

Procedure for histones

(1) Hydrolyze in 5% trichloracetic acid (TCA) either 15 min at 90°C or 3 hr at 60°C.
(2) Rinse 3 × 10 min in ethanol 70%.
(3) Rinse in water 5 min.
(4) Rinse in Tris buffer pH 8.1, 2 min (Gomori buffer, diluted 10×).
(5) Stain 30 min in a solution of 0.1% brilliant sulfaflavine in buffer pH 8.1.
(6) Differentiate in buffer pH 8.1 for 5 min.
(7) Dehydrate in ethanol 95% and 100%.
(8) Mount from xylene in Fluormount.

Fixation should be carried out in formalin (4%). Excitation at 360–400 nm.

2. Ninhydrin

Rosselet (21) has adapted a biochemical method using ninhydrin in alkaline solution for the cytofluorometric determination of protein-bound arginine. In the modified procedure the reaction is performed in a mixture of glycerol, methanol, and sodium hydroxide. The fluorescence originates from an intermediate product of the transformation of the ninhydrin anion to o-carboxymandelic acid through the formation of an addition compound with the guanidine group of arginine. The reaction has been tested quantitatively on protein solutions, protein fibers, and isolated cell nuclei. In very closely packed objects, such as sperm, only a part of the arginine is stained.

Procedure for arginine

(1) Add a freshly made mixture of reagent A and reagent B, 1:1, to the preparation and store for 5–24 hr in the dark. (Reagent A: 0.5% ninhydrin in pure methanol. Reagent B: 1 N NaOH plus pure glycerine 1:1.) Since the fluorescence increases slowly between 5 and 24 hr a constant reaction time is important and a standard object should be used on each slide (i.e., isolated calf thymus nuclei).

(2) Add a coverslip to the preparation and take the measurements. For excitation the range 360–400 nm is used.

3. Dansyl Chloride

According to Rosselet and Ruch (22) dansyl chloride may be used for the determination of lysine. By means of model experiments on protein films and isolated rat liver nuclei it could be shown that the reaction, if carried out in ethanol, is specific and linear for amino groups. The lysine content may be measured in absolute values by comparison with polylysine films. In the case of sperm cells only 50% of the lysine reacts.

Procedure for lysine

(1) Stain in a 0.1% solution of dansyl chloride (puriss.) in 95% ethanol saturated with $NaHCO_3$ (pure) for 4 hr. The maximum fluorescence is reached after 3 hr and remains practically constant for a further 10 hr.

(2) Rinse in ethanol 95% for 30 min.

(3) Mount from absolute ethanol and xylene in Fluormount.

Use ethanol or ethanol–acetic acid for fixation. Excitation at 360–400 nm.

REFERENCES

1. Agrell, I., and Berggvist, H. A., Cytochemical evidence for varied DNA complexes in the nuclei of undifferentiated cells. J. Cell. Biol. 15, 604–606 (1962).

2. Alfert, M., and Geschwind, I. I., A selective staining method for the basic proteins of cell nuclei. *Proc. Natl. Acad. Sci. U.S.* **39**, 991–999 (1953).
3. Bahr, G. F., and Wied, G. L., Cytochemical determination of DNA and basic protein in bull spermatozoa. Ultraviolet spectrophotometry, cytophotometry, and microfluorometry. *Acta Cytol.* **10**, 393–412 (1966).
4. Bosshard, U., Fluoreszenzmikroskopische Messungen des DNS Gehaltes von Zellkernen. *Z. Wiss. Mikroskopie* **65**, 391–408 (1964).
5. Caspersson, T., "Cell Growth and Cell Function." Norton, New York, 1950.
6. Caspersson, T., Lomakka, G., and Rigler, R., Jr., Registrierender Fluoreszenzmikrospektrograph zur Bestimmung der Primär- und Sekundärfluoreszenz verschiedener Zellsubstanzen. *Acta Histochem.* Suppl. **6**, 123–134 (1965).
7. Caspersson, T., Hillarp, N. Å., and Ritzén, M., Fluorescence microspectrophotometry of cellular catecholamines and 5-hydroxytryptamine. *Exptl. Cell Res.* **42**, 415–428 (1966).
8. Deitch, A. D., Microspectrophotometric study of the binding of the anionic dye, naphtolyellow S by the tissue sections and by purified proteins. *Lab. Invest.* **4**, 324–351 (1955).
9. Kasten, F. H., Schiff-type reagents in cytochemistry. I. Theoretical and practical considerations. *Histochemie* **1**, 466–509 (1959).
10. Killander, D., and Rigler, R., Jr., Initial changes of deoxyribonucleoprotein and synthesis of nucleic acid phytohemagglutinine stimulated human leucocytes *in vitro*. *Exptl. Cell Res.* **39**, 701–704 (1965).
11. Leemann, U., and Ruch, F., Unpublished observations (1969).
12. Leemann, U., Ruch, F., and Sträuli, P., Characterization of the nuclei of Ehrlichascites carcinoma by means of quantitative cytochemistry. *Acta Cytol.* **12**, 381–394 (1968).
13. Leuchtenberger, C., Vendrely, R., and Vendrely, C., A comparison of the content of DNA in isolated animal nuclei by cytochemical and chemical methods. *Proc. Natl. Acad. Sci. U.S.* **37**, 33–38 (1951).
14. Loeser, C. N., and West, S. S., Cytochemical studies and quantitative television fluorescence and absorption spectroscopy. *Ann. N.Y. Acad. Sci.* **97**, 329–526 (1962).
15. Perrin, F., Loi de décroissance du pouvoir fluorescent de la concentration. *Compt. Rend.* **178**, 1978–1980 (1924).
16. Ploem, J. S., The use of a vertical illuminator with interchangeable dichroic mirrors for fluorescence microscopy with incident light. *Z. Wiss. Mikroskopie* **68**, 129–142 (1967).
17. Prenna, G., Qualitative and quantitative application of fluorescent Schiff-type reagents. *Mikroskopie* **23**, 150–154 (1968).
18. Rigler, R., Microfluorometric characterization of intracellular nucleic acids and nucleoproteins by acridine orange. *Acta Physiol. Scand. Suppl.* **267**, 1–122 (1966).
19. Ritzén, M., Quantitative fluorescence microspectrophotometry of catecholamine-formaldehyde products. *Exptl. Cell Res.* **44**, 505–520 (1966).
20. Ritzén, M., Quantitative fluorescence microspectrophotometry of 5-hydroxytryptamine-formaldehyde products in models and in mast cells. *Exptl. Cell Res.* **45**, 178–194 (1966).
21. Rosselet, A., Mikrofluorometrische Argininbestimmung. *Z. Wiss. Mikroskopie* **68**, 22–41 (1967).
22. Rosselet, A., and Ruch, F., Cytofluorometric determination of lysine with dansyl chloride. *J. Histochem. Cytochem.* **16**, 459–466 (1968).
23. Rost, F. W. D., and Pearse, A. G. E., Quantitative studies of fluorescence fading in tissue sections. *Proc. 3rd Intern. Cong. Histochem. Cytochem. New York, 1968*, pp. 226–227. Springer, Berlin, *1968*.

24. Ruch, F., Ultraviolett-Mikrospektrographie. *Leitz Mitt. Wiss. Technol.* **1**, 250–255 (1961).
25. Ruch, F., Fluoreszenzphotometrie. *Acta Histochem.* Suppl. **6**, 117–121 (1964).
26. Ruch, F., Determination of DNA content by microfluorometry. *In* "Introduction to Quantitative Cytochemistry" (G. L. Wied, ed.), pp. 281–294. Academic Press, New York, 1966.
27. Ruch, F., Dichroism and difluorescence. *In* "Introduction to Quantitative Cytochemistry" (G. L. Wied, ed.), pp. 549–555. Academic Press, New York, 1966.
28. Ruch, F., and Bosshard, U., Photometrische Bestimmung von Stoffmengen im Fluoreszenzmikroskop. *Z. Wiss. Mikroskopie* **65**, 335–341 (1963).
29. Salisburg, G. W., and Vandemark, N. L., "Physiology of Reproduction and Artificial Insemination of Cattle." Freeman, San Francisco, California, 1961.
30. Sandritter, W., Bosselmann, K., Rakow, L., and Jobst, K., Untersuchungen zur Feulgenreaktion. Die Lagnzeithydrolyse bei verschiedenen Zelltypen. *Biochim. Biophys. Acta* **91**, 645–647 (1964).
31. Saurer, W., Eigenfluoreszenz von Proteinen und Nukleinsäuren im Zellkern. Doctoral Dissertation No. 3834. Swiss Federal Institute of Technology, Zurich (1966).
32. Schmidt-Weinmar, H. G., Quantitative Auswertung von Fluoreszenz-Mikrophotographien mit dem Zeiss Integrations-Photometer nach Zeitler und Bahr. *Z. Wiss. Mikroskopie* **69**, 80–93 (1968).
33. Thaer, A. A., Instrumentation for microfluorometry. *In* "Introduction to Quantitative Cytochemistry" (G. L. Wied, ed.), pp. 409–426. Academic Press, New York, 1966.
34. Vendrely, R., and Vendrely, C., La teneur du noyau cellulaire en acide desoxyribonucléique à travers les organes, les individus et les espèces animales. *Experientia* **4**, 434–436 (1948).
35. West, S. S., Fluorescence microspectroscopy of mouse leucocytes superavitally stained with acridine orange. *Acta Histochem.* Suppl. **6**, 135–152 (1965).
36. West, S. S., Loeser, C. N., and Schoenberg, M. D., Television spectroscopy of biological fluorescence. *IRE (Inst. Radio Engrs.), Trans. Med. Electron.* **7**, 138–142 (1960).

OPTICAL ROTATORY DISPERSION AND THE MICROSCOPE*

Seymour S. West

UNIVERSITY OF ALABAMA IN BIRMINGHAM MEDICAL CENTER, BIRMINGHAM, ALABAMA

Recent development of a new single-beam method for measuring rotation of the plane of polarization of linearly polarized light (*34*) has provided a means for greatly increasing the sensitivity of the polarizing microscope. The fundamental limitations on the sensitivity of the conventional polarizing microscope have been analyzed by Inoué (*15*) and Inoué and Hyde (*17*). These authors showed that the depolarized light, which is the cause of the limited sensitivity, is a constant for a given set of optical conditions. This being so, West (*32*) suggested that the new single-beam method of measuring optical rotation could circumvent the limitations of the microscope and permit its use as a spectropolarimeter, i.e., an instrument that measures optical rotation as a function of wavelength. Such an instrument has been constructed. Optical rotatory dispersion (ORD) spectra from unfixed cells, both stained and unstained, have been obtained. The new instrument can be termed a *microspectropolarimeter*. It should be of interest in quantitative cytochemistry as a biophysical method for investigating the chemistry of the cell with localization limited only by the resolution of the microscope. Operation is possible over a wide spectral range limited only by the spectral emission of the light source, the transmission or reflection properties of the optical elements, the spectral transmission of the optical path, and the spectral sensitivity of the photoelectric detector. Sensitivity to small quantities of substance is approximately the same as that of fluorescence microspectrophotometry.

In the visible region of the spectrum the microspectropolarimeter makes possible cytochemical studies on particular intracellular constituents in intact cells. This can be greatly aided by supravital staining with appropriate dyes.

* This work was supported, in part, by U.S. Public Health Service Grant DE-2670 and Jet Propulsion Laboratory (NASA) Contract 951925.

Direct correlation between the ORD spectra from cells and those obtained from solution studies is possible. Where the dyes used are also fluorochromes, additional insight into the dye–cell interaction can be provided by fluorescence microspectrophotometry of single unfixed cells. Corrected fluorescence spectra produced by supravitally stained cells may also be compared with fluorescence spectra obtained from solution studies. ORD spectra and fluorescence spectra are two independent determinations on the same dye–substrate complex. Possibilities for additional independent determinations such as phosphorescence spectra, decay time, and response to alterations in environmental conditions also exist. All of these studies can be performed on unfixed cells, and can result in quantitative data directly comparable with similar data from solution studies. Each of these additional determinations not only adds to the certainty of identification, but also provides further insight into the physical chemistry and the biologic role of a particular intracellular substance. ORD measurements on cells can serve to identify and localize a particular intracellular constituent, provide some information on its molecular conformation, and study the interaction of intracellular macromolecules.

Before proceeding further it may be well to define a few terms. A number of literature sources which are more comprehensive are also recommended. These include Lowry (20), Partington (26), Djerassi (12), Velluz et al. (30), Mislow (22). Recent review articles by Beychok (9, 10) and Eyring et al. (13) will also be found very useful.

Linearly polarized light is composed of two circularly polarized components equal in amplitude but rotating in opposite directions (Fig. 1). In the figure I_L and I_R represent the left- and right-circularly polarized components,

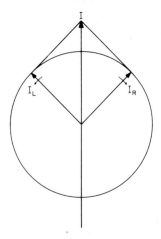

FIG. 1. Circularly polarized components of linearly polarized light.

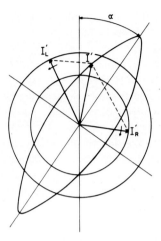

FIG. 2. Effect of circular birefringence and circular dichroism on linearly polarized light.

respectively. The direction of propagation of the light is out of the plane of the paper and orthogonal to it. The vector sum of the two circularly polarized components describes a plane (plane of polarization) along the direction of propagation of the light. The vector **I** in Fig. 1 is the trace of the plane of polarization in the plane of the paper and indicates its orientation. Conventionally, the directions of rotation are assigned with the observer looking toward the light source. Substances which are *optically active* rotate the plane of polarization of transmitted linearly polarized light. This *optical activity* is a physical optical property of dissymmetric substances which results from their ability to refract and absorb left- and right-circularly polarized light to different extents. There are two indices of refraction, one for left-circularly polarized light and one for right-circularly polarized light. Hence the term *circular birefringence* to describe this behavior. In traversing an optically active medium one of the circularly polarized components is retarded with respect to the other. The two circularly polarized components combine to form plane-polarized light once again upon leaving the optically active substance, but the phase shift that one component has experienced with respect to the other results in a rotation of the plane of polarization. The acute angle through which the plane of polarization has been rotated is the *optical rotation* produced by the optically active substance. It is the purpose of all polarimeters to measure this angle.

Optical rotation varies as a function of wavelength. The dependence of optical rotation on wavelength is termed *optical rotatory dispersion* (ORD).

Similarly, an optically active substance has two extinction coefficients, one for left-circularly polarized light and the other for right-circularly polarized light. These extinction coefficients are appreciably different from each other only in the neighborhood of an absorption band. The unequal absorption of left- and right-circularly polarized light is called *circular dichroism* (CD).

In the neighborhood of an absorption band, linearly polarized light passing through an optically active substance emerges elliptically polarized. Figure 2 illustrates this situation. The plane of polarization has been rotated through an angle α by the circular birefringence of the substance. The left- and right-circularly polarized components, I_L' and I_R' respectively, are of unequal amplitude due to the circular dichroism. The vector I', the resultant of I_L' and I_R', describes an ellipse. Optical rotation and circular dichroism are not dependent upon the orientation of the specimen with respect to the plane of polarization, provided the optical pathlength is held constant.

Though not treated here, mention should nevertheless be made of *linear birefringence* and *linear dichroism*. Both of these phenomena are dependent on the orientation of the specimen with respect to the plane of polarization of impinging linearly polarized light. Substances which are linearly birefringent divide the impinging linearly polarized light into two beams which are linearly polarized at right angles to each other. There is a different index of refraction for each of these beams so that one is delayed with respect to the other in traversing the specimen. Upon emerging the two beams recombine producing, in general, elliptically polarized light. For a given thickness of material the ellipticity of the emerging polarized light depends upon the orientation of the specimen with respect to the plane of polarization of the impinging light. Linearly polarized and circularly polarized emerging light are simply special cases of elliptical polarization. Linear dichroism is the variation of absorption of linearly polarized light as a function of the orientation of the specimen. The familiar Polaroid filter polarizer is an example of linear dichroism.

The polarizing microscope, in the past, has been devoted exclusively to linear birefringence and linear dichroism. The biologist has made use of these phenomena to gain information on the submicroscopic structure of anisotropic biologic objects (*1–8, 14, 16, 18, 23, 24, 27–29*). Although the microspectropolarimeter is sensitive to both circular and linear birefringence these two effects can be separated. The microspectropolarimeter is insensitive to ellipticity and senses only the rotation of the major axis. In the case of circularly polarized light no optical rotation is detected.

A biologic macromolecule may participate in a number of optical phenomena. Many of these phenomena are interrelated, can be measured in cytological preparations, and constitute independent experiments. These relationships are shown in Fig. 3 for a model substance. The absorption spectrum has a peak at a wavelength λ_0. There may be a fluorescence band associated with this absorption band as shown on the right. The ORD spectrum is an

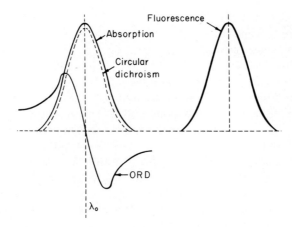

FIG. 3. Optical phenomena.

S-shaped curve with an inflection point between the peak and the trough lying at λ_0. The CD spectrum has the same shape as the absorption spectrum with its peak also occurring at λ_0. Circular dichroism is not measurable outside the region of the absorption spectrum. The S-shaped part of the ORD curve and the CD spectrum together are called the *Cotton effect*. The region of the spectrum they occupy is termed the *Cotton region*. Both CD and ORD are signed. CD may be positive or negative, lying above or below the baseline. When the peak of the ORD spectrum appears on the long wavelength side of the Cotton region it is termed a positive Cotton effect.

With the exception of CD all the phenomena depicted in Fig. 3 have been measured in unfixed cells. For a variety of reasons, absorption microspectrophotometry finds its greatest use on fixed material. But it is relatively insensitive and the results are difficult to interpret in molecular terms. Fluorescence is several orders of magnitude more sensitive than absorption. Truly vital fluorescent dyes (fluorochromes) are available and methodology has been developed for obtaining fluorescence emission spectra from single, unfixed cells (*31, 33*). Correlation with solution studies and with ORD is possible. Although CD measurements through the microscope are possible they have not yet been attempted. Circular dichroism and optical rotatory dispersion are related phenomena and it is possible, at least in principle, to calculate one if the other is known.

The Cotton region is the most desirable portion of the ORD spectrum in which to obtain data. Most biologic macromolecules exhibit Cotton effects in the ultraviolet. This region of the spectrum is lethal and precludes the use of unfixed cells. However, all biologic macromolecules have the characteristic ability of inducing asymmetry in small molecules such as fluorescent dyes with which they form complexes (11). These small molecules are symmetric and do not display optical activity in the unbound state. Complexes of fluorescent vital dyes with cells may thus be studied by at least two independent means, ORD and fluorescence microspectroscopy.

It is important to bear in mind that the microspectropolarimeter measures net optical rotation. In a situation as complex as the interior of the cell a number of constituents may simultaneously contribute to the ORD spectrum. This could result in a complex ORD spectrum difficult to interpret. Suitable vital fluorochromes can serve to identify a particular intracellular constituent if the cytochemistry of the dye–cell interaction is known. Induced ORD in bound dyes displays Cotton effects in the neighborhood of the dye absorption band which is in the visible region of the spectrum. This permits the use of unfixed cells for ORD measurements. When the dyes are fluorochromes visual observation can localize the dye within the cell and fluorescence spectroscopy can quantitate the amount of dye and qualitatively determine the molecular species present.

I. Theory of Operation

The fundamentals of polarimetry are shown in Fig. 4. For simplicity ordinary light is depicted as linearly polarized in only two mutually perpendicular planes, whereas it actually is polarized in all possible planes. The polarizer is an optical element which defines a plane of vibration for

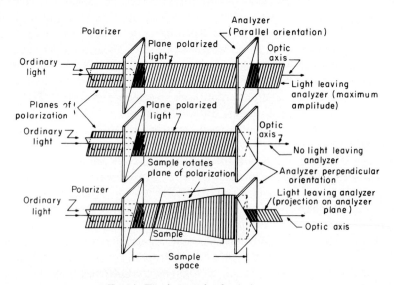

FIG. 4. Fundamentals of polarimetry.

light transmitted through it. The emerging light is linearly polarized. The analyzer is also an optical element with one plane of vibration. The upper diagram in Fig. 4 shows that light linearly polarized in a plane parallel to the plane defined by the analyzer is transmitted with no attenuation, neglecting losses due to reflection and absorption. Light polarized in a plane perpendicular to the plane of the analyzer is (middle diagram) completely blocked. An observer looking through the analyzer toward the polarizer sees a black field under these conditions. The bottom diagram in the figure shows the effect of an optically active medium upon the plane of polarization of linearly polarized light passing through it. The plane of polarization of the light is continuously rotated as it passes through the optically active sample. The total optical rotation experienced by the linearly polarized light depends on the nature of the sample, the wavelength of the light, and the pathlength through the sample. Environmental factors also play a role. Due to the presence of the optically active sample the plane of polarization of the light impinging on the analyzer is no longer orthogonal to the plane of vibration of the analyzer. Light now passes through the analyzer, with an amplitude which is determined by the angle between the plane of vibration of the analyzer and the plane of polarization of the linearly polarized light impinging on it. The acute angle between the incident and emerging radiation is given by Eq. (1) which is the familiar law of Malus

$$I^2 = I_0{}^2 \cos^2\phi \tag{1}$$

where I_0 is the amplitude of linearly polarized light incident on the analyzer; I, the amplitude of plane-polarized light transmitted by the analyzer; and ϕ, the azimuthal angle between the plane of polarization of the incident light and the plane of polarization defined by the analyzer. The quantities in Eq. (1) are squared because the eye or a photoelectric detector is sensitive to power. The square of the amplitude is the *intensity* of the light.

In an ideal system which obeys Eq. (1) exactly, no light leaves the analyzer when ϕ is equal to 90°. Sensitive measurements of optical rotation require a fairly good approximation to the ideal. The conventional measuring procedure is to set the analyzer to maximum extinction with a reference sample between the polarizers. The position of the analyzer is recorded from the angular scale with which it is equipped. An optically active sample is then inserted. An increase in the intensity of light leaving the analyzer is observed. The analyzer is rotated through the acute azimuthal angle which restores maximum extinction. The position is recorded. The difference between the two readings is the optical rotation produced by the optically active sample. The sensitivity of this measurement, i.e., how small an optical rotation can be detected, depends markedly on how closely complete extinction is approached. This is limited by the nature of the polarizers and the amount of depolarization scattering that occurs in the system.

As mentioned in the introduction, the microscope introduces a considerable amount of depolarization scattering which severely limits its sensitivity. Unfortunately, most biologic objects of interest are weakly birefringent and the optical rotations produced are not measurable by the above method. A further complication arises from the fact that circular birefringence and linear birefringence may be present simultaneously. The conventional methodology cannot readily separate these two. The methodology presented below circumvents the limitation on sensitivity, can separate linear from circular birefringence, and offers some new possibilities for ORD studies of interest in cytochemistry. Though the discussion below is confined to the microscope, it is equally applicable to solution studies using standard polarimeter cells and appropriate optics.

That the polarizing microscope obeys Eq. (1) is shown in Fig. 5. Curve a is the amplitude of the light leaving the analyzer as a function of the azimuthal angle. Curve b, which was made from actual measurements, shows the variation in intensity of the light leaving the analyzer as a function of the azimuthal angle. This is the $\cos^2$ function given in Eq. (1). Curves a and b are linear plots with the labeling of the ordinate given on the left side of the figure. The troughs of Curve b thus appear to go to zero. However, if

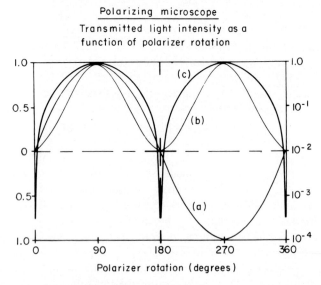

FIG. 5. Transmission of light through the microscope as a function of the azimuthal angle. Analyzer fixed, polarizer rotated. Amplitude (a), intensity (b), and logarithm of intensity (c).

the experimental data is plotted logarithmically one readily sees that the extinction is far from complete. This is illustrated by Curve c. The ordinate for Curve c is labeled to the right. Similar results were given by Köhler (19) in his extensive discussion of the polarizing microscope. The degree of failure to achieve complete extinction varies directly with numerical aperture. The higher the numerical aperture the less complete the extinction. The failure to achieve complete extinction can also be observed visually by examining the back focal plane of the objective when the polarizer and analyzer are crossed. A black cross on a lighter field is observed, particularly at high numerical aperture.

Since the depolarization in the microscope is a constant for a given set of operating conditions (15), it appeared likely that the new method of measuring optical rotation developed by West et al. (34) which is insensitive to depolarization could serve to circumvent this difficulty in the microscope. In the new methodology the analyzer is caused to rotate at a constant speed. Equation (1) may then be expressed as

$$I^2 = I_0{}^2 \cos^2 (\omega t + \theta) \tag{2}$$

where t is the time in seconds; ω, the radian frequency, $2\pi f$; f, the rotational frequency; and θ, the phase angle. Since $2 \cos^2\theta = 1 + \cos 2\theta$, Eq. (2) can be transformed to

$$I^2 = (I_0{}^2/2) + (I_0{}^2/2) \cos (2\omega t + 2\theta) \tag{3}$$

Neglecting any losses due to absorption, scattering, and other causes Eq. (3) represents the wave shape of the intensity of the light as a function of time as it emerges from the rotating analyzer. Note that the right-hand side of Eq. (3) contains a constant term and a time-varying term. A linear photoelectric detector is placed in the path of the beam leaving the rotating analyzer. The electric signal produced by the photoelectric detector is the same as that described by Eq. (3) except for a constant factor. Thus the photoelectric current is given by

$$i_S = (kI_0{}^2/2) + (kI_0{}^2/2) \cos (2\omega t + 2\theta) \tag{4}$$

where k is the transducer constant in μamp/μwatt.

The phase angle 2θ depends only upon the azimuthal change that has been imparted to the linearly polarized light traversing the sample space. Two measurements are required to measure a phase angle since a fiduciary point must be established. This is readily accomplished in a double-beam system where reference and sample signals are simultaneously present. The system under consideration, however, is a single-beam system. The fiduciary point from which the phase angle will be measured must, therefore, be stored for later comparison with the succeeding signal. This can be accomplished by housing the analyzer in a drum with a magnetic track on the perimeter. The entire assembly rotates as a unit. The fiduciary point is obtained by choosing a point on the waveform that can be recovered with great accuracy. A fast pulse generated at that instant in time is recorded on the magnetic track around the rotating analyzer. The index from which phase shift is to be measured is thus physically fixed to the optical phenomenon that gave rise to it. Ordinarily the recording of the fiduciary point will be made with an optically inactive or reference sample on the stage of the microscope. For the reference signal, the phase angle may be taken as zero. If an optically active sample is now placed on the stage of the microscope the phase of the sample signal with respect to the phase of the reference signal, represented by the recorded index, can be determined. The optical rotation (OR) produced by the sample is given by

$$OR = 2\theta \tag{5}$$

The depolarized light, produced by the microscope, does not affect the argument of the cosine term on the right side of Eq. (4). Only the magnitudes of the constant term and the coefficient of the cosine term are affected. The constant term can be removed by AC-coupling the photoelectric signal. Also, phase measurements can be largely insensitive to variations in signal amplitude. Thus Eq. (4) shows that the depolarization, which is a constant, can be separated from the azimuthal change effected by the sample and the latter detected and measured.

A biologic object such as a cell introduces additional optical factors not included in Eq. (4) as it stands. These include depolarization scattering, nonspecific absorption, and circular dichroism. Scattering which causes depolarization and nonspecific absorption affect the intensity of light striking the photoelectric detector, but do not change the phase angle, θ. Circular dichroism in the sample will, in general, cause the light incident on the analyzer to be elliptically polarized. These factors can be introduced as shown in Eq. (6).

$$I^2 = (P/2) \exp(-Kd) + (Q/2) \exp(-Md) \cos(2\omega t + 2\theta) \tag{6}$$

where the terms are defined as follows:

$P = B^2 + C^2 + S^2$
$K = k_A c_A$
$M = (k_A c_A + k_S c_S)$
B, Amplitude of the major axis of the ellipse
C, Amplitude of the minor axis of the ellipse
$Q = B^2 - C^2$
S, Scattered, depolarized light which passes through the analyzer
k_A, Absorption coefficient of nonoptically active material

k_S, Scattering coefficient
c_A, Concentration of absorbing material
c_S, Concentration of scattering material
d, Cell length
B and C are given by Eqs. (7) and (8):

$$B^2 = I^2[\exp(-k_R dc/2) + \exp(-k_L dc/2)]^2 \qquad (7)$$

and

$$C^2 = I^2[\exp(-k_R dc/2) - \exp(-k_L dc/2)]^2 \qquad (8)$$

where k_R is the absorption coefficient of right-circularly polarized light, and k_L is the absorption coefficient of left-circularly polarized light, and c is the concentration of dichroic substance

The quantities shown in Eqs. (7) and (8) are wavelength-dependent. B and C have values of consequence only in the neighborhood of an absorption band.

Examination of Eq. (6) shows that the presence of circular dichroism, depolarization scattering, and nonspecific absorption modifies the first term on the right and affects only the coefficient of the time-dependent second term. The first term on the right thus remains a constant for a given set of conditions and is discarded by passing the signal through the AC-coupled circuit. Since only the amplitude and not the phase angle of the second term is altered, the system is relatively insensitive to variations in signal amplitude and can operate in the presence of considerable amounts of circular dichroism, depolarization scattering, and nonspecific absorption. For example, optical rotation measurements have been obtained from sucrose solutions with a neutral density filter having an optical density of 5.5 interposed. These measurements were made without the microscope, but should be equally true for it.

The extent to which the optical density of a sample can be increased while still retaining useful accuracy depends, in part, on the spectral purity of the light and, in part, on the absorption spectrum of the substance under study. Instrumental factors also play a role, but need not become important until the spectral purity of the light is very high. For the measurement on sucrose above, the limitation was the spectral purity of the source, not other instrumental factors. A sample with a narrow absorption band and high optical density will enhance the effect of stray light on the measurements when measurements are taken in the neighborhood of the absorption band.

In the case of the microscope there are additional sources of stray polarized light. One of these, the inhomogeneous distribution of optically active chromophore, is already well known from absorption microspectrophotometry as "distribution error." Another appears when the cytologic area of interest does not completely fill the photometric aperture.

Stray polarized light, whatever its source or wavelength, results in a measured or apparent optical rotation which is different from the true optical rotation produced by the substance of interest. The effect of stray polarized light is shown graphically in the vector diagram of Fig. 6. The vectors represent the amplitude and phase of the sinusoidal modulation of the light intensity produced by the rotation of the analyzer. The x axis coincides with the reference from which optical rotation is measured and represents the spatial position of the plane of polarization in the absence of an optically active sample. The desired beam and the stray polarized light can be considered as independent components as they enter the rotating analyzer. The rotating analyzer performs the vector addition. The beam leaving the rotating analyzer and impinging on the photodetector is intensity-modulated at the frequency determined by the rotating analyzer, which is held constant. The amplitude and phase, however, are altered. More than one stray polarized light vector may be present simultaneously. Only stray polarized light due to spectral impurity is considered in Fig. 6. In Fig. 6

A_0 Intensity at wavelength λ_0

λ_0 Desired wavelength in the neighborhood of absorption maximum

ϕ_0 Optical rotation of λ_0 by the sample (true rotation)

A_1 Intensity at wavelength λ_1

λ_1 Wavelength at which sample has weak absorption

ϕ_1 Optical rotation of λ_1 by the sample

A_A Resultant intensity

ϕ_A Apparent optical rotation

For the case shown in Fig. 6 the apparent optical rotation is less than the true rotation. However, the converse could also occur depending on the particular situation. The importance of the stray effects will vary with the nature of the material and the problem. For cytochemistry the stray polarized light effects due to strong absorption may not constitute the largest source

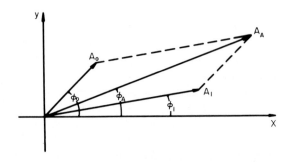

FIG. 6. The effect of stray polarized light, vector representation.

of error. Of greater probable importance are the stray polarized light effects due to inhomogeneous distribution of optically active chromophore (distribution error). Depolarization scattering is of no consequence. Thus, consideration need not be given to differences in refractive index between the cell and the mounting medium or between intracellular bodies and the cytoplasm or nucleoplasm in which they are embedded.

The conditions for obtaining a true reading with the microspectropolarimeter are thus quite similar to those which apply to absorption microspectrophotometry, i.e., (1) high spectral purity and (2) the smallest possible photometric aperture. However, the requirement for spectral purity is, in general, much more stringent, especially if measurements are to be made in the neighborhood of a strong absorption band. The conservative use of light in this methodology makes it possible to use filters to reduce the stray light produced by the monochromator.

The requirement for reducing the photometric aperture can be satisfied down to the same limits which are true for absorption microspectrophotometry, approximately 1-μ diameter area in the object plane. Detection and measurement of optical rotation is not limited to this area. It depends only on the amount of rotation present. But, localization of the source of optical rotation is limited to the spatial resolution of the microscope. It should be pointed out that there is no knowledge, at present, of the importance of distribution error in microspectropolarimeter measurements of optical rotatory dispersion from cytological samples. This is a new technique and a great deal more information is required than is presently available to assess potential sources of error in cytochemical terms.

II. Instrumentation

Figure 7 is a block diagram of the microscope arranged for spectropolarimetry. Two possible modes of operation are shown. For the purposes of explaining the principles of operation, the arrangement with the monochromator lying between the light source and the polarizer will be considered. This is also the easiest arrangement to implement and subjects the specimen to the minimum amount of radiation.

The monochromatic radiation leaving the monochromator is linearly polarized by the polarizer (part of the microscope condenser) and then caused to transilluminate the specimen by the substage condenser. The microscope objective collects the transmitted radiation and forms an image of the object plane in the plane of the photometric aperture. This aperture may be placed either before or after the analyzer. In either case, the light passes through the rotating analyzer before impinging on the photosensitive surface of the photoelectric detector. The photoelectric signal is amplified by an AC-coupled preamplifier which removes the constant term shown on the right side of Eq. (6). The sinusoidal signal now oscillates about a pre-determined baseline which can be taken as zero volts for the balance of this discussion. The zero-crossing point on the waveform has the largest slope and is chosen, therefore, as the point on the waveform that will be used to measure its phase. It is the function of the zero-crossing detector to locate this point on the waveform and generate a fast pulse at the instant zero-crossing occurs.

Two measurements are required to measure a phase angle. The polarimeter under consideration here is a single-beam device so that these two measurements cannot be made simultaneously. One has to be stored. This is accomplished by recording the pulse produced by the zero-crossing detector on the magnetic recording track on the drum which houses the rotating analyzer. This recording serves as the reference from which changes in phase angle (optical rotation) are measured.

After the reference pulse is recorded the sample is brought into the field of view. The signal produced in the presence of the sample is transmitted to the zero-crossing detector where again a fast pulse is produced at the instant of zero-crossing. This pulse is not recorded, although it can be, if desired. The recorded reference pulse and the pulse representing the sample are both simultaneously transmitted to an electronic counter which serves as a phase comparator. The electronic counter contains a stable crystal

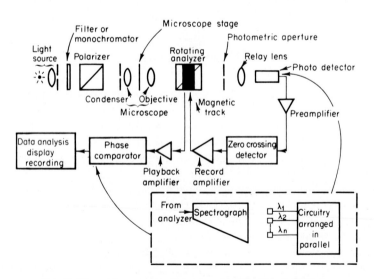

Fig. 7. The microspectropolarimeter, simplified block diagram.

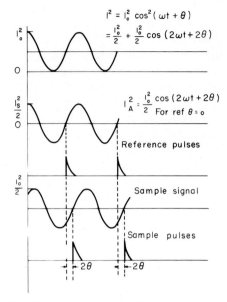

FIG. 8. Waveforms showing method of measuring phase angle.

oscillator. Pulses from the oscillator are counted during the time interval between the recorded reference pulse and the sample pulse. The waveforms are shown in Fig. 8. The time interval representing the phase difference, designated 2θ in the figure, is the desired measurement of optical rotation. The number of oscillator pulses counted is thus a direct measure of the optical rotation exhibited by the sample.

The rotational frequency of the rotating analyzer must be precisely known for the time-interval measurement to represent a particular phase angle. This is provided for by synchronizing the motor driving the rotating analyzer with the quartz crystal oscillator in the counter. The period of rotation of the rotating analyzer is measured with the counter and an accurate relationship obtained between the time-interval measurement and the phase angle it represents. The measurements of phase angle are absolute when the accuracy and stability of the quartz crystal oscillator are traceable to the Bureau of Standards. This is the case for the counter employed in the author's experimental setup. The measurements appear on the counter display and are also printed by a printer driven by the counter.

All spurious rotations, such as those produced by optics which are not strain-free, are self-canceling since they are common to the reference and sample signals. The requirements for strain-free optics are thus greatly relaxed. In addition, since stray depolarized light is tolerable, the specifications on the efficiency of polarizing elements can also be relaxed considerably.

A mode of operation which differs from the one just described is possible. The arrangement is shown in the dashed box in Fig. 7. Here the microscope is illuminated by white light which is polarized, as in the previous case, before being transmitted through the microscope. A spectrograph is placed behind the photometric aperture. The spectrum produced can be examined for optical rotation as a function of wavelength by positioning a single photodetector at each wavelength of interest, in turn. Each wavelength will have impressed on it an optical rotation which is a function of the nature of the sample. There is probably not much advantage and perhaps less convenience to making the measurements this way.

However, if the interest lies in kinetics, then the single photodetector may be replaced with a battery of photodetectors operating in parallel. Each photodetector is positioned to detect a particular wavelength in the ORD spectrum. With operation in parallel as shown, an ORD spectrum could be recorded in half a revolution of the rotating analyzer. At the present rotational speed (2 revolutions per second) an ORD spectrum could be recorded in 0.25 sec and be repeated in the next quarter second, and so on. Higher rotational speeds are possible, if faster kinetics are to be studied. It should be possible, by this means, to study the kinetics of particular macromolecular interactions as they participate in the metabolic processes in living cells.

III. Results

The microspectropolarimeter is a new instrument designed to measure ORD in microscopic samples. To the best of the author's knowledge the literature provides no report of ORD measurements with the microscope. There is, however, a large, rapidly growing literature on ORD and CD dealing with solution studies on biologic macromolecules. It, therefore, seems reasonable to expect that cells are also circularly birefringent. The question remains, however, as to whether the intensity of this phenomenon in cells is sufficient to permit detection and measurement.

The first studies with the microspectropolarimeter reported here were addressed to (1) determining whether experimental behavior of the instrumentation agreed with the theoretical predictions, and then (2) determining whether sensitivity was sufficient to obtain ORD spectra from single, supravitally stained cells. ORD spectra from solution studies on acridine orange (AO)–DNA complexes are available in the literature. If the microspectropolarimeter has sufficient sensitivity, ORD spectra from cells stained supravitally with acridine orange should, in some measure, resemble the ORD spectra from solution studies. Also, since only *net* optical rotation is measured,

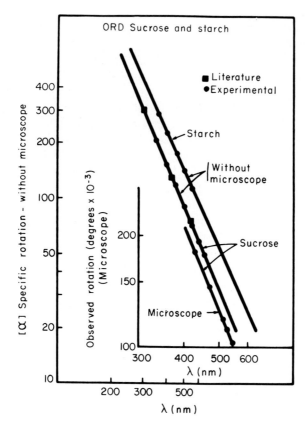

FIG. 9. ORD spectra of starch and sucrose solutions. Upper two curves taken without the microscope. Inset, sucrose solution with the microscope. (See text.)

ORD spectra from unstained cells must also be obtained and subtracted from the ORD spectra from the stained cells.

Operation of the microspectropolarimeter was determined experimentally by first investigating the instrumental methodology itself without the microscope. Standard polarimeter cells and sucrose solutions were used to see if the measurements are absolutely accurate, as predicted. The results are shown in Fig. 9. Agreement with the literature (21) values is excellent. The symbol [α], usually found in tables, is the *specific rotation* which is given by

$$[\alpha] = (\alpha/lc) \qquad (9)$$

where $\alpha = 2\theta$ in the notation used here. It is the observed rotation in degrees. Also, l is the pathlength in decimeters and c, the concentration in grams per ml. The observed rotation α can also be obtained from the circular birefringence. Thus

$$\alpha_l = (1800/\lambda)(n_L - n_R) \tag{10}$$

n_L and n_R are the indices of refraction for left- and right-circularly polarized light, respectively, and λ is the wavelength of light in vacuum.

Some appreciation of the sensitivity of ORD measurements can be obtained by considering that for an observed rotation of 100 millidegrees over a pathlength of one decimeter with $\lambda = 500$ nm, $(n_L - n_R)$ is approximately equal to 2.8×10^{-9}. The specific rotation may vary with concentration. This is not the case for sucrose, which makes it a good test substance.

Also shown in Fig. 9 is an ORD spectrum for an opalescent starch solution. The experimental points all fall on a straight line, demonstrating that depolarization scattering has little or no effect on the measurement of optical rotation. This is of special importance for cytochemical studies on unfixed cells suspended in aqueous media. The use of such mounting media cannot be avoided if cells are to be kept in a physiologically functional state. But such mounting media do increase the amount of depolarization scattering produced by the cells. Hence, insensitivity to depolarized light is of prime importance to biophysical cytochemical studies employing polarized light.

Performance with the microscope incorporated into the system is shown in the insert in Fig. 9. The ordinate is expressed as "observed rotation" rather than specific rotation because polarimeter cells with accurately known pathlength were not available for the microscope. The slope of the curve, however, is the quantity of importance. It is identical with the slope of the curve produced without the microscope indicating that the microspectro-polarimeter is also capable of absolute measurements. Indeed, using estimates of the pathlength of the microscope preparation values of specific rotation in good agreement with the true values were obtained.

The studies on cells were conducted with mouse Ehrlich's ascites tumor cells. Three aliquots of a suspension of ascites cells in physiologic saline, buffered to pH 7.0 with Mac Ilvaine's buffer were treated as follows: One aliquot was left untreated; the second and third aliquots were supravitally stained with acridine orange. The procedure consisted of centrifuging the cells, decanting the supernatant, and then resuspending the cells in acridine orange dissolved in a physiologic saline solution buffered to pH 7.0 with Mac Ilvaine's buffer. One of these cell populations [stained cell (1) in Figs. 10 and 11] was stained with a dye concentration of approximately 10^{-16} moles of dye per cell. The other cell population [stained cell (2) in Figs. 10 and 11] was stained with a dye concentration of approximately 10^{-15} moles

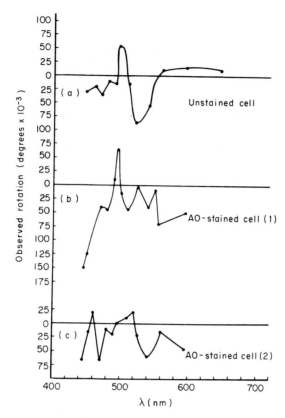

FIG. 10. ORD spectra from Ehrlich's mouse ascites tumor cells suspended in physiologic saline and saline–AO solutions, pH 7.0. Raw data from unstained cell (a) and AO stained cells (b) and (c). (See text.)

of dye per cell. Staining was allowed to continue until equilibrium between the cells and the dye solution was reached.

Slides were then prepared from each of the three aliquots by placing a drop of the cell suspension on a microscope slide, covering it with a cover-slip, and sealing with paraffin. Note that the stained cells were suspended in the staining solution on the slide and that equilibration of the staining reaction was permitted to occur before slides were prepared. Both of these procedural steps must be observed if reproducible data is desired.

Slides used for ORD measurements with the microspectropolarimeter were examined for fluorescence after the ORD measurements had been completed. One of the molecular species formed when acridine orange is complexed with cells is labile and phototoxic. This occurs only at the higher values of intracellular dye content (*31, 33*). Therefore, as a precaution against possible denaturation and cell injury, slides to be used for ORD measurements were not examined with the fluorescence microscope until after the ORD measurements were completed. Under the fluorescence microscope cells from population (1) showed no cytoplasmic staining and green nuclei. Cells from population (2) also showed no cytoplasmic staining and the nuclei were green only very slightly tinged with yellow. No differences were observed in the fluorescence microscope between cells examined immediately after the slides were prepared, and those examined after the ORD measurements were completed. In so far as visual fluorescence observations

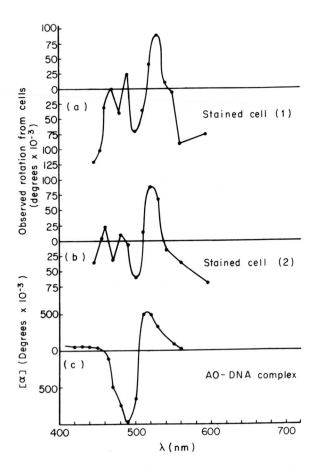

Fig. 11. Comparison between corrected ORD spectra from AO-stained cells, (a) and (b), and ORD spectrum from AO–DNA complex in solution (c). Curve (c) redrawn from Neville and Bradley (25).

apply no apparent damage to the cells resulted from the ORD measurements.

The photometric aperture for the ORD measurements was circular and slightly smaller than an ascites tumor cell nucleus. ORD spectra from three cells are shown in Fig. 10. Curve a is from an unstained cell. Curves b and c are curves from individual cells from populations (1) and (2), respectively. Curve b represents the smaller intracellular dye content. Curves b and c are the raw data as produced by the microspectropolarimeter and represent the net optical rotation. Curve a is surprising in that the large rotations and clearly shown Cotton effect were unexpected. The relatively large rotations produced by the unstained cell clearly demonstrate the need for such measurements to clarify the data from stained cells.

The net optical rotation observed at a given wavelength is the linear, algebraic sum of the rotations contributed by all the optically active chromophores present. It is, therefore, permissible to obtain the optical rotation produced by one component by simple subtraction of the optical rotations introduced by the interfering components.

The raw data shown in Fig. 10 were subjected to this process by subtracting Curve a from Curve b and from Curve c. The results are given in Fig. 11 where Curve a is cell (1) and Curve b is cell (2). Notice the similarity of the two curves and the appreciable Cotton effect with point of inflection at approximately 500 nm. The Cotton effect region for induced optical rotation should lie in the neighborhood of the absorption maximum of the dye. For acridine orange the absorption maximum of the monomeric form of the dye and also of the dye–nucleic acid complex, at low dye to nucleic acid ratios, lies at 502 nm. The dye alone in solution is not optically active. Thus the Cotton effect centered at approximately 500 nm appears to indicate that the relatively large Cotton effects shown in Curves a and b are due to acridine orange–nucleic acid complexes.

The possibility of making direct comparisons with the results from solution studies is an important aspect of microspectropolarimeter measurements in unfixed cells. Curve c in Fig. 11 is the ORD spectrum of an AO–DNA complex in solution obtained by Neville and Bradley (25). The similarity between the ORD spectra from stained cells and from the solution study is clearly seen in the neighborhood of the Cotton effect region. At the shorter wavelengths the ORD spectra from cells show detail not present in the solution data. If true, this would not be surprising since the cell is very much more complex than the simple solution. However, this is still early data, and with no prior knowledge to fall back on, additional experimental work is required to establish the validity and significance of the detail in the short wavelength portion of the ORD spectra from AO-stained cells.

IV. Discussion

The results presented clearly demonstrate that the microscope with suitable modification can function as a microspectropolarimeter. The microspectropolarimeter produces absolute measurements limited only by the knowledge of the material under examination. The instrument is extremely sensitive with regard to sample size and should find application wherever the amount of sample is limited. The other attributes of the methodology such as insensitivity to unpolarized light (depolarization scattering), and ability to make measurements in the presence of optical densities greater than 4 or 5, to name a few, should make it very useful for the study of biopolymers. Adaptation of the instrument for operation in the ultraviolet can be accomplished with relative ease. It is also a relatively simple matter to replace the microscope with an optical system adapted for use with standard polarimeter cells.

The data presented also demonstrates conclusively that it is possible to obtain ORD spectra from intact, unfixed cells, both stained and unstained. Cells can be examined under essentially physiologic conditions. The use of vital dyes not only permits operation in the visible region of the spectrum and avoids ultraviolet damage to the cells, but can serve to identify a particular intracellular constituent, even when present in a complex mixture, without prior chemical separation. The latter is equally true for intrinsic spectra, but may be practically impossible to apply to a system as complex as a cell. Dyes, therefore, constitute very powerful molecular probes.

ORD spectra are indicative of the conformation of molecules. The stereochemistry of biologic macromolecules is of very great importance for developing fundamental understanding of intracellular events. Microspectropolarimetry provides a means, though not the only one, of carrying out stereochemical investigations at the cellular level. The data produced is identical, in kind, with that obtained from simplified model systems. As an example of this latter point, the data presented here indicate that the conformation (double helix) of intranuclear nucleic acids is the same as has been found in solution and by X-ray diffraction. While there has not been much argument on this score, all the evidence for the existence of intranuclear DNA in the form of the Watson-Crick double helix has been indirect. The ORD spectra from AO-stained cells are direct evidence of the intranuclear existence of the double helix form. The implications for further study of intracellular DNA, RNA, and other constituents of the cell are obvious. Such studies may be of particular importance for macromolecules more labile than the nucleic acids.

One other point worth mentioning is the potential ability of the microspectropolarimeter to follow the kinetics of intracellular reactions. Although considerable additional instrumentation is required to make this possible, such data would have important bearing on the understanding of control mechanisms in cellular processes.

Finally, it should be realized that microspectropolarimetry is only one means for obtaining from cells data which can be useful to the theoretical chemist. Other possibilities, such as fluorescence microspectrophotometry, already exist. No single method or phenomenon is likely to provide all the information that is needed. The cytochemist should freely employ a variety of techniques and phenomena. But in every instance experiments should be designed in such fashion that the data obtained from cells is subject to the same restrictions and is expressed in the same terms as data from experiments carried out in solution. The data presented here do not quite satisfy these criteria. The shapes of the ORD curves from cells in Fig. 11 can readily be compared with the shape of the ORD spectrum obtained from solution. But the quantity of nucleic acid present in the cells cannot be deduced from the ORD spectra, in contrast to the solution data which is expressed in absolute terms (specific rotation). Quantitating the amount of an intracellular substance responsible for a given ORD spectrum presents a difficult problem because of both distribution error and the difficulties entailed in accurate measurement of microscopic distances. Future work will be devoted to this problem, but it may be that the simplest solution is to use an independent means for quantitating the amount of substance. In the case of intranuclear AO–NA complexes the amount of AO–NA complex present can be measured by fluorescence microspectrophotometry. Under proper conditions even a simple filter fluorescence microphotometer can be used. Despite the difficulty of quantitating the intracellular amount of a particular substance, microspectropolarimetry offers considerable promise for biophysical cytochemistry and should find important applications over a broad spectrum of biologic problems.

REFERENCES

1. Allen, R. D., and Francis, D. W., Cytoplasmic contraction and the distribution of water in the amoeba. *Symp. Soc. Exptl. Biol.* **19**, 259–271 (1965).
2. Allen, R. D., and Nakajima, H., Two exposure, film densitometric method measuring phase retardations due to weak birefringence in fibrillar of membrane cell constituents. *Exptl. Cell Res.* **37**, 230–249 (1965).
3. Allen, R. D., and Rebhun, L. I., Photoelectric measurement of small fluctuating retardations in weakly birefringent, light-scattering biological objects. *Exptl. Cell Res.* **29**, 583–592 (1962).
4. Allen, R. D., Brault, J., and Moore, R. D., A new method of polarization microscopic analyses. I. Scanning with a birefringence detection system. *J. Cell Biol.* **18**, 223–235 (1963).
5. Allen, R. D., Francis, D. W., and Nakajima, H., Cyclic birefringence changes in pseudopods of Chaos Carolinensis revealing the localization of the motive force in pseudopod extension. *Proc. Natl. Acad. Sci. U.S.* **54**, 1153–1161 (1965).

6. Allen, R. D., Brault, J. W., and Zeh, R. M., Image contrast and phasemodulated light methods in polarization and interference microscopy. *In* "Advances in Optical and Electron Microscopy" (V. E. Cosslett and R. Barer, eds.), Vol. 1, pp. 77–114. Academic Press, New York, 1966.

7. Bajer, A., and Allen, R. D., Structure and organization of the living mitotic spindle of Haemanthus endosperm. *Science* **151**, 572–574 (1966).

8. Bartels, P. H., Principles of polarized light. *In* "Introduction to Quantitative Cytochemistry" (G. L. Wied, ed.), pp. 519–538. Academic Press, New York, 1966.

9. Beychok, S., Circular dichroism of biological macromolecules. *Science* **154**, 1288–1299 (1966).

10. Beychok, S., Rotatory dispersion and circular dichroism. *Ann. Rev. Biochem.* **37**, 437–462 (1968).

11. Blout, E., and Stryer, L., Anomalous optical rotatory dispersion of dye: polypeptide complexes. *Proc. Natl. Acad. Sci. U.S.* **45**, 1591–1593 (1959).

12. Djerassi, C., "Optical Rotatory Dispersion." McGraw-Hill, New York, 1960.

13. Eyring, H., Liu, H. C., and Caldwell, D., Optical rotatory dispersion and circular dichroism. *Chem. Rev.* **68**, 525–540 (1968).

14. Frey-Wyssling, A., "Submicroscopic Morphology of Protoplasm," pp. 82 et seq. Elsevier, Amsterdam, 1953.

15. Inoué, S., Studies on depolarization of light at microscope lens surfaces. *Exptl. Cell Res.* **3**, 199–208 (1951).

16. Inoué, S., and Dan, K., Birefringence of the dividing cell. *J. Morphol.* **89**, 423–455 (1951).

17. Inoué, S., and Hyde, W. L., Studies on depolarization of light at microscope lens surfaces. *J. Biophys. Biochem. Cytol.* **3**, 831–838 (1957).

18. Inoué, S., and Sato, H., Arrangement of DNA in living sperm: A biophysical analysis. *Science* **136**, 1122–1124 (1962).

19. Köhler, A., Die Verwendung des Polarizationsmikroskops fur biologische Untersuchungen. *In* "Handbuch der biologischen Arbeitsmethoden" (E. Abderhalden, ed.), Sect. II, Part 2, pp. 907–1108. Urban & Schwarzenberg, Munich, 1928.

20. Lowry, T. M., "Optical Rotatory Power." Longmans, Green, New York, 1935. (Dover, New York, 1964, reprint).

21. Meites, L., ed., "Handbook of Analytical Chemistry," pp. 6–245. McGraw-Hill, New York, 1963.

22. Mislow, K., "Introduction to Stereochemistry." Benjamin, New York, 1966.

23. Missmahl, H. P., Birefringence and dichroism of dyes and their significance in the detection of oriented structures. *In* "Introduction to Quantitative Cytochemistry" (G. L. Wied, ed.), pp. 539–547. Academic Press, New York, 1966.

24. Nakajima, H., and Allen, R. D., The changing pattern of birefringence in plasmodia of the slime mold, Physarum Polycephalum. *J. Cell Biol.* **25**, 361–374 (1965).

25. Neville, D. M., Jr., and Bradley, D. F., Anomalous rotatory dispersion of acridine orange-native deoxyribonucleic acid complexes. *Biochim. Biophys. Acta* **50**, 397–399 (1961).

26. Partington, J. R., An advanced treatise on physical chemistry. "Physico-chemical Optics," Vol. 4. Longmans, Green, New York, 1953.

27. Ruch, F., Birefringence and dichroism of cells and tissues. *In* "Physical Techniques in Biological Research" (G. Oster and A. W. Pollister, eds.), Vol. 3, pp. 149–176. Academic Press, New York, 1956.

28. Ruch, F., Dichroism and difluorescence. *In* "Introduction to Quantitative Cytochemistry" (G. L. Wied, ed.), pp. 549–555. Academic Press, New York, 1966.

29. Swann, M. M., and Mitchison, J. M., Refinements in polarized light microscopy. *J. Exptl. Biol.* **27**, 226–237 (1950).
30. Velluz, L., Legrand, M., and Grosjean, M., " Optical Circular Dichroism." Academic Press, New York, 1965.
31. West, S. S., Fluorescence microspectroscopy of mouse leukocytes supravitally stained with acridine orange. *Acta Histochem.* Suppl. **6**, 135–156 (1965).
32. West, S. S., A proposed microspectropolarimeter. *Abstr. 11th Ann. Meeting Biophys. Soc., Houston* **7**, p. 11, 1967.
33. West, S. S., Fluorescence microspectrophotometry of supravitally stained cells. *In* "Physical Techniques in Biological Research" (A. W. Pollister, ed.), Vol. 3C, pp. 253–321. Academic Press, New York, 1969.
34. West, S. S., Liskowitz, J., Usdin, V. R., and Welch, E. A., A polarimeter for detection of extraterrestrial life. *Proc. 18th Ann. Conf. Eng. Med. Biol.* **7**, 176 (1965).

PRINCIPLES OF QUANTITATIVE AUTORADIOGRAPHY

Ronald J. Przybylski*

DEPARTMENT OF ANATOMY, SCHOOL OF MEDICINE, CASE WESTERN RESERVE UNIVERSITY,
CLEVELAND, OHIO

I. Introduction and Perspective

Upon the successful completion of a descriptive study of a biologic system, the morphologist often begins to inquire into the possibility of further investigating the various dynamic aspects of the system. Questions arise such as: (1) Are the various parts of this system in a constant state of turnover or are there periods of relative inactivity? (2) Does the system operate at the same rate continuously and further, does the average rate accurately reflect the rates of the individual components? (3) Are all components of the biologic system constantly present or do they undergo cyclic changes of formation, stabilization and disappearance? (4) What is the channel of inflow and outflow of newly synthesized components? Any attempt to answer these questions with purely morphologic observations of preserved specimens without accompanying experimentation usually produces a myriad of subjective postulates. Therefore, some means of analysis must be employed which gives greater insight into the dynamics of morphology. Autoradiography provides a useful approach to such analyses.

Biochemical and physiologic investigations are often concerned with obtaining the answers to the questions posed above. However, quite frequently the biologic systems under investigation in these studies are considered (for the sake of simplicity) to be homogeneous. If the biologic system is suspected to be heterogeneous, then autoradiographic techniques in conjunction with biochemical and physiologic experiments utilizing radioisotopic tracers can be employed. This approach can detect heterogeneity

* This study was supported with funds obtained from the Heart Association of Northeast Ohio, Inc.

because the morphologic parameters remain intact and their differential activities are preserved and perceived via the autoradiographic image.

Autoradiography utilizes the photographic action of ionizing radiation for the detection of radioactive substances. Thus, the isotopic compound is presented to the biologic system and is then incorporated into its metabolic pathways, hopefully in the same way as the same chemical without the isotopic carrier present. Once inside the system, one hopes to (1) detect its presence, (2) assess its amount and thereby distinguish between homogeneous and heterogeneous uptake, and (3) be confident of the identity of the molecule containing the isotope which is producing the autoradiographic image.

The autoradiographic detection of an isotopic molecule in a biologic system is obtained by the application of a sensitive photographic emulsion to the preserved specimen. Decay of the radioactive compound will produce a silver deposit, the latent image, in the photographic emulsion which is subsequently enlarged by photographic developing procedures. The objectives of our experimental design dictate the means of application of the photographic emulsion and the limits of autoradiographic resolution desired. Thus, if we are interested in regional or organ level detection of the isotopic compound, then we should expect to be able to define its location to, at best, centimeters. If our interest lies at the tissue level of detection, then the resolution limits are reduced to millimeters. However, if localization of the isotope is sought at the cellular or intracellular level then the limits of definition are reduced to microns or angstroms. The type of isotope employed will also determine the means of detection and the limits of definition.

Initially, we are concerned only with the presence or absence of the isotope. Therefore, we need only to document the kind and amount of tissues or cells that have incorporated the precursor as opposed to those that have not. Subsequently, we may attempt to gain some information on the differential utilization of the precursor under varied experimental conditions and/or in a time study, so we attempt to quantitate the system. In autoradiography, quantitation can be relative or absolute. Relative quantitation is based on the correspondence between the amount of the radioactivity in the morphological structure and the number of silver grains it has produced. These values for a particular morphologic structure are then compared to the amounts assessed for similar structures in the experimental system. The actual counting of the number of silver grains is usually done visually. However, numerous investigators have mechanized the quantitation. Each approach has its limits and advantages. Absolute quantitation, which is difficult to obtain and is seldom employed, utilizes the number of observable silver grains to calculate the exact number of isotopic compounds present in the structure. This type of quantitation necessitates a series of standards of known isotopic dilutions and the photographic response of the sensitive emulsion to the isotopes employed. For most investigations, relative quantitation of the data is sufficient to assess differences in biologic activity on a comparative basis provided that the conditions of experimentation, processing, and means of autoradiographic detection have been standardized.

A final factor which is of prime importance in quantitative autoradiography is the degree of confidence that exists with regard to the molecular species containing the isotopic chemical. In other words, does the structural component which shows an autoradiographic label actually contain the isotopically labeled molecule or has the precursor been degraded and resynthesized as another metabolite which has been incorporated into a molecular species completely unlike the one under study? One should always bear in mind that the autoradiographic image detects the isotopic marker and does not indicate the actual molecule to which it is attached. Adequate extraction procedures must be employed to identify the molecular species carrying the isotope. Controls of this type necessitate numerous arduous steps in the experimental design. However, they are extremely important and provide a firm basis for quantitative determinations. Unfortunately, this aspect is often overlooked or at times impossible to achieve.

Since it is impossible to cover all the factors that influence quantitation in autoradiography, I suggest that the reader refer to several excellent articles on the topic of autoradiography and use their bibliographies. An excellent text on the topic has appeared by W. A. Rogers (37). Feinendegen (12) has considered tritium-labled molecules in his text which contains an excellent supplementary bibliography. Perry (31) has considered the various factors involved in quantitative autoradiography in an excellent article. Other articles on the methodology, theory, and limitations of the autoradiographic technique can be found in Volumes I and II of "Methods in Cell Physiology." The articles by Caro (6) and Salpeter (39) are especially helpful toward understanding autoradiography with the electron microscope. Stevens (45) has presented an excellent article on the methodology of autoradiography at the levels of the light and electron microscopes.

II. Methods of Obtaining Relative Measurements of Radioactivity

Two general approaches have been utilized to obtain relative measurements of radioactivity: (1) visual counting, and (2) densitometry. In visual counting of single grains one assesses the number produced by decay of low energy isotopes. Higher energy isotopes such as ^{14}C and ^{32}P produce tracks of silver grains and therefore a single track of several grains is counted as one isotopic decay. Since this approach involves continued use of the light microscope for long periods and is in itself tedious, it has considerable subjective element, and grain counts should be expected to vary among observers and with the same observer after protracted observation. Brief periods of 1–2 hr produce more reliable data. Reliability in counting is also dependent on the observer's ability to recognize a silver grain and distinguish it from extraneous contamination. It may appear unnecessary to comment on this aspect. However, recognition of silver grains is very important to quantitation of single grain and track counting in the light microscope and it is imperative for isotopic localization at the level of the electron microscope. Trained investigators can show a reproducibility in visual counts of 5–10%. Special methods can be employed to increase the precision of grain counting and reduce the fatigue. These include (1) projecting the autoradiographic image on a screen; (2) employing a series of photographs utilizing optical systems with depths of field large enough to include all the grains; or (3) a focal series of photographs [see refs. (25) and (29) for details].

Track counting necessitates an expertise in recognition and a knowledge of the behavior of β- and δ-particles as they travel through the emulsion. Specimens containing low levels of radioactivity are most useful in that they facilitate easy recognition of the source. It is customary to define a β-particle track as four or more silver grains in a row (20), and an autoradiographic image of less than 50 tracks per 100 μ^2 can lead to a precision of about 4% (21). δ-Tracks are produced by electrons that have been ejected by particles passing through the parent atom. The tracks produced by these particles are difficult to assess since they often fork at an angle of approximately 90 degrees and then terminate in large, closely spaced silver grains. The two branches often are misinterpreted as two distinct tracks rather than a continuous one.

Practically all the quantitation of autoradiographic experiments at the present time are based on visual grain counts. Attempts to automate grain counting have employed photometric devices. Several such devices have been described in the literature. Dudley and Pelc (11) described the use of a "flying spot" microscope. Mazia et al. (24) utilized a microdensitometric trace across the autoradiograph. Tolles (47) combined a cytoanalyzer and nuclear track scanner. A promising approach has been to measure photometrically the light reflected by the silver grains when subjected to incident dark field illumination (9, 14, 36). The silver grains must be of a uniform size to obtain accurate quantitation. Miniature silver grains and extraneous particulates will reflect unequal amounts of light and hinder the accuracy of the measurements. Current technology for photometrically counting silver grains and the cost of setting up such equipment prohibits its universal acceptance. Consequently, this technique is usually used to supplement visual counting rather than to replace it. A further limitation is the necessity of having large areas of the specimen to count ($10^4 \mu^2$). Studies of single cells cannot be done. However, within its limits the technique is extremely useful and provides a reasonable amount of accuracy, as seen from the studies of

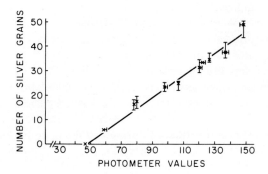

Fig. 1. A graph showing the photometric measurement of light reflected by silver grains under incident dark field illumination related to the number of silver grains in the field of measurement. In the absence of silver grains, a positive deflection is obtained (48 units in this situation). Above this value, direct linearity between photometric readings and visual grain counts is obtained. Vertical marks: mean, highest, and lowest values from three photometric readings. From Rogers (*37*) with permission of the author and Elsevier Publishing Company.

Rogers (*37*) (Fig. 1). For further details of the instrumentation necessary for photometric quantitation see Rogers (*37*).

III. Collection and Presentation of Autoradiographic Data

To obtain a relative quantitation of silver grains by visual counting it is necessary to count several samples of the same specimen. If the distribution of radioactivity among the specimens is uniform, then variations in the grain count reflect the randomness of radioactive decay. However, it is obvious that usually a sizable biologic variation is present in all studies of living systems. Consequently, a number of samples from numerous specimens obtained from several organisms must be counted. The degree of confidence which we can rest in our average count is very closely tied up with the size of the sample taken and with the standard error of the average grain count. (The standard error is that deviation from the true average count which a given count has a 31.7% probability of exceeding.) The actual statistic parameters can be found in any statistics text.

Inherent in a determination of the average number of silver grains in our specimen is the necessity of correcting for background. This is done by visual counting of the silver grains present in the emulsion that overlies the microscope slide adjacent to the tissue of interest, and subtracting this count from that obtained for the silver grains located over the radioactive specimen.

A more refined approach to background counting is to utilize a section of tissue which contains no isotope and is attached to the microscope slide adjacent to the radioactive tissue. This approach provides a barrier between any extraneous radiation present on the microscope slide and the sensitive emulsion. It also has the advantage of controlling for positive or negative chemography of the tissue section itself. Positive chemography is the formation of silver grains in the emulsion over the tissue by chemicals other than the incorporated isotopes. Such chemicals may be naturally present in the tissue or may have been introduced during histologic preparation. Negative chemography is the removal of formed silver grains from the emulsion by chemicals present in the tissue section which are able to oxidize the metallic silver and thus cause an exaggerated fading of latent images.

Background silver grains are those formed in the emulsion by factors other than the decay of the isotope being counted. Background grains can be present because of a variety of reasons. Among the common causes are the following:

(1) Contaminating chemicals may be present in the emulsion, obtained from items used in preparing the emulsion by the manufacturer or by the investigator doing autoradiography. Common sources of contamination are metallic ions contributed by impure water, spoons or forceps, photographic developer and fixer dust, and residual deposits of emulsion left on glassware used for autoradiography.

(2) Environmental radiation will produce background. Common sources are cosmic rays, X-ray machines, and high energy isotopes in the vicinity of the container used to store the autoradiographic emulsion; to some degree the glass of the microscope slide may contain potassium-40 and traces of α-emitting isotopes. The gelatin itself may contain traces of radioactive material.

(3) Spontaneous background can be the result of a reaction of the highly sensitized emulsion to melting temperature and stress induced by drying.

(4) Other factors producing background usually include abrasion produced by rubbing microscope slides together during processing, or a careless finger placed on the emulsion. Small scintillations emitted from silica gel can be a minor cause of background. Silica gel is often used as a drying agent during storage of the autoradiographs while the sensitive emulsion is exposed to the isotopic decay.

An examination of background can be quite informative (Fig. 2). Obviously, the complete absence of background radiation during the entire experiment is the ideal situation for autoradiography. A second best situation is depicted by Curve A, obtained from Rogers (37, p. 96). Upon application of the emulsion to the isotopically labeled specimen, the background is

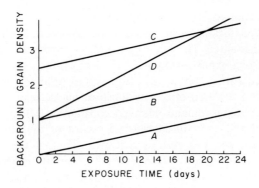

EXPOSURE TIME (days)

FIG. 2. A graph illustrating the accumulation of background silver grains with increasing exposure times and minimal fading of latent images. (A) The ideal situation, with no background at the start of exposure, and slow buildup during exposure, due to environmental radiation. (B) The more usual situation where background levels are appreciably low at the start of the exposure. (C) High levels of background at the start of the exposure, suggesting old emulsion or faulty techniques in the preparation of the autoradiogram. (D) Appreciably low background at the start of the exposure, but rapid increase during exposure possibly due to environmental radiation. From Rogers (37) with permission of the author and the Elsevier Publishing Company.

extremely low. During the process of exposure and photographic development there is a gradual but minimal increase in background grain density. More usually (Curve B), some background is present initially and builds up as exposure time increases. Curve C shows that a high level of background density was present initially. Such results usually suggest that the emulsion had been exposed to some source of radiation, abnormally high temperatures, or has surpassed its shelf life, which is 2–3 months. Curve D shows us that background was initially low and built up inordinately fast during a relatively short exposure. High levels of environmental radiation should be expected. It should be obvious that the level of background grain density should be known before the emulsion is applied to the radioactive specimen. If high grain density is present initially (Curve C) then the emulsion should not be used. Some authors have proposed the use of H_2O_2 to eradicate background. Treatment of 5–6 hr by H_2O_2 vapors decreases background by 95% without affecting the sensitivity of the emulsion (8). In several laboratories, including my own, storage of the autoradiographs during exposure at $-70°C$ has been extremely beneficial in keeping background to a minimum. Over a period of 3–4 weeks there is no noticeable increase in grain density.

Having obtained a relative quantitation of silver grain density by visual counting which has been corrected for background, it is necessary to express the data in some manner which is meaningful to the experimenter. The grain count can be expressed as the number of grains per cell structure, per cell, or per tissue type. Grain density can also be expressed as a function of unit area of the field counted. The field can include entire tissues, cells, or portions of cells. Thus, one expresses number of silver grains per cell, per nucleus, or per unit area, 1 μ^2 or 100 μ^2, of cytoplasm or total cell. When expressing grain density as a function of an entire cell or nuclei, it must be ascertained whether this parameter remains relatively constant or is fluctuating. It should be noted that the silver grain density is a record of the number of labeled molecules present per unit area of specimen. Thus, if the total number of molecules remains constant then volume fluctuations in the structure will dilute or concentrate these molecules. Since the photographic emulsion chiefly sees the cross-sectional area of a structure and a minor portion of its third dimension (1–2 μ in the case of tritium) then the grain density above the structure will be variously decreased or increased. Thus, if we present grain density per square micron of a particular structure we must ascertain whether volume fluctuations have occurred. If they have, it may be more beneficial to express the data as total number of grains per nucleus. Quite often it is not realized that changes in cell or cell organelle size do occur during experimental manipulation or during a time study of a developmental process. Similarly, changes incurred during the preservation and preparation of biologic material may produce sizable changes in the volume of cells or cellular organelles. Shephard (42) has measured the nuclear volume of living melanophores during differentiation and has demonstrated very nicely that the nuclei of developing *Fundulus* melanophores increase in volume during differentiation (Fig. 3). Further, he fixed the developing embryo in acetic acid–alcohol and formalin and showed that in all stages of development the nuclear volume of the fixed preparation was less than the living. There was

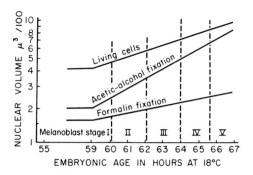

FIG. 3. A comparison of nuclear volumes determined from living cells and from cells preserved with two different fixatives. The roman numerals show the division of *Fundulus* melanoblast differentiation into five stages and related to embryonic age. Obtained from Shephard (42) with permission of the author and Academic Press.

a consistent and constant decrease in nuclear volume due to formalin fixation which approximated a factor of 2.7. However, the decrease in nuclear volume after acetic acid–alcohol fixation was not uniform across the developmental time span. In the melanoblast stages, the decrease in volume due to fixation approximated a factor of 2.5 whereas this decrease in differentiated cells approximated 1.2.

IV. Factors Affecting the Quantitation of Autoradiographs

The technique of autoradiography is relatively simple operationally and is easily incorporated into laboratory routines. However, as with most techniques, the correct use of the technique at both the light and electron microscope level and the proper assessment and interpretation of the data depends upon the investigator's awareness of the numerous variables affecting the autoradiographic image. It is sufficient to itemize the steps involved in the technique to obtain an initial listing of the potential variables.

(1) Administration and Incorporation of the Isotope
  (a) Amount administered
  (b) Method of administration—inhalation, ingestion, injection: subcutaneous, vascular system, peritoneum, intramuscular, diffusion across the epidermis, nutrient media for *in vitro* studies
  (c) Vascular supply of the tissue in an intact organism and the size of the circulating precursor pool
  (d) Size and turnover of the precursor pool, metabolic pathways of utilization, biologic half-life of the synthesized molecule
  (e) Effect of physical and biologic environment on cellular activity: temperature, pH, hormonal effects, etc.
(2) Histologic or Cytologic Preparation
  (a) Fixation and nonspecific absorption of the isotope
  (b) Dehydration, embedding, sectioning
  (c) Extraction controls
  (d) Prestaining
(3) Application of the Autoradiographic Emulsion
  (a) Stripping film
  (b) Liquid emulsion—dipping vs. loop application
(4) Storage Conditions of Autoradiographs during Exposure
(5) Photographic Processing
  (a) Type of developer
  (b) Time and temperature of developing
  (c) Clearing

(6) Physical Factors Affecting the Autoradiographic Density
    (a) Staining
    (b) Background
    (c) Thickness and size of specimen
    (d) Density of specimen
    (e) Emulsion thickness and sensitivity
    (f) Energy spectrum of the isotope

The biologic variables in autoradiographic experiments are usually the most difficult to overcome. Administration of isotopically labeled precursors to *in vitro* systems has the tremendous advantage of being able to administer accurate amounts of the precursor and to present it for a known period of time. Similarly, a "chase" of the unlabeled precursor can be easily administered. On the contrary, tracer experiments *in vivo* are more difficult to control. The isotopes can be administered in a variety of ways and in various amounts (one rule of thumb dictates 1–10 μc/gm body weight). Usually, one can obtain from the literature the dosage, method of administration, etc., and length of autoradiographic exposure for the experimental animal of choice. Peterson and Baserga (*34*) have shown that the means of administration does affect the amount of the isotope incorporated by various tissues. Intravascular injections are useful for rapid distribution of the isotopes and for determining the time of clearance. Most isotopes are immediately removed by tissues from the circulating system. Feinendegen and Bond (*13*) showed that thymidine-^{3}H passed from the bloodstream and into replicating DNA of bone marrow cells within 15 sec after injection. In the time that it took the blood to circulate twice through the organism, there was a 50% decrease in the amount of the isotope in the bloodstream. Similar observations of rapid clearance have been noted for RNA precursors and labeled amino acids (*7, 35*). In developing chick embryos, leucine-^{3}H in the circulation had a half-life of about 20 min (Fig. 4). Borsook *et al.* (*3*) reported 97% clearance of ^{14}C-labeled amino acids in the blood of rats 10 min after injection.

Information of this type is essential because it dictates the presentation time of the isotopic precursor in many experiments. It should be noted that the vasculature and hemodynamics of the circulatory system at the tissue or cells of interest will drastically affect passage of the precursor into the cells. Obviously, cells near a capillary will have the first opportunity for incorporation. However, the presence of "soluble pools" within the cells and their constant interchange with their environment often minimizes large discrepancies in precursor distribution. However, the size of the pool and its turnover rate usually differs with the cell type. It would be extremely beneficial to be able to assay these parameters for "soluble pools." However, where heterogeneous populations of cells are involved this is usually an

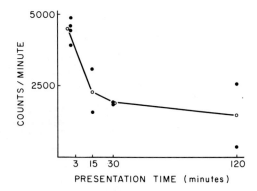

FIG. 4. A measure of the amount of leucine-^{3}H remaining in the blood plasma with time after a single injection expressed as counts per minute. Procedure: 10 μ of leucine-^{3}H (0.567 c/mmole specific activity) were injected into a yolk sac vein of a day 8.5 chick embryo. At various times, approximately 0.1 ml of blood was withdrawn and centrifuged to remove intact cells. Subsequently, 10 μl of the sample diluted with 0.1 ml of distilled water were plated onto planchets. Counts were obtained from a Nuclear-Chicago gas-flow counter. Closed symbols represent the average of a duplicate count on individual samples. Open symbols represent the average of the individual samples. From Przybylski (37).

impossibility at the present time. Labeled precursors are not usually assessed in autoradiographic experiments since they are usually water-soluble and therefore lost during commonly used preparative procedures, or steps are employed to remove them such as treating the tissue with cold trichloracetic acid (TCA). Several approaches have been utilized to preserve water-soluble substances. These techniques of necessity use freezing and " dry " mounting of emulsions to prevent movement. The reader is directed to Miller, Stone, and Prescott (26) for further details.

The preservation of biologic specimens by chemical fixation has presented problems to cytologists for many years. Ideally, accurate morphology and complete retention of cellular components at the sites present in the living cell is desired. However, numerous studies have shown that one type of chemical fixative provides better preservation than another and does not necessarily provide the most appealing morphology. For example, Carnoy's fixative is poor for morphology but is excellent for nucleic acids. For our purpose, it is also good for autoradiography since it does not affect the emulsion and can be used for enzymatic extraction with nucleases.

Kopriwa and Leblond (16) have adequately demonstrated the usefulness of various fixatives for preserving cytidine-^{3}H in tissue sections for autoradiography (Table I). Carnoy proved the best fixative with Bouin and formalin showing a close second and third. However, Zenker's fluid prepared with mercury salts presented a negative chemography and poor preservation

TABLE I

INFLUENCE OF VARIOUS FIXING AGENTS ON THE RETENTION OF LABEL IN MOUSE
LIVER AFTER INJECTION OF CYTIDINE-^{3}H[a]

Fixative	Number of grains/1000 μ^2 [b]
Carnoy	124.4 $\pm$ 3.8
Bouin	116.6 $\pm$ 5.1
Formalin	105.5 $\pm$ 1.9
Zenker	3.1 $\pm$ 0.2
Zenker plus iodine and cysteine[c]	75.1 $\pm$ 2.1

[a] With permission of Kopriwa and Leblond (16). © 1962, The Histochemical Society, Inc. Permission granted by the Williams & Wilkins Company acting for the copyright owner.

[b] Unstained sections.

[c] To remove mercury salts after Zenker fixation, liver sections were placed in Lugol's solution ($I_2:KI:H_2O = 1:2:100$) for 5 min; 5% thiosulfate, 5 min; running water, 10 min; 0.2 M cysteine, 2 hr; running water, 20 min; and distilled water before drying.

of the cellular components. Schneider and Maurer (41) have shown (Table II) that neutral formalin or 0.5% TCA–5% formalin proved a better fixative than Carnoy for several types of tissue labeled with cytidine-^{3}H. Unfortunately, some difficulty has been encountered in getting nucleases to extract after formalin fixation; RNA is more easily extracted than DNA. However, it can be done if conditions of fixation are monitored closely (46). Methanol has been shown to be quite useful for nucleic acid studies (12, p. 189).

Autoradiography with the electron microscope necessitates both excellent preservation of morphology and the isotopic compounds administered. The two most commonly employed fixatives have been glutaraldehyde and osmium tetroxide, which present acceptable morphologic preservation. Glutaraldehyde has also proven effective in enzyme cytochemistry (38). However, some problems have arisen because of the loss of labeled components during fixation and tissue preparation.

Few studies have been made in which glutaraldehyde has been compared to other fixatives for the preservation of isotopic compounds. One such study (33) employing leucine-^{14}C shows that equal amounts of label are preserved with glutaraldehyde, osmium tetroxide, and formaldehyde in protein as determined on trichloroacetic acid precipitates of liver slices. These liver slices had protein synthesis inhibited by puromycin and therefore this value represents residual protein synthesis. Retention of label in normal protein-synthesizing liver slices was not done. The actual point of the study was to show that glutaraldehyde, unlike osmic acid and formaldehyde, will bind free labeled amino acid to bovine serum albumin and to tissue slices.

TABLE II

INFLUENCE OF VARIOUS FIXING AGENTS ON THE RETENTION OF LABEL IN MOUSE TISSUES AFTER INJECTION OF CYTIDINE-^{3}H^a

Fixative	pH	Liver epithelium		Renal tubule epithelium		Adrenal cortex		Adrenal marrow		Colon epithelium		Colon ganglion cells		Small intestine muscle		Small intestine ganglion cells	
		%	g	%	g	%	g	%	g	%	g	%	g	%	g	%	g
η-Formalin	7.0	100	230	100	190	100	570	100	180	100	220	100	700	100	110	100	360
Carnoy-chloroform	3.4	39	90	80	170	79	450	72	130	73	160	14	100	36	40	47	170
0.5% TCA–5% formalin	~1.3	70	160	69	130	51	290	78	140	77	170	20	140	91	100	140	500
5% TCA–6% formalin	~0.9	22	50	—	—	—	—	—	—	—	—	—	—	—	—	—	—
20% Glacial acetic acid– concentrated ethanol	2.8	70	160	—	—	—	—	—	—	—	—	—	—	—	—	—	—
Methanol	—	22	50	—	—	—	—	—	—	—	—	—	—	—	—	—	—

^{a} Data taken 20 min after injection. With permission of Schneider and Maurer (41) and Fischer Verlag.

In the case of the model system (BSA) six washings with 5% TCA after glutaraldehyde fixation left more than 25% of the free labeled amino acid fixed to the protein. Formaldehyde fixation was equivalent to the TCA control, i.e., 0.5% of free amino acid remaining after six washes. The distressing revelation for autoradiographic studies was that up to 5% of the free labeled leucine was bound to the liver slices due to fixation with glutaraldehyde. Thus, a real artifact in the autoradiographs can be introduced by the fixation of the free amino acids to the tissue.

Osmium tetroxide fixation and processing for electron microscopy results in the loss of some protein and low molecular weight peptides. The protein loss as measured by TCA precipitation can be as great as 10% (7). The buffer system employed also affects the amount of protein extracted (up to 15%) (22) and changes the geometry of the cells (48). Osmium tetroxide, unlike acetic acid–alcohol, is a poor fixative for nucleic acids (2) and so it is surprising to find differences in incorporation pattern with labeled nucleic acid precursors. Figure 5 (35) shows an example of such a difference in developing pancreatic acinar cells of the chick after cytidine-^{3}H administration. The obvious differences lie in the amount of radioactivity preserved in the tissue and the pattern of incorporation. The former difference leads to a misinterpretation of the time of peak radioactivity. The difference in the pattern of incorporation may reflect a chemical difference wherein some types of RNA are preserved better than others. The point to be made is that several investigators have attempted to localize the sites of nucleolar RNA synthesis utilizing the techniques of electron microscope autoradiography after fixation with osmium tetroxide without assessing the effect of the fixative on their data and interpretations.

Current interests in the field of lipid synthesis and absorption related to cellular events have resulted in several analyses of the influence of fixation and processing for electron microscopy on the retention of isotopic lipid precursors. Generally, the results of these studies on amoeba (17), lung (27), liver (43), heart (44), and intestinal mucosa (4, 40) show that fixation of labeled precursor and the subsequent loss in the washes and organic solvents varies with the type of tissue being studied and the chemical structure of the intracellular lipid. In amoeba, Korn and Weisman (17) assayed the retention of palmitic acid-^{3}H in specimens processed for electron microscopy after OsO_4, permanganate, and glutaraldehyde fixation. None of the fixatives were able to retain the neutral lipids in the amoebae. Assays for phospholipid showed that OsO_4 was able to retain most of these components, permanganate fixation showed 75% retention, and glutaraldehyde fixation alone was not able to preserve phospholipids in amoebae. Guinea pig lungs contain large amounts of lipid and show very large losses of injected choline-^{3}H during processing for electron microscopy (27). Fixation with OsO_4 in

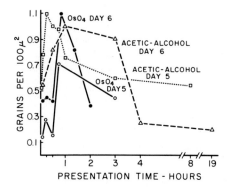

FIG. 5. Comparison of the incorporation of cytidine-³H into nucleoli from day 5 and day 6 pancreatic acinar tissue fixed either in 2% osmic acid or acetic alcohol (1:3). Incorporation is expressed as the number of silver grains per square micron. Autoradiographic exposure time for the acetic acid–alcohol fixed and paraffin embedded tissue was 2 weeks; for osmic acid fixed and plastic embedded material it was 40 days. From Przybylski (35).

s-collidine buffer showed a 66.9% loss, 4% formalin showed a 48.5% loss while a primary fixation with glutaraldehyde followed by a secondary fixation with OsO_4 showed the least loss, 39.0%, which is still a sizable amount. These studies considered only the total amount of label extracted and did not distinguish between label that had been incorporated into cellular structure and that which occurred in the precursor pools. Rat jejunum (4) showed a 26% loss of linoleic acid-1-^{14}C when fixed with OsO_4 and processed for electron microscopy. Most of this loss occurred in the fixative (4%) and in the propylene oxide solvent (10%). In rat heart muscle (4) a large retention of 9,10-oleic-acid-³H occurred after glutaraldehyde fixation. Total loss of label in the processing approximated 12%. Saunders et al. (40) have shown that unsaturated, but not saturated, lipids formed reaction products in jejunal mucosa when fixed with osmium tetroxide and assessed by thin-layer chromatography. Palmitate and cholesterol-^{14}C were retained at the 50% level whereas 85% of mucosal oleate and linoleate-^{14}C were retained.

Thus, all of these studies point out the fact that when one attempts to assess the cellular pathways of lipid absorption or the synthesis of lipid-containing structures, one must recognize that loss of label is occurring and that this loss must be assayed to determine if crucial steps in the biologic process are being lost due to extraction of the labeled lipid. It is remarking the obvious to say that a fair amount of cellular structure is being removed and that the final electron micrograph does not accurately record morphology.

V. Staining

Relating the autoradiographic grain density to morphology necessitates staining of the tissue and cell components for visualization, or using phase-contrast microscopy. With light microscope autoradiographs, staining of the tissue before the application of the emulsion must be done with a stain that does not affect the grain density. Prestaining offers the risk of introducing components that will react with the emulsion and the possibility of removing the labeled material. Prestaining with Feulgen has been shown to reduce the grain count after thymidine-^{3}H administration (18). Positive chemography has been noted when sections are prestained with celestin blue (19) and other dyes which are able to reduce silver halides. Prestaining sections for electron microscope autoradiography presents the possibility of introducing radioactive heavy metals or a negative chemography due to bound osmium (39).

Staining of the tissue after the autoradiograph has been obtained offers the possibility of displacing the silver grains due to swelling and shrinkage of the emulsion. The stain itself may remove the reduced silver grains. In my experience, 0.25% azure B staining for 2–4 hr at 37°C can result in the removal of all the silver grains from an autoradiograph. Background grains appear to be removed first. In doing autoradiography with the electron microscope, it is desirable to remove the overlying gelatin to enhance resolution and visualization of cellular fine structure. Various methods to achieve this end have been utilized, among which are included trypsin digestion, acid and alkali digestion, and dissolution with hot water. All of these techniques run the risk of loss of silver grains and possible minor displacement during gelatin removal. The reader is directed to the articles by Stevens (45) and Salpeter (39), and Leblond et al. (19) and their bibliographies for further discussions of staining.

VI. Preparation of the Autoradiographs

Various factors in the preparation of autoradiographs for light and electron microscopy definitely affect the final grain density. Initially, the type of emulsion and its method of application to the tissue section must be decided. Two types of emulsion are available for light microscope studies: liquid emulsion and stripping film. The latter consists of a silver halide emulsion applied to a gelatin layer adhering to a glass plate. The emulsion and base are peeled off of the glass plate as a thin film, floated on water, and applied to the histologic section so that emulsion and tissue are juxtaposed. The thickness of the emulsion varies from 5 to 25 μ. Thinner layers can be obtained by allowing the emulsion and gelatin to stretch on the water bath.

The important aspect to note is that the emulsion is applied as an intact film. It is closely applied to the tissue section except where deep defects in the surface are encountered. At these points, the emulsion and gelatin film will bridge the gap. Its lack of contact results in inadequate recording of the underlying radioactivity.

Autoradiographic emulsion in the gelled form is liquified upon heating and applied to the tissue preparation in the sol state. All interstices are filled and some penetration of the tissue proper can occur. Thus, all portions of the tissue can be assessed for radioactivity. Emulsion thickness is regulated by dilution of the stock emulsion with water or by forming a thin film by dipping a wire loop (~10-cm diameter) into the liquified emulsion and applying this film directly to the tissue sections mounted on glass slide.

TABLE III

GRAIN SIZE AND RELATIVE SENSITIVITY OF VARIOUS NUCLEAR EMULSIONS[a]

Emulsion	Grain diameter[b] (μ)	Relative sensitivity
Ilford L.4	0.12	132
Demers	0.12	61[c]
Ilford K.5	0.18	100
Kodak V-1055	0.17	51
Kodak AR-10	—	57
Kodak NTB-3	0.23	48
Kodak NTB	0.27	—
Ilford G.5	0.32	—

[a] From Caro and van Tubergen (8) with permission of the authors and The Rockefeller University Press.

[b] The grain diameter was measured in the electron microscope and is accurate to ±10%. The relative sensitivity was measured by taking the average grain count per cell in autoradiographs of bacteria uniformly labeled with tritiated leucine. Development was for 2 min, in D-19 at 20°C, except for L.4 (4 min). The differences in autoradiographic response reflect, in part, differences in grain size, since an emulsion with large silver halide crystals has few of them per unit volume.

[c] This value was obtained without sensitization. When the emulsion was sensitized with triethanolamine the sensitivity became equivalent to that of L.4. On the basis of these results, L.4 was chosen as the most suitable emulsion for electron microscopic autoradiography. For light microscopic autoradiography, it has the disadvantage of showing frequent tracking (several grains caused by one decay) when used in thick layers with tritium. This complicates grain counts in quantitative work. (This tracking largely disappears for reasons which will appear later, when the emulsion is applied as a monolayer of crystals, such as is obtained in electron microscopic preparations.) For this reason we have usually selected K.5 for light microscopic preparations on the basis of its high sensitivity, low tracking, and similarity in composition with L.4.

For electron microscopic studies, emulsion thickness is kept to a minimum as dictated by the penetrability of the electron beam for the visualization of structure and maximum resolution. Numerous methods of emulsion application have been devised with an ultimate goal of obtaining a single layer of silver halide crystals uniformly distributed over the sections and embedded in minimal amounts of supporting gelatin.

The choice of emulsion is further dictated by the grain size. Obviously, the grains must be large enough to see with the light microscope. With the electron microscope, this is not a problem. Rather, an attempt is made to keep grain size small so that the underlying structure containing the labeled compound can be seen and photographed.

Another factor to be considered in the choice of an emulsion is its relative sensitivity. The sensitivity is usually measured by the number of silver grains obtained per unit distance in the tract of particles at minimum ionization. Such information has been difficult to obtain for tritium. However, Caro and van Tubergen (8) have provided an estimate of relative sensitivity for some nuclear emulsions and their grain size (Table III). Rogers (37) has compiled a table in which the crystal grain size and sensitivity to various ionizing particles is given. Using tritium-labeled compounds it is best to choose among Ilford L.2, K.2, and G.2 or Kodak NTB, NTE, and AR-10. Emulsions which are useful for ^{14}C and ^{35}S are Ilford L.3, K.3, and G,3 and Kodak NTB-2. These can also be used for low energy emitters.

VII. Storage Conditions and Photographic Development

The formation of the autoradiographic image occurs by the action of ionizing radiation on the silver halide crystals embedded in the gelatin of the emulsion. Electrons in the crystals are mobilized by the passage of the charged or uncharged particle emitted from the radioactive source. These electrons reduce Ag^+ to metallic silver atoms and this silver appears to be deposited in "sensitivity specks" present in each crystal. This combination results in the formation of a latent image. Those crystals containing the latent image are now more susceptible to the reducing action of photographic developers. However, it should be realized that those crystals without latent images are also acted upon in time and that an overdeveloped autoradiograph can appear completely black as it is fogged by room light. Thus, two important points should be noted: (1) latent image fading should be minimized and (2) photographic developing procedures should be closely regulated to minimize the formation of grains from crystals which do not contain latent images. Photographic grains obtained from crystals containing the latent image constitute a record of the isotopic decay, whereas those crystals

TABLE IV

EFFECT OF VARIOUS STORAGE CONDITIONS ON THE OVERALL AUTORADIOGRAPHIC RESPONSE OF K.5 AND L.4 EMULSIONS[a,b]

	Storage conditions				
Temperature	$-20°C$	$4°C$	$4°C$	$4°C$	$20°C$
Atmosphere	Air	CO_2	Air	Air	Air
Drying agent	Drierite	Drierite	Drierite	$ZnCl_2$[c]	Drierite
	Emulsion and exposure time				
K.5 (2 days' exposure)	36	60	100	112	136
K.5 (5 days' exposure)	83	80	100	100	116
K.5 (7 days' exposure)	100	106	100	—	121
L.4 (7 days' exposure)	85	114	100	—	140

[a] From Caro and van Tubergen (8) with permission of the authors and The Rockefeller University Press.

[b] The grain count per bacterial cell has been normalized with respect to the value found for storage in air at $4°C$ over Drierite (set arbitrarily at 100). The cells used in the various experiments had different amounts of label and the average grain counts varied from 1.5 to 3 grains per cell. In each series all slides were developed together in D-19 at $20°C$ for 2 min, except for the L.4 emulsion which was developed 4 min. Each point represents the average of 300 cells on three different slides. The fact that, in the K.5 experiments, the differences between various storage conditions are less pronounced with longer exposure times might indicate that the length of the exposure is also an important factor in determining the best storage conditions.

[c] Excess $ZnCl_2$ in contact with saturated solution, giving approximately 10% relative humidity.

which do not contain latent images contribute to background density which we employ in our calculations as a correction factor.

Table IV (8) shows the effect of storage conditions on the autoradiographic response of some emulsions. Storage conditions have been varied with regard to temperature, humidity, and atmosphere. These conditions affect the sensitivity of the emulsion, the retention of the latent image, and background. Storage in CO_2 does not seem to markedly affect sensitivity, while temperature has a definite effect. Salpeter (39) has shown that Kodak NTE emulsion requires storage in an atmosphere of dry nitrogen or helium to impede fading of the latent image. In my experience, storage at temperatures approximating $-70°C$ greatly decreases background and latent image fading.

The photographic developing process must be so regulated that a maximum of latent images and a minimum of background grains are developed. The resultant silver grains should be large enough to facilitate recognition without hindering resolution and recognition of underlying structures. Figures 6 and 7 show the efficiency of the developer D-19 on two nuclear track emulsions, Ilford L.4 and K.5 (8). The L.4 emulsion reached a plateau of grain count after 2 min of development during which background grains remain low in number. Prolonged development increases grain count due to isotopic decay with developing time, while background grains are few until 3 min, at which time they increase logarithmically. Obviously, temperature is critical. With increasing temperature the reaction proceeds faster and timing becomes difficult, leading to an increase in background grains and occasional lifting and displacement of the emulsion. Below 20°C the efficiency of the developer decreases. Thus, in all quantitative studies the photographic processing of autoradiographs should be completely standardized and optimal conditions for the emulsions employed should be used.

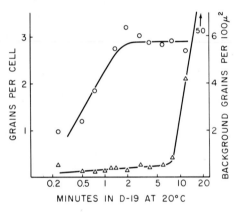

FIG. 6. Development curve for Ilford L.4. The grain count per cell reaches a plateau after 2 min in D-19 at 20°C, while the background does not rise significantly until 6–8 min. From Caro (8) with the permission of the author and The Rockefeller University Press.

VIII. Physical Factors in the Specimen and the Emulsion

The ideal situation for autoradiographic studies and its relative quantitation entails several factors. The structure containing the isotopic compounds should have boundaries that are discrete enough to allow us to correctly assess its dimensions. This structure should be spatially separate from adjacent structures so that a cross-fire from emitted particles does not present a problem. The amount of radioactive material within the structure should be totally preserved and sufficiently large to produce an autoradiographic image in a reasonable length of time. Ideally, the photographic emulsion should be sensitive enough and latent image fading should be so

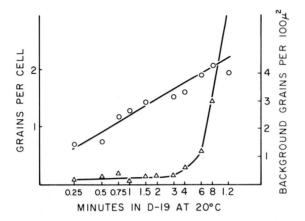

FIG. 7. Development curve for Ilford K.5. Grain counts made on preparations of *E. coli* labeled with leucine-³H show that an increase occurs with development time. The background remains low for the first 3 min of development and then rises suddenly. Since the slope of the grain count per cell curve has a low value (notice that the time scale is logarithmic), reproducible results can be obtained. From Caro (*8*) with permission of the author and The Rockefeller University Press.

negligible that an absolute minimum of isotopic decays will render the silver halide crystal developable. Since latent image formation and growth of the silver grain is initiated by "sensitivity specks," only one such sensitivity speck per silver halide crystal is desired. The resultant developed silver grain should be sufficiently small to allow its identification with the structure in question. The geometric factors of specimen and emulsion thickness and the degree of scattering should be minimal. A thin specimen is necessary to allow the structure in question to lie immediately adjacent to the sensitive emulsion. In this way, radiation can be more accurately assessed due to a minimal loss of emitted particles by scattering and absorption within the structure. Obviously, the ideal situation cannot be realized. I have outlined many of the variables introduced by the organism, the histologic preparative procedures, the preparation of the autoradiographs and their subsequent processing. The critical investigator must also be aware of the various factors which exist in the specimen itself and the spatial relationship of specimen to overlying emulsion and ultimately the overlying silver grains.

The purpose of the autoradiographic technique is the detection of radioactive material in a biologic structure. The extension of the technique to relative quantitation involves a comparison of grain densities related to the structure under varied experimental conditions. Thus, the investigator must have confidence that the silver grains present over a biologic structure in the autoradiographic preparation did indeed result from the emission of ionizing radiation from the radioactive components in the structure instead of adjacent structures. Consequently, much concern should be given to resolution in radioautography and the factors influencing it. Several authors have considered these problems (*1, 5, 28, 30, 31*). It is generally agreed that the factors affecting the localization of the site of radioactive decay and thus limiting autoradiographic resolution include specimen thickness, the size of the developed grain, and the size of the silver halide crystal. These factors are independently variable and statistic and contribute to a final error. Bachmann and Salpeter (*1*) point out that the final error is not a simple sum of the individual limiting errors but is more like a root mean square. They propose the following formula for the total error:

$$E_t = (E_g + E_x + E^2{}_{dg})^{\frac{1}{2}}$$

where E_g is the error caused by geometric factors; E_x is the error caused by the size of the silver halide crystal; and E_{dg} is the error caused by the size of the developed grain. In their discussion E_t is considered the radius of a circular zone around the midpoint of a developed silver grain.

The geometric factors to be considered are the thickness of the specimen and the emulsion, the degree of scattering and cross-fire of the emitted particles, and self-absorption within the specimen. If the radioactive source

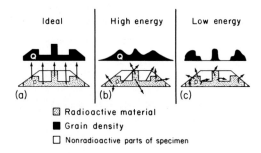

FIG. 8. Schematic illustration of the relationship between the amount of radioactive material incorporated in a specimen (outlined below) and the autoradiographic grain density (in black above). The emulsion is in contact with the upper surface of the specimen. In the ideal case, (a), there is a perfect image of the distribution of radioactivity so that grains Q bear an identical relation to the whole as do their source points, P. In the high energy case, (b), electrons emitted at P may give rise to grains, Q, which are at other points in the image. This leads to image broadening and loss of contrast. Note the diminution in the narrow central peak compared to the broader peaks at each end. In the low energy case, (c), many electrons, especially those emitted from the part of the specimen distant from the emulsion, would be absorbed in the specimen and hence not reach the emulsion. This will greatly improve the resolution, but it will result in a distribution which is distorted in a way which depends on the proximity of different regions to the emulsion. From Perry (*31*) with permission of the author and Academic Press.

lies immediately adjacent to the photographic emulsion then geometric effects due to the specimen do not exist. However, if the specimen deviates from infinite thinness we must consider the effect of specimen scattering and the absorption on the emitted particle, the degree of spread due to the greater distance from the emulsion, and the energy of the particle being emitted. Perry (*31*) has diagrammatically presented these relationships. Figure 8 shows the ideal situation where the grain density corresponds exactly to the distribution and amount of radioactive material in the specimen. If particles of higher energy are being recorded, such as ^{32}P (maximum energy, 1.7 MeV; average range in water of 1250 μ), ^{35}S (0.167 MeV; 50 μ), and ^{14}C (0.155 meV; 40 μ) then the image spread is considerable and self-absorption by the specimen is negligible. Resolution in this situation is poor. It can be increased by decreasing cross-fire if the specimen and emulsion thickness are decreased. However, decreasing the emulsion thickness would reduce the number of emitted particles recorded, since a sizable number of them would have sufficient energy to pass through the emulsion without forming a latent image.

Using radioactive sources such as ^{3}H (0.018 MeV; 1 μ) presents a situation where absorption effects by the specimen are sizable. However, the range of the particles is small and contributes to an increase in resolution. Consequently, the thickness of the specimen should be kept to 1 or 2 μ. If this is technically difficult as in the case of monolayer tissue culture, then it should be realized that the silver grains being counted in this type of preparation result from emitted particles located within the top 1–2 μ of the specimen. The possibility exists that a nuclear grain density count can be recorded as near zero even though a fair amount of radioactivity is present. If the cytoplasm is present between the emulsion and the nucleus and the nucleus is not a flat structure approximating 1 μ in thickness, then the possibility exists that nuclear density would approximate zero even though a fair amount of isotope is present.

If section thickness is kept to 1 μ the probability of forming silver grains by all the emitted particles is greatly enhanced. Those originating at the surface will be unaffected by the specimen, whereas those particles emitted 1 μ from the emulsion must contend with the dry mass of the specimen and thus suffer some absorption. It cannot be assumed that all portions of a cell or various types of cells contribute equally to absorption of emitted particles. Similarly, it cannot be assumed that the densities of cells and cellular components in living preparations can be used to determine their effects on absorption since several studies have shown that chemical fixation, dehydration, and embedding and sectioning can drastically change their densities in the desiccated cells. Within the cell, the nucleolus has been shown to have the greatest density and is usually 2-3 times greater than the other cell components. One can easily realize that histologic preparations of various cell types all differ in their respective densities. Since dry mass is roughly approximated by protein content, it is not unusual to suspect that a muscle cell in the early stages of development has less dry mass than one which has synthesized massive amounts of contractile proteins. Similarly, a pancreatic acinar cell containing large amounts of secretory zymogen has more dry mass than one that has released its cytoplasmic stores of enzymes.

Maurer and Primbsch (23) have presented the results of their studies relating grain density to section thickness and self-absorption (Fig. 9). Using methacrylate-embedded mouse liver labeled with cytidine-^{3}H and tyrosine-^{3}H, these authors cut sections of varying thickness, applied them to glass slides, removed the methacrylate, and produced autoradiographs with AR-10 stripping film. Grain density per square micron of nucleolus, karyoplasm, and cytoplasm was determined and related to section thickness. Grain density increased initially with section thickness to a point where saturation was evident due to section thickness. This saturation was observed to occur

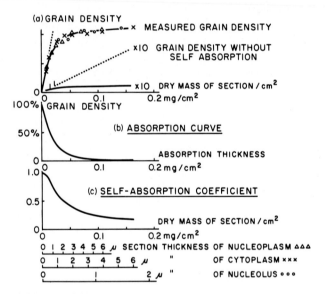

Fig. 9. (a) Grains per unit area (ordinate) vs. dry mass of sections in mg/cm² (abscissa). Experimental points: ○, nucleolus; △, nucleoplasm, × cytoplasm. All points fall on the single curve (solid line) which is valid for desiccated biologic material. Dotted curve is extrapolation to an infinitely thin section and gives the grains per unit area which would obtain if there were no self-absorption. (b) Absorption curve, obtained from the solid curve of (a) by differentiation. (c) Self-absorption coefficient, obtained by plotting the ratio of the solid to the dotted curves of (a). Scales at bottom illustrate the thickness in microns of each cell compartment which corresponds to the abscissa values of dry mass in (a), (b), and (c). From Mauer and Primbsch (23) with permission of the authors and Academic Press.

at 0.25 μ for nucleoli and 1 μ for karyoplasm and cytoplasm. The dry mass of the cellular compartments in 1 μ sections was determined with an interference microscope and gave values for the nucleolus equal to 0.095 mg/cm^2; karyoplasm, 0.018 mg/cm^2; and the cytoplasm, 0.025 mg/cm^2. With these data, they were able to calculate by differentiation an "absorption curve" which indicates the fraction of tritium atoms that will pass through tissue to produce silver grains. These data give some insights to the effect of overlying tissue on the production of silver grains. The calculated self-absorption coefficient indicates the effect of dry mass upon the production of silver grains (calculated as the ratio of the measured to the ideal number of silver grains). Thus, we see that a measure of 50% of the ideal activity in material with a dry mass of 0.03 mg/cm^2 can be obtained from a 2 μ section of nucleoplasm, a 1.5 μ section of cytoplasm, and a 0.25 μ section of nucleolus. The reader is directed to the original article for details and to the article by Perry *et al.* (*32*) for an application of these principles in a quantitative analysis of RNA synthesis in HeLa cells using autoradiographic techniques.

The range of β-particles is determined by their energy and density of the matter they traverse. Tritium is extremely useful in autoradiography because in biologic material 80% of the tritium electrons have a range below 1 μ and 99% do not reach 2 μ and only 2.5% reach 2.5 μ (*5, 31*). The maximum range of tritium β-particles is less than 1 μ since the halide crystals are about six times more dense than water. Hence, the contribution to image spread is negligible. Hill (*15*) using tritium embedded in Araldite sections has shown that the grain density of Ar-10 film descends to one-half its peak 0.21–0.30 μ from the source. This corresponds to a resolution of about 0.6 μ (grain size AR-10 is 0.2–0.3 μ) which is adequate for most light microscope autoradiography.

Little advantage can be obtained by decreasing the thickness of the emulsion when using tritium. Some advantage is recognised by using silver halides with smaller grain sizes. On the contrary, when using isotopes which emit higher energy β-particles a distinct increase in resolution by reducing emulsion thickness can be realized. Such a reduction minimizes cross-fire and thus the perceived silver grains more closely correspond to their source. Since the grain density is proportional to the radiation intensity which varies inversely as the square of the distance, it should be realized that a reduction in emulsion thickness results in the recording of only the lower energy portion of the spectrum of emitted particles. In many cases, the increased resolution is more beneficial than the possible detriment of decreased grain density.

For studies on subcellular structures with the electron microscope, considerations of resolution are of extreme importance in the identification of the cellular structure prior to an assessment of grain density. Emulsion thickness and the size of the silver halide crystal will clearly affect resolution when the number of grains per source is high. If we must identify two point sources 0.1 μ apart, then, it is necessary to have the crystal size reduced to an absolute minimum so that a number of grains can be present in between the two sources. Similarly, numerous grains are desired so that as the exposure increases a sufficient number of grains are present to record the emitted particles by forming numerous latent images. This can be achieved by decreasing crystal size and increasing emulsion thickness. A second layer of crystals can record rather efficiently since the crystals in the first layer do not form a completely dense layer. Numerous spaces between adjacent crystals are evident in all the micrographs of monolayer emulsions.

Three factors definitely contribute to the accurate identification of the labeled source. Of the numerous crystals present over a radioactive source, those immediately above the point of emission have a greater probability of being hit and forming a latent image than those further away because they offer a greater cross-sectional area to the source. Thus, 50% of the developed silver grains will lie within the area of a single crystal. In order for the emitted particle to encounter adjacent crystals, they must travel through the section (density 1.1) at an acute angle and through a portion of a silver halide crystal (density 6.7) to reach the second crystal where it is stopped. Caro (5) has stated that in his experience the passage of a tritium β-particle through a crystal results in absorption or scattering of the particle out of the plane of the emulsion. This occurred with a probability close to one. Thus, several crystals in a monolayer cannot be exposed. A second factor is that the latent image produced in a crystal due to passage of a particle occurs at "sensitivity specks" which may not coincide with the path of particle entry. An extension of this situation leads to a third factor wherein the developed silver grain may not coincide with the position of the latent image, although it does contact it at some point. This phenomenon is obvious if one considers the growth of the whorled silver grain seen in the electron microscope autoradiographs. This factor can be minimized by using techniques of physical development which allow the identification of the latent image. However, the diameter of the crystal must still remain the radius of the circle encompassing the possible source of radioactivity. To date, the minimal crystal size available is approximately 0.05 μ. The possibility exists of reducing this size to 0.01 μ and still having a reasonably sensitive crystal.

REFERENCES

1. Bachmann, L., and Salpeter, M. M., Autoradiography with the electron microscope. A quantitative evaluation. *In* "Quantitative Electron Microscopy" (G. F. Bahr and E. H. Zeiter, eds.), pp. 303–315. Williams & Wilkins, Baltimore, Maryland, 1965.

2. Bahr, G. E., Osmium tetroxide and ruthenium tetroxide and their reactions with biologically important substances. *Exptl. Cell Res.* **7**, 458–479 (1954).

3. Borsook, H., Deasy, C. L., Hagen-Smit, A. J., Kieghley, G., and Lowy, P. H., Metabolism of C^{14}-labeled glycine, L-histidine, L-leucine and L-lysine. *J. Biol. Chem.* **187**, 839–848 (1950).

4. Buschmann, R. J., and Taylor, A. B., Extraction of absorbed lipid (linoleic acid-1-C^{14}) from rat intestinal epithelium during processing for electron microscopy. *J. Cell Biol.* **38**, 252–255 (1968).

5. Caro, L. C., High resolution autoradiography. II. The problem of resolution. *J. Cell Biol.* **15**, 189–199 (1962).

6. Caro, L. C., High-resolution autoradiography. *Methods Cell Physiol.* **1**, 327–364 (1964).

7. Caro, L. C., and Palade, G. E., Protein synthesis storage and discharge in the pancreatic exocrine cell. An autoradiographic study. *J. Cell Biol.* **20**, 473–495 (1964).

8. Caro, L. C., and van Tubergen, R. P., High-resolution autoradiography. I. Methods. *J. Cell Biol.* **15**, 173–188 (1962).

9. Dendy, P. P., A method for the automatic estimation of grain densities in autoradiography. *Phys. Med. Biol.* **5**, 131–137 (1960).

10. Deuchar, E. M., Staining sections before autoradiographic exposure: Excessive background caused by celestine blue. *Stain Technol.* **37**, 324 (1962).

11. Dudley, R. A., and Pelc, S. R., Automatic grain counter for assessing quantitative high-resolution autoradiographs. *Nature* **172**, 992–993 (1953).

12. Feinendegen, L. E., "Tritium-Labeled Molecules in Biology and Medicine." Academic Press, New York, 1967.

13. Feinendegen, L. E., and Bond, V. P., Differential uptake of H^3-thymidine into the soluble fraction of single bone marrow cells determined by autoradiography. *Exptl. Cell Res.* **27**, 474–482 (1962).

14. Gullberg, J. E., A new change-over optical system and a direct recording microscope for quantitive autoradiography. *Exptl. Cell Res. Suppl.* **4**, 222–230 (1957).

15. Hill, D. K., Resolving power with tritium-autoradiographs. *Nature* **194**, 831–832 (1962).

16. Kopriwa, B. M., and Leblond, C. P., Improvements in the coating technique of radioautography. *J. Histochem. Cytochem.* **10**, 269–284 (1962).

17. Korn, B. D., and Weisman, R. A., Loss of lipids during preparation of amoebae for electron microscopy. *Biochim. Biophys. Acta* **116**, 309–316 (1966).

18. Lang, W., and Maurer, W., Zur Verwendbarkeit von Feulgen-gefärbten Schnitten für quantitative Autoradiographie mit markierten Thymidin. *Exptl. Cell Res.* **39**, 324 (1962).

19. Leblond, C. P., Kopriwa, B. M., and Messier, B., Radioautography as a Histochemical tool. *In* "Histochemistry and Cytochemistry" (R. Wegmann, ed.), p. 1. Macmillan, New York, 1963.

20. Levi, H., A discussion of recent advances towards quantitative autoradiography. *Exptl. Cell Res. Suppl.* **4**, 207–221 (1957).

21. Levi, H., and Nielsen, A., Quantitative evaluation of autoradiograms on the basis of track or grain counting. *Lab. Invest.* **8**, 82–93 (1959).

22. Luft, J. H., and Wood, R. L., The extraction of tissue protein during and after fixation with osmium tetroxide in various buffer systems. *J. Cell Biol.* **19**, 46A (1963).

23. Maurer, W., and Primbsch, E., Grösse der beta-Selbstabsorption bei der ^{3}H-Autoradiographie. *Exptl. Cell Res.* **33**, 8–18 (1964).

24. Mazia, D., Plaut, W., and Ellis, G., A method for the quantitative assessment of autoradiography. *Exptl. Cell Res.* **9**, 305–312 (1955).

25. Micou, J., and Goldstein, L., A simple method to reduce the strain in manual grain counting of autoradiographs. *Stain Technol.* **34**, 347–348 (1959).
26. Miller, O. L., Jr., Stone, G. E., and Prescott, D. M., Autoradiography of water-soluble materials. *Methods Cell Physiol.* **1**, 371–380 (1964).
27. Morgan, T. E., and Huber, G. L., Loss of lipid fixation for electron microscopy. *J. Cell Biol.* **32**, 757–760 (1967).
28. Moses, M. J., Application of autoradiography to electron microscopy. *J. Histochem. Cytochem.* **12**, 115–130 (1964).
29. Ostrowski, K., and Sawicki, W., Photomicrographic method for counting photographic grains in autoradiograms. *Exptl. Cell Res.* **23**, 625–628 (1961).
30. Pelc, S. R., Theory of electron microscope autoradiography. *J. Roy. Microscop. Soc.* [3] **81**, 131–139 (1963).
31. Perry, R. P., Quantitative autoradiography. *Methods Cell Physiol.* **1**, 305–326 (1964).
32. Perry, R. P., Ererra, M., Hell, E. A., and Durwald, H., Kinetics of nucleoside incorporation into nuclear and cytoplasmic RNA. *J. Biophys. Biochem. Cytol.* **11**, 1–13 (1961).
33. Peters, T., Jr., and Ashley, C. A., An artefact in radioautography due to binding of free amino acids to tissue by fixatives. *J. Cell Biol.* **33**, 53–60 (1967).
34. Peterson, R. O., and Baserga, R., Route of injection and uptake of tritiated precursors. *Arch. Pathol.* **77**, 582–586 (1964).
35. Przybylski, R. J., Cytodifferentiation of the chick pancreas. III. The content and synthesis of ribonucleic acid, and proteins in developing acinar cells. *J. Morphol.* **123**, 173–190 (1967).
36. Rogers, A. W., A simple photometric device for the quantitation of silver grains in autoradiographs of tissue sections. *Exptl. Cell Res.* **23**, 228–239 (1961).
37. Rogers, A. W., "Techniques of Autoradiography." Elsevier, Amsterdam, 1967.
38. Sabatini, D. D., Bensch, K., and Barnett, R. J., Cytochemistry and electron microscopy. The preservation of cellular ultrastructure and enzymatic activity by aldehyde fixation. *J. Cell Biol.* **17**, 19–58 (1963).
39. Salpeter, M. M., General area of autoradiography at the electron microscope level. *Methods Cell Physiol.* **2**, 229–254 (1966).
40. Saunders, D. R., Wilson, J., and Rubin, C. E., Loss of absorbed lipid during fixation and dehydration of jejunal mucosa. *J. Cell Biol.* **37**, 183–187 (1968).
41. Schneider, G., and Maurer, W., Autoradiographische Untersuchung über den Einbau von ^{3}H-Cytidin in die Kerne einiger Zellarten der Maus und über den Einfluss des Fixationsmittels auf die ^{3}H-Activität. *Acta. Histochem.* **15**, 17–181 (1963).
42. Shephard, D. C., An approach to the study of cytodifferentiation in melanophores. *Develop. Biol.* **6**, 311–332 (1963).
43. Stein, O., and Stein, Y., Lipid synthesis, intracellular transport, storage and secretion. I. Electron microscopic radioautographic study of liver after injection of tritiated palmitate or glycerol in fasted and ethanol treated rats. *J. Cell Biol.* **33**, 319–340 (1967).
44. Stein, O., and Stein, Y., Lipid synthesis, intracellular transport and storage. III. Electron microscopic radioautographic study of the rat heart perfused with tritiated oleic acid. *J. Cell Biol.* **36**, 63–77 (1967).
45. Stevens, A. R., High resolution autoradiography. *Methods Cell Physiol.* **2**, 255–310 (1966).
46. Swift, H., The quantitative cytochemistry of RNA. *In* "Introduction to Quantitative Cytochemistry" (G. L. Wied, ed.), pp. 355–386. Academic Press, New York, 1966.

47. Tolles, W. E., Methods of automatic quantitation of microautoradiographs. *Lab. Invest.* **8**, 99–112 (1959).
48. Tooze, J., Measurements of some cellular changes during the fixation of Amphibian erythrocytes with osmium tetroxide solutions. *J. Cell Biol.* **22**, 551–563 (1964).

TISSUE PREPARATION FOR THE AUTORADIOGRAPHIC LOCALIZATION OF HORMONES*

Walter E. Stumpf†

DEPARTMENT OF PHARMACOLOGY, UNIVERSITY OF CHICAGO, CHICAGO, ILLINOIS

I. Introduction

Information about the sites of action of hormones may be obtained by following their routes in the body after exogenous application. Distribution patterns related to time can be optimally obtained by procedures which (a) maintain the *in situ* conditions of hormone distribution, (b) provide cellular and subcellular resolution, and (c) simultaneously inform about the topographic relationship of their sites of deposition.

Autoradiography had the potential to provide this information; however, progress in techniques was necessary in order to fulfill the listed requirements. Classical histologic preparations did not permit the study of diffusible, that is, noncovalently bound, substances. Liquid fixatives; dehydrating, rehydrating and clearing fluids; embedding media; section floating on water and wet section mounting; and thawing of frozen sections are all reported sources of translocation, redistribution, or extraction of tissue constituents and exogenously applied material as well (reviewed in *32, 45*).

The high sensitivity of the autoradiographic procedure, which is due to the use of radioactive tracers and information storage in photographic emulsion, not only permits the study of low and physiologic doses of labeled hormones with, for instance, a concentration of estradiol below its saturation at about 2×10^{-8} mole/kg rat uterus (*16*), but simultaneously records information degradation related to tissue preparation as well. It is for this reason that the attractive pictorial approach often provides the ambiguous and

* Supported by U.S. Public Health Service Grant AM-12,649 and Berlin Laboratories Fellowship Grant.

† Present address: Departments of Anatomy and Pharmacology, University of North Carolina, Chapel Hill, North Carolina.

conflicting results which have been published in the literature. The latter can be demonstrated in the elucidation of the deviating reports on the auto-radiographic localization of estradiol-^{3}H in the uterus (*32, 33, 43*), pituitary (*35*), and brain (*34, 37*). This has been reviewed by the author and related to results obtained by different technical approaches such as centrifugal fractionation (*36*) as well as his own comparative studies with six different autoradiographic techniques (*32, 43*). In these autoradiographic experiments the degree of diffusion, redistribution, and leaching of the radioactive label —estradiol-^{3}H and mesobilirubinogen-^{3}H were used—was related to the extent to which liquids, including embedding media and thawing of frozen sections, were employed. Only when all of such invalidating steps were avoided, were results obtained which agree with our present knowledge about the pattern of distribution, the locus of action, or excretion as derived from various other approaches. It is noteworthy that in the comparative study, the data obtained from each autoradiographic technique were reproducible within the same technique, although results from the different techniques contrasted with each other.

Attempts to localize hormones date back to the time when radioiodide became available (*13*) and later radioiodine-labeled thyroid hormones were used (*10, 15*). As yet, however, there is no consensus about the site of thyroid hormone synthesis in the thyroid follicles between pictorialists and bio-chemists or even among the pictorialists themselves. The latter is not sur-prising, since all the reported studies were done with wet autoradiographic techniques, and Nadler *et al.* (*22*), for example, admit loss of 10–15% of the total radioactivity in the thyroid during their tissue preparation. Localization of other hormones such as progesterone (*17, 27*), estrone and deoxycorti-costerone (*17*), testosterone (*21*), and insulin (*2*) has been reported, but all of the data must be critically evaluated since they are likely to include varying degrees of artifacts due to inadequate autoradiographic technique. Bogoroch (*4*) localized aldosterone-^{3}H in cell nuclei of the toad bladder with an auto-radiographic technique which excludes nearly all of the known sources of extraction and translocation. Her data are not in contrast with the reported biochemical action of aldosterone.

II. Adverse Effects of Fixation

Although liquid fixation proved generally useful for the preservation of tissue morphology, its limitations for histochemistry have been recognized by many investigators. Table I is a compilation of selected references from the literature in which loss of tissue mass during commonly used fixation procedures was measured. It has been stated therefore that "fixation" should not be mistaken for immobilization of tissue constituents and that the term "fixation" is misleading, since liquid fixatives are actually denaturing solvents. The denaturing effect, which may lead to varying enzyme inactivation; the threat of translocation and leaching; the possibility of binding and trapping *in situ* unbound material (*26*); as well as the uncontrolled and delayed quenching of metabolic processes by the time required for penetration of the fixative into the tissue, may render this procedure inadequate for more delicate histologic and histochemical studies. Therefore, quenching by freezing followed by freeze-drying (*11*, *44*), and freeze-substitution (*30*) have been introduced; and in order to also circumvent embedding, freeze-drying of frozen sections (*1*, *18*, *39*) and frozen section freeze-substitution (*6*) have been recommended.

It is in line with these efforts to eliminate or minimize artifacts occurring in tissue preparation that in recent years freeze-etching (*31*) has been increasingly applied. The use of frozen tissue sections in the hydrated state would exclude the need for freeze-drying and thus eliminate another potential source of artifact.

In Table II an attempt is made to outline the general interrelationship between *in situ* preservation and possible information degradation as related to tissue preparation.

III. Freeze-Dried Section Preparation for the Autoradiogram

After reviewing the literature on the autoradiography of diffusible substances (*41*), it became apparent that a new technical approach was needed and a method had to be designed which would provide cellular and subcellular resolution and yet avoid all known sources of diffusion during tissue preparation and mounting. Theoretically three approaches are available: (a) use of frozen sections, nondehydrated; (b) use of frozen sections, freeze-dried; or (c) use of frozen tissue blocks, freeze-dried and vapor-fixed, with the fixative forming a specific link between the labeled substance and the tissue in order to obtain immobilization and to prevent shifting and loss during subsequent embedding and other treatment. The first approach, although optimal, would require maintenance of low temperatures during sectioning and film mounting in order to prevent thawing or disruptive recrystallization of the ice. For high resolution light microscopy this would probably mean maintaining a temperature below $-30°C$, optimally below $-60°C$ (*44*), until the end of the photographic exposure of the autoradiograph. Our own efforts to mount frozen sections on emulsion-coated slides at low temperatures without deleterious rise in temperature during handling were not successful. While the third approach of finding a specific fixative may be realizable in exceptional cases, it would still require controls without

TABLE I

Loss of Tissue Constituents by Fixation

Authors	Fixatives	Duration	Temperature	Labeled molecules	Loss of tissue or radioactivity
Sylvén (46)	(1) Saline 0.9% (2) Formaldehyde soln. 5% (3) Abs. alcohol (4) Carnoy solution	24 hr	Room temp.	—	10–30% Loss of the total tissue mass of fresh organs Slices from perfused rabbit liver, lung, skeletal muscle, and spleen were used
Merriam (19)	(1) 10% Formalin (2) Acetic acid–alcohol	19–20 hr	Room temp.	—	With formalin: 25% (liver), 22% (muscle) loss of total dry weight, 5–8% further loss during subsequent dehydration and infiltration, 0.2–0.6% of total dry weight protein loss in water after embedding With acetic acid–alcohol similar loss
Dallam (7)	OsO_4 buffered	20 hr	—	—	10.2–16.9% Loss of protein from rat heart and kidney pieces (50–150 mg) during whole tissue preparation 29.2–31.8% Loss from isolated mitochondria and 13.8–14.9% loss from isolated liver microsomes

Author	Fixative	Time	Temperature	Isotope	Loss	Remarks
Ostrowski et al. (24)	Formalin or Carnoy's fluid; Absolute alcohol or acetone	24 hr	Room temp.	—	1% / 2.3%	Protein elution from rat liver or kidney pieces of fresh tissue weight
Droz and Warshawsky (8)	Bouin's fluid and treatment with ethanol and dioxane	48 hr	Room temp.	1-leucine-^{14}C	10–50%	loss of total radioactivity from tissue of mice sacrificed and exsanguinated at 30 min or 24 hr after intraperitoneal injection
Morgan and Huber (20)	s-Collidine-buffered OsO_4; 4% Buffered formalin followed by OsO_4; 2% Glutaraldehyde followed by OsO_4; Tricomplex fixation	2 hr / 3 hr and 2 hr / 24 hr and 2 hr / 2 hr	Room temp. / Room temp. / Room temp. / Room temp.	Choline-methyl-^{3}H Cl	66.9% / 46.5% / 39% / 51.3%	Loss of total radioactivity from 1-mm cubes of lung tissue, excised 0.5–30 hr after i.p. injection, fixed in 10 volumes of fixative; 9.9–37.8% was lipid-soluble radioactivity
Schneider and Schneider (29)	Formol 10%; Formol 40%; Carnoy solution	180 days	Room temp.	—	21% dry weight / 40% dry weight	21% dry weight loss from 2-cm brain slices (including loss of protein, DNA and RNA) / (90% of the extractable lipoids)

TABLE II

MICROSCOPIC TISSUE PREPARATION AND INFORMATION DEGRADATION[a]

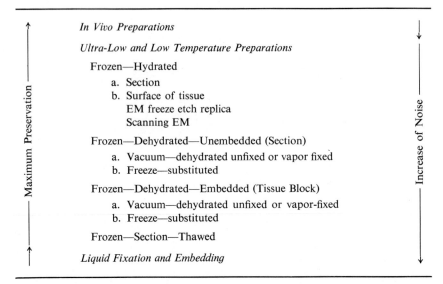

Maximum Preservation ←		Increase of Noise →
	In Vivo Preparations	
	Ultra-Low and Low Temperature Preparations	
	Frozen—Hydrated	
	a. Section	
	b. Surface of tissue	
	EM freeze etch replica	
	Scanning EM	
	Frozen—Dehydrated—Unembedded (Section)	
	a. Vacuum—dehydrated unfixed or vapor fixed	
	b. Freeze—substituted	
	Frozen—Dehydrated—Embedded (Tissue Block)	
	a. Vacuum—dehydrated unfixed or vapor-fixed	
	b. Freeze—substituted	
	Frozen—Section—Thawed	
	Liquid Fixation and Embedding	

[a] In biological experiments the general objective is to obtain information about the *in vivo* state. Instead of searching for the "right" fixative or "right" embedding material it may be more useful to perfect minimal-treatment procedures. The sequence in the table may vary, depending on the subject of study.

fixative. The second procedure, that is, the use of freeze-dried frozen sections, appeared to be the most promising. Anfinsen, Lowry, and Hastings (*1*) in 1942 prepared freeze-dried and unembedded sections for the use in quantitative histochemistry. The histologic quality of freeze-dried sections, prepared at $-40°C$ in a cryostat, had been compared with those fixed with formalin and embedded in paraffin as well as with thawed sections by Bloom *et al.* (*3*) and an apparatus had been designed by Grunbaum *et al.* (*12*) to facilitate the preparation of freeze-dried sections. However, except for quantitative histochemical studies (*18*), freeze-dried sections have only on exceptional occasions been used for histomorphologic preparations. This was apparently due to the difficulties experienced in obtaining undamaged and undistorted sections, since ice crystal formation, shrinkage after incomplete freeze-drying, and autolysis may occur during inadequate preparation and handling of unfixed freeze-dried sections. Fitzgerald (*9*) in 1961 proposed the use of freeze-dried sections for autoradiography, but apparently did not succeed in obtaining useful preparations in his laboratory. More successful preparation of freeze-dried sections and their application in autoradiography became possible after progress in low temperature cutting (40, 42), vacuum freeze-drying (44), and section handling (32) had been made. Rapid freezing, maintenance of low temperatures until the end of freeze-drying, and conditions of low relative humidity for storage and handling of the freeze-dried sections are prerequisites if optimal results are to be obtained.

Cryostatic cutting of 0.5 to 1 μ thin sections, required for high autoradiographic resolution, has been achieved by utilizing lower temperatures ($-50°$ to $-70°$C) than commonly employed. Cryostatic sectioning at these low temperatures was originally investigated by the author in order to minimize ice crystal artifacts from recrystallization. Although cutting at temperatures below $-45°$C was believed to be impossible (25), it has been found that cutting of thin sections is facilitated at lower temperatures. There exists an interrelationship between temperature and section thickness which must be observed in order to be successful. While, for instance, 0.5 to 1.0 μ sections can be cut easily at $-60°$C knife temperature, 2 or 3 μ sections are difficult or impossible to cut since the tissue fractures. At $-20°$C, however, 5 to 10 μ sections are easily cut, but 1 μ sections are difficult or impossible to cut since compression of the sections occurs.

Knife temperatures as low as $-50°$ to $-70°$C may be desirable, or even required, in order to obtain optimal tissue preparation for critical light microscopic studies such as phase contrast or autoradiography of ion distribution. Moreover, for many light microscopic autoradiographic studies 1–4 μ thick sections, prepared at knife temperatures between $-25°$ to $-45°$C, apparently provide satisfactory conditions. This temperature can economically be supplied by a single-stage refrigerator (Harris Manufacturing Co., Cambridge, Massachusetts). Figure 1 depicts such a cryostat, which is equipped with a rotary microtome (International Equipment Co., Needham Heights, Massachusetts) with a cutting range of 2 to 0.1 μ. Sections thicker than 2 μ can be cut by readvancement without cutting. This type of an open top Low Temperature Wide Range Cryostat has an advantage over other commercially available cryostats in that it (a) permits cutting in a wider range of section thickness, including sections thinner than 1 μ and (b) provides safer conditions for tissue preservation, since a larger compartment with colder and heavier air reduces the threat of inadvertent warming of the tissue.

The microtome may be lubricated with molybdenum disulfide (Molykote Z powder, Alpha-Molykote Corp., Stamford, Connecticut) applied dry or in absolute alcohol as a vehicle, or by a Teflon coating. Sectioning is performed while observing through a dissection microscope which is mounted at the top ledge of the cryostat. As a light source for cutting, fiber optic light pipes (Iota Cam Model 150-T, Fiber Optic Corp., Wakefield, Massachusetts) which provide cold light are used. Cutting of thin frozen sections requires

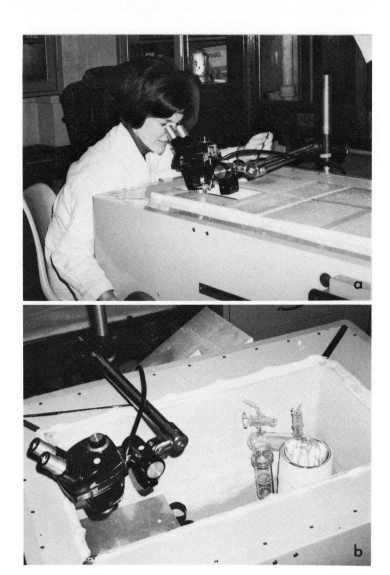

FIG. 1. Wide-range low temperature cryostat for cutting of frozen sections and freeze-drying. (a) Cutting is performed under observation through a dissecting microscope, illuminated by a fiber optic light pipe providing cold light. The sections are transferred from the knife with a fine brush held in the left hand. The cryostat is covered during sectioning, except for two small openings for the dissecting microscope and the left hand. (b) After cutting of the sections, the freeze-drying apparatus is assembled within the cryostat. The cryosorption chamber of the vacuum pump (see Fig. 2) is immersed into a Dewar with liquid nitrogen. Freeze-drying is concluded after 12 hr. The Dewar with the pump is then removed from the cryostat and, after allowing the specimen chamber to warm to room temperature, the vacuum is broken with dry gas.

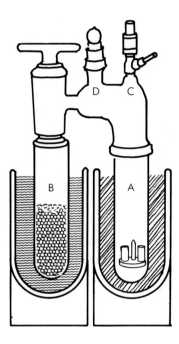

Fig. 2. Cryosorption pump for freeze-drying of tissue. The portable pump is assembled within the cryostat with the help of a mechanical forepump connected to stopcock C. A, Sample chamber with tissue-containing vials, immersed in a Dewar with alcohol–dry ice slush; B, cryosorption chamber with molecular sieve (Linde 5A) immersed in a Dewar with liquid nitrogen; D, port for ion gauge. (The Cyro-Pump may be obtained from Thermovac Industries Corp., Copiague, Long Island, New York.)

the development of skill and, in general, difficulties are initially encountered in obtaining flat sections. The use of an antiroll device for small 0.5–2 μ sections, however, is not needed. Rolling up may be prevented by assuring homeothermic conditions, guiding the section with the bristles of a fine brush during cutting, and pausing for a few seconds before the end of the cutting stroke while the section is supported at both ends—at the one by the brush, while the other is still adhering to the tissue block. Sharpness of the knife, cutting angle, cutting speed, and differences in tissue density may further influence the process; all of these factors must be considered if the cutting is not successful.

Since it is important that the temperature of the tissue is kept low until the end of freeze-drying, the frozen sections are transferred to the precooled sample chamber of the cryosorption pump within the cryostat. The freeze-drying apparatus is also assembled and freeze-drying initiated within the cryostat by help of a forepump. This procedure is easily done in an open-top cryostat (Fig. 1). The forepump action facilitates assembly of the sample chamber and cryosorption pump through a self-seating O-ring joint and provides for a faster and better final vacuum by brief, approximately 5 min, evacuation and activation of the molecular sieve. The compact U-shaped cryosorption pump (Fig. 2) was designed to facilitate assembly within the cryostat and to provide a low enough vacuum (44) for efficient freeze-drying at temperatures of about $-70°C$ with a vapor pressure of ice of about 10^{-3} mm Hg at this temperature.

Premature rise in temperature before the end of freeze-drying may result in tissue disruption due to recrystallization, or shrinkage of the nondried portion of the tissue (23), which would produce autoradiographic artifacts. Improper freeze-drying procedures seem to be widely employed and probably are a frequent source for nonsatisfactory results. In order to determine the time required for freeze-drying under the conditions described in Section V, a Cahn electrobalance was included in the vacuum system and the drying time was followed by continuous registration of weight, temperature, and vacuum (44). It was found that (a) drying of tissue proceeds nonlinearly in an exponential course, and that (b) different tissues dry at different rates. For instance, the half-drying time $(T/2)$ for a 10-mg cube of liver was approximately 30 hr, of kidney 50 hr, and of brain 100 hr. Seven half-drying times were arbitrarily selected for the termination of freeze-drying. Extrapolating from the information obtained from the weighing experiments with tissue blocks between 2 and 29 mg initial weight, it was calculated that freeze-drying of sections may be finished after 12 hr drying time and safely terminated at 20–24 hr, considering delay in obtaining the initial optimal vacuum and providing a safety margin for variations of tissues and in the activation of the molecular sieve (44).

IV. Handling of Unfixed Freeze-Dried Sections

After freeze-drying it is important that the tissue be kept in a low humidity environment in order to prevent enzymes from autolyzing the tissue. Therefore the vacuum should be broken with a dry gas instead of ambient air. When ambient air was used to break the vacuum, the hygroscopic tissue showed increase in weight in the previously mentioned weighing experiments, which was dependent on the relative humidity of the air. The weight increase reached a plateau about 15 min after breaking the vacuum with a gain of about 1% of the dry weight for 10% relative humidity at room temperature. The highest gain registered was a 7% increase of the dry weight at 85% relative humidity. These experiments indicate that a rapid equilibration with ambient humidity occurs within the freeze-dried tissue, which may introduce diffusion and destruction of tissue constituents. Therefore, unfixed freeze-dried sections are stored in a desiccator, the sections mounted at a relative humidity of 20% to 40%, and the mounted slides stored over Drierite at −15°C for the photographic exposure. These precautions are apparently sufficient for light microscopic studies as judged by the histologic quality obtainable (44). If vapor fixation is required, this can be done easily by exposing the freeze-dried sections to the fixative in a separate container or breaking the vacuum by introducing a vapor for fixation and/or staining. Heating of the freeze-dried tissue to above 60°C before breaking the vacuum may also be used in order to reduce or eliminate the threat of enzymic autolysis. The freeze-drying approach excludes, however, the study of volatile substances, especially when combined with terminal heating.

V. The Method of Dry-Mounting of Freeze-Dried Sections

(1) *Simultaneous freeze-quenching and mounting.* A tissue block of 1–2 mm^3 is excised and placed on a tissue holder using minced liver as an adhesive. Tissue with tissue holder are immersed in liquified propane cooled by liquid nitrogen to about −180°C. If larger tissue blocks are used, slower cooling or the use of a different coolant is required in order to avoid fracturing of the specimens. The mounted tissue is stored in a liquid nitrogen refrigerator until further use.

(2) *Cryostatic sectioning.* The tissue is transferred from the storage tank to the cryostat in a liquid nitrogen Dewar, mounted on the microtome, and trimmed for cutting. The sections are transferred from the knife with a fine brush to a vial in the vicinity of the knife. The vial is covered with a fine punched out wire mesh which permits wiping the section from the brush and prevents section loss during freeze-drying and breaking of the vacuum. Several vials may be contained in a tissue carrier which is transferred by a long and precooled forceps to the sample chamber of the cryosorption pump (Thermovac Industries Corp., Copiague, Long Island, New York) within the cytostat.

(3) *Freeze-drying.* Sample chamber and cryosorption pump (Fig. 2) are assembled within the cryostat with the help of a vacuum produced by an outside forepump, which facilitates self-seating of the O-ring joint. After brief evacuation the assembled cryosorption pump is transferred without delay to two outside Dewar flasks, containing dry ice slush as a coolant for the sample chamber and liquid nitrogen as a coolant for the molecular sieve contained in the cryosorption pump. A vacuum better than 10^{-5} torr is obtained. Freeze-drying may be terminated safely after about 20 hr by breaking the vacuum, that is, introducing nitrogen gas (Fig. 2, stopcock C)

after the sample chamber has been removed from the coolant, and allowed to equilibrate to room temperature. The tissue carrier is removed and stored in a desiccator until further use. The cryosorption pump is then removed from the liquid nitrogen and the molecular sieve will desorb the trapped water and reactivate within the pump at room temperature or in an oven at about 80°C for 2 hr.

(4) *Dry-mounting of sections.* Histologic slides which have been coated with liquid photographic emulsion (Kodak NTB 3) are used. They were air-dried, and stored over Drierite. At room temperatures of 20°–22°C and under conditions of a relative humidity between 20% and 40% for the avoidance of electrostatic discharge artifacts, if too low, and diffusion artifacts, if too high, sections are placed on clean pieces of Teflon (Crane Packing Co., Morton Grove, Illinois) using a fine forceps. Although handling of sections should be minimal, checking under a dissecting microscope may be especially

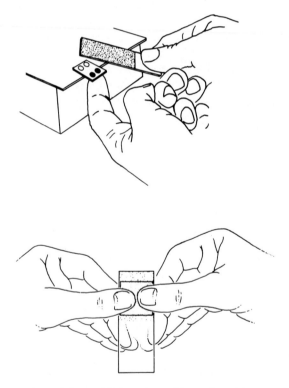

FIG. 3. Dry-mounting of freeze-dried sections: The radioactivity-containing sections, ●, and control sections without radioactivity, ○, are placed on a Teflon piece. Under safe-light a dried emulsion-coated slide is placed over the Teflon (upper picture) and both are pressed together (lower picture).

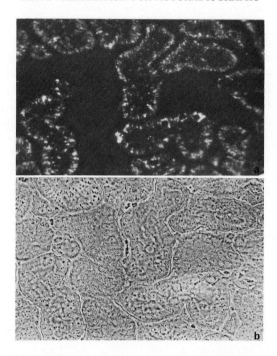

Fig. 4. Freeze-dried section of rat kidney showing autofluorescence (a) and phase contrast (b) of the same area; $2\,\mu$, unfixed and unembedded section, dry-mounted on a gelatin-coated slide, viewed under mineral oil. $341.7\times$ (a) was photographed under ultra-violet light with Zeiss excitation filter 1 and barrier filters 41 and 50, showing yellow and bluish green autofluorescence in the cytoplasm of proximal tubules, while the distal tubules (without the brush border) are nearly devoid of fluorescence.

important for judging the quality of thin and small sections, their positioning, and in order to remove folds. Under safe-light an emulsion-coated slide is placed over the Teflon (Fig. 3, upper picture) and Teflon and slide are pressed together between forefinger and thumb (Fig. 3, lower picture). After release of the pressure the Teflon falls off with the tissue adhering to the emulsion. The mounted slides are stored in a black box (modified Clay-Adams box) with a Drierite compartment and exposed at $-15°C$.

(5) *Photographic processing and staining.* At the end of exposure the slide-containing box is allowed to adapt to room temperature before opening. A slide is removed and the section area is breathed at, once or twice, to briefly moisten the emulsion for better adherance of the section to the emulsion. If the breathing step is not included, loss of sections may occur during photographic processing or staining. The slide is developed with Kodak D 19 developer for about 1 min at $21°C$, briefly rinsed in tap water, fixed in Kodak fixer for 5–7 min, gently rinsed in tap water for 5–10 min,

and stained. Immediate staining with a single-step staining procedure such as methyl green–pyronin is optimal for most autoradiographs, since it does not introduce silver grain fading and provides good histologic differentiation without obscuring of silver grains. After a few seconds of staining and brief rinsing, the slide is air-dried and a coverslip is mounted with Permount. It is important during photographic processing and staining to keep the temperature constant in order to avoid reticulation of the emulsion.

VI. Applications

Freeze-dried and unembedded sections prepared in the described manner have been used successfully for different histochemical applications. Optimal preservation of structure has been demonstrated in high resolution phase-contrast photomigrograms of the kidney, parotid gland, and heart (44). Figures 4a and 4b show a phase-contrast picture of the rat kidney with a corresponding picture of the autofluorescence of the tubules. Catecholamine

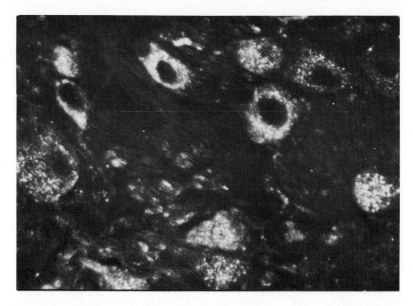

FIG. 5. Fluorescence photomicrograph of the superior cervical ganglion of the rat showing catecholamine fluorescence and autofluorescence, 1-μ freeze-dried section, treated for 2 hr at 80°C with paraformaldehyde vapor. The paraformaldehyde was exposed over a saturated calcium nitrate solution with a relative humidity of 51% at 24.5°C for 24 hr prior to use. The unembedded section was dry-mounted on cleared photographic emulsion after the vapor treatment and viewed under mineral oil. 820×. The same filters were used as for Fig. 4a.

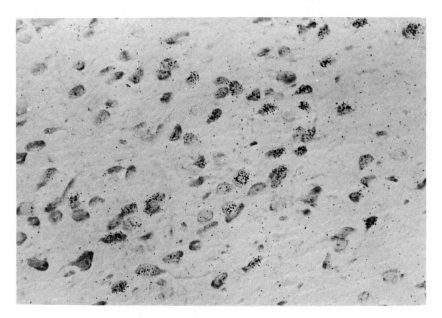

FIG. 6. Autoradiograph of the nucleus ventromedialis tuberis cinerei (frontal plane) prepared 2 hr after subcutaneous injection of 0.4 μg of 17β-estradiol-6,7-³H into mature female rat, ovariectomized 36 hours before the injection. Radioactivity is concentrated in nuclei of neurons accumulated in the pars lateralis of the nucleus ventromedialis (*34*); 2 μ, exposed 117 days, stained with methyl green–pyronin. 152×.

fluorescence can be studied in 1-μ freeze-dried sections by reducing the freeze-drying process to 1 day and eliminating the embedding step, thus permitting examination of the section immediately after the exposure to paraformaldehyde (Fig. 5).

Two micron thin sections of undecalcified bone have been obtained by the described low temperature tissue preparation and used for the study of tetracycline deposition (*28*). Several applications of this method in the auto-radiography of hormones and other diffusible substances such as digoxin-³H (*28*), sorbitol-³H (*32*), and other extracellular space indicators (*5*) have been reported. Estradiol-³H localization has been studied most extensively in the uterus *in vivo* (*36*) and *in vitro* (*14, 32*). Penetration and uptake of the radio-active material *in vitro* could be followed by autoradiography and was found to be limited to a margin of about 200 to 300 μ of the outer tissue which is in contact with the incubation media. This is in agreement with the limits of tissue thickness useful for *in vitro* studies, as reported by Warburg and McIlwain (cf. Stumpf, *32*). Estradiol-³H localization has also been reported in the vagina, oviduct, ovary, induced mammary tumor, liver, adrenal (*38*), pituitary (*35*), diencephalon (*34*, and amygdala *37*). A distinct pattern of cellular and subcellular distribution of radioactivity was found to be typical for target tissues, while nontarget tissues such as liver and adrenal showed different results. The tissues which showed a nuclear concentration in the autoradiographs, except for brain, had been reported in the literature to respond to prolonged estrogen treatment with hyperplasia or tumor formation. This includes interstitial cells of the testis (38). Pituitary anterior lobe cells showed nuclear concentration of estradiol-^{3}H in 50% to 85% of all cells. Acidophiles, basophiles (castration cells), and chromophobes all concentrated and retained to varying degree estradiol-^{3}H, which has been interpreted as being suggestive of a pluripotentiality of anterior pituitary cells (35). In addition, in the posterior lobe, single cells or clusters of cells which are located at the border to the intermediate lobe and stain like anterior lobe cells with methyl green–pyronin, concentrated radioactivity in the same manner, while other posterior lobe cells and intermediate lobe cells did not concentrate the label (35).

In the hypothalamus of female and male rats the topography of estradiol-^{3}H-concentrating neurons (Fig. 6) was determined (34) and found to be in general agreement with the reported topography of terminations of nerve fibers of the stria terminalis, a nerve-fiber bundle which originates in the nuclei of the amygdala. This was interpreted as supporting evidence for an amygdaloid–hypothalamic interrelationship in ovulation control and sexual behavior (34). In most of the above-mentioned tissues the chemical nature of the radioactivity had been identified as being 17β-estradiol-^{3}H.

Freeze-dried section preparations may also be useful for other studies in which translocation artifacts are to be avoided, for instance, microincineration or electron-probe determination of metal-ion distribution. Electron microscope autoradiography with hormones may become possible by utilizing freeze-dried sections. However, tissue preparation for electron microscopic resolution requires considerably lower temperatures for tissue sectioning and freeze-drying than those sufficient for light microscopic studies, and much technical work has to be done in order to obtain acceptable morphologic preservation, while dry application of photographic emulsion desirable for the avoidance of diffusion may even be more difficult to accomplish.

APPENDIX

Methyl Green–Pyronin (MGP) Stain for DNA and RNA as Used in Freeze-Dried Section Autoradiographs*

* The procedure, originally used as a stain for plasma cells, was provided by Dorothy Cummings, M. T., Department of Pathology, University of Chicago, and modified for the staining of dry-mounted freeze-dried sections.

Solutions

Pyronine Y (C. I. 45005)	0.25 gm
Methyl green (C. I. 42590)	0.75 gm
Phosphate buffer, pH 5.3	100.0 cc
0.5 M Disodium phosphate (Na_2HPO_4)	52.5 cc
0.1 M Citric acid	47.5 cc

    (Both substances are dissolved in 25 % methyl alcohol in order to prevent mold growth; correct for pH 5.3. Stock solutions are stable.)

To this add:

0.5 % Phenol soln.	0.5 cc
1.0 % Resorcinol soln. (fresh)	2.5 cc

The above MGP soln. reagent requires 2–3 days to age, and will remain stable up to 6 months at room temperature. Can be reused. Filter before use.

Procedure

1. The mounted freeze-dried sections (in autoradiography after photographic processing) are stained in MGP reagent for approximately 10 sec to 1 min depending on the type of tissue and section thickness.
2. Wash gently in 2 changes of tap or distilled water.
3. Air-dry and mount with synthetic resin.

Results

Cytoplasmic and nucleolar RNA red; DNA blue (the blue color is due to the presence of methyl violet in commercial methyl green).

REFERENCES

1. Anfinsen, C. B., Lowry, O. H., and Hastings, A. B., The application of the freeze drying technique to retinal histochemistry. *J. Cellular Comp. Physiol.* **20**, 231–235 (1942).
2. Beck, L. V., and Fedynskyi, N., Evidence from combined immunoassay and radioautography procedures that intact insulin-^{125}I molecules are concentrated by mouse kidney proximal tubule cells. *Endocrinology* **81**, 475–485 (1967).
3. Bloom, D., Swigart, R. H., Scherer, W. G., and Glick, D., Studies in histochemistry. XXX. A study by phase contrast microscopy of cytological effects of freeze-drying procedures on cultured fibroblasts and guinea pig tissues. *J. Histochem. Cytochem.* **2**, 178–184 (1954).
4. Bogoroch, R., Studies on the subcellular localization of tritiated steroids. *In* "Autoradiography of Diffusible Substances" (L. J. Roth and W. E. Stumpf, eds.), pp. 99–112. Academic Press, New York, 1969.
5. Brown, D. A., Stumpf, W. E., Diab, I. M., and Roth, L. J., Assessment of autoradiographic resolution using soluble extracellular fluid indicators. *In* "Autoradiography of Diffusible Substances" (L. J. Roth and W. E. Stumpf, eds.), pp. 161–182. Academic Press, New York, 1969.
6. Chang, J. P., and Hori, S. H., The section freeze-substitution technique. I. Method. *J. Histochem. Cytochem.* **9**, 292–300 (1961).

7. Dallam, R. D., Determination of protein and lipid lost during osmic acid fixation of tissues and cellular particulates. *J. Histochem. Cytochem.* **6**, 178–181 (1958).

8. Droz, B., and Warshawsky, H., Reliability of the radioautographic technique for the detection of newly synthesized protein. *J. Histochem. Cytochem.* **11**, 426–435 (1963).

9. Fitzgerald, P. J., "Dry"-mounting autoradiographic technic for intracellular localization of water-soluble compounds in tissue reactions. *Lab. Invest.* **10**, 846–856 (1961).

10. Ford, D. H., and Gross, J., The localization of I^{131}-labeled triiodothyronine and thyroxine in the pituitary and brain of the male guinea pig. *Endocrinology* **63**, 549–560 (1958).

11. Gersh, I., The Altmann technique for fixation by drying while freezing. *Anat. Record* **53**, 309–337 (1932).

12. Grunbaum, B. W., Geary, T. J. R., and Glick, D., Studies in histochemistry. XLIII. The design and use of improved approaches for the preparation and freeze-drying of fresh-frozen sections of tissue. *J. Histochem Cytochem.* **4**, 555–560 (1956).

13. Hamilton, J. G., Soley, M. H., and Eichorn, K. B., Deposition of radioactive iodine in human thyroid tissue. *Univ. Calif. (Berkeley) Publ. Pharmacol.* **1**, 339–367 (1940).

14. Jensen, E. V., Suzuki, T., Kawashima, T., Stumpf, W. E., Jungblut, P. W., and DeSombre, E. R., A two-step mechanism for the interaction of estradiol with rat uterus. *Proc. Natl. Acad. Sci. U.S.* **59**, 632–638 (1968).

15. Jensen, J. M., and Clark, D. E., Localization of radioactive l-thyroxine in the neurohypophysis. *J. Lab. Clin. Med.* **38**, 663–670 (1951).

16. Jungblut, P. W., Hätzel, I., DeSombre, E. R., and Jensen, E. V., Über Hormon-"Receptoren." Die oestrogenbindenden Principien der Erfolgsorgane. *Colloq. Ges. Physiol. Chem.* **18**, 58–82 (1968).

17. LeBlond, C. P., Metabolism of ^{14}C-labelled steroids. *In* "Ciba Foundation Conference on Isotopes in Biochemistry" (G. E. W. Wolstenholme, ed.), pp. 4–13. McGraw-Hill (Blakiston), New York, 1951.

18. Lowry, O. H., The quantitative histochemistry of the brain. *J. Histochem. Cytochem.* **1**, 420–428 (1953).

19. Merriam, R. W., Standard chemical fixations as a basis for quantitative investigations of substances other than deoxyribonucleic acid. *J. Histochem. Cytochem.* **5**, 43–51 (1957)

20. Morgan, T. E., and Huber, G. L., Loss of lipid during fixation for electron microscopy. *J. Cell Biol.* **32**, 757–760 (1967).

21. Mosebach, K.-O., Jühe, H., and Dirscherl, W., Initiale Verteilung von Radio-C in unreifen Ratten nach Infusion physiologischer Mengen von Testosteron-4-^{14}C. *Acta Endocrinol.* **54**, 557–567 (1967).

22. Nadler, N. J., Benárd, B., Fitzsimons, G., and LeBlond, C. P., A radioautographic technique to demonstrate inorganic radioiodide in the thyroid gland. *In* "Autoradiography of Diffusible Substances" (L. J. Roth and W. E. Stumpf, eds.), pp. 121–130. Academic Press, New York, 1969.

23. Naidoo, D., and Pratt, O. E., Enzyme histochemistry of the brain. I. A critical study of a freeze drying method for nervous tissue. *Acta Histochem.* **3**, 85–103 (1956).

24. Ostrowski, K., Komender, J., and Kwarecky, K., Quantitative investigations on the solubility of proteins extracted from tissues fixed by different chemical and physical methods. *Experientia* **17**, 183–184 (1961).

25. Pearse, A. G. E., "Histochemistry," 3rd ed., Little, Brown, Boston, Massachusetts, 1968.

26. Peters, T., Jr., and Ashley, C. A., Binding of amino acids to tissue by fixatives. *In* "Autoradiography of Diffusible Substances" (L. J. Roth and W. E. Stumpf, eds.), pp. 267–278. Academic Press, New York, 1969.
27. Rogers, A. W., Thomas, G. H., and Yates, K. M., Autoradiographic studies on the distribution of labelled progesterone in the uterus of the rat. *Exptl. Cell Res.* **40**, 668–670 (1965).
28. Roth, L. J., and Stumpf, W. E., Autoradiography of drugs at histologic levels. *In* "Radioactive Isotopes in Pharmacology" (P. J. Waser and B. Glasson, eds.), pp. 135–148. Wiley (Interscience), New York, 1969.
29. Schneider, G., and Schneider, G., Qualitative und quantitative Untersuchungen über Stoffverluste bei Formol- und Carnoy-Fixierung von menschlichem Hirngewebe. *Acta Histochem.* **28**, 227–242 (1967).
30. Simpson, W. L., An experimental analysis of the Altmann technic of freezing-drying. *Anat. Record* **80**, 173–189 (1941).
31. Steere, R. L., Electron microscopy of structural detail in frozen biological specimens. *J. Biophys. Biochem. Cytol.* **3**, 45–60 (1957).
32. Stumpf, W. E., High resolution autoradiography and its application to *in vitro* experiments. Subcellular localization of ^{3}H estradiol in rat uterus. *In* "Radioisotopes in Medicine: *In vitro* Studies" (R. L. Hayes, F. A. Gosvitz, and B. Murphy, eds.), At. Energy Comm. Symp. Ser. No. 13 (CONF-671111), pp. 633–660. Oak Ridge, Tennessee, 1968.
33. Stumpf, W. E., Too much noise in the autoradiogram? *Science* **163**, 958–959 (1969).
34. Stumpf, W. E., Estradiol concentrating neurons: Topography in the hypothalamus by dry-mount autoradiography. *Science* **162**, 1001–1003 (1968).
35. Stumpf, W. E., Cellular and subcellular ^{3}H estradiol localization in the pituitary by autoradiography. *Z. Zellforsch/Mikroskop. Anat.* **92**, 23–33 (1968).
36. Stumpf, W. E., Subcellular distribution of ^{3}H estradiol in rat uterus by quantitative autoradiography. A comparison between ^{3}H estradiol and ^{3}H norethynodrel. *Endocrinology* **83**, 777–782 (1968).
37. Stumpf, W. E., and Sar, M., Distribution of radioactivity in hippocampus and amygdala after injection of ^{3}H-estradiol by dry-mount autoradiography. *Physiologist* **12**, 368 (1969).
38. Stumpf, W. E., Nuclear concentration of ^{3}H-estradiol in target tissues. Dry-mount autoradiography of vagina, oviduct, ovary, testis, mammary tumor, liver and adrenal. *Endocrinology,* **85**, 31–37 (1969).
39. Stumpf, W. E., and Roth, L. J., Vacuum freeze-drying of frozen sections for dry-mounting, high-resolution autoradiography. *Stain Technol.* **39**, 219–223 (1964).
40. Stumpf, W. E., and Roth, L. J., Frozen sectioning of tissues at ultra-low temperatures. *Cryobiology* **1**, Suppl. 1, 22 (1964).
41. Stumpf, W. E., and Roth, L. J., Dry-mounting high-resolution autoradiography. *In* "Isotopes in Experimental Pharmacology" (L. J. Roth, ed.), pp. 133–143. Univ. of Chicago Press, Chicago, Illinois, 1965.
42. Stumpf, W. E., and Roth, L. J., Thin sections cut at temperatures of $-70°C$ to $-90°C$. *Nature* **205**, 712–713 (1965).
43. Stumpf, W. E., and Roth, L. J., High resolution autoradiography with dry-mounted freeze-dried frozen sections. Comparative study of six methods using two diffusible compounds ^{3}H estradiol and ^{3}H mesobilirubinogen. *J. Histochem. Cytochem.* **14**, 274–287 (1966).
44. Stumpf, W. E., and Roth, L. J., Freeze-drying of small tissue samples and thin frozen sections below $-60°C$. A simple method of cryosorption pumping. *J. Histochem. Cytochem.* **15**, 243–251 (1967).

45. Stumpf, W. E., and Roth, L. J., High-resolution autoradiography of H³-estadiol with unfixed, unembedded 1.0 μ freeze-dried frozen sections. *Advan. Tracer Methodol.* **4**, 113–125 (1968).

46. Sylvén, B., On the advantage of freeze-vacuum dehydration of tissues in morphological and cytochemical research. *Acta, Unio. Intern. Contra Cancrum* **7**, 708–712 (1951).

AUTHOR INDEX

Numbers in parentheses are reference numbers and indicate that an author's work is referred to although his name is not cited in the text. Numbers in italics show the page on which the complete reference is listed.

A

Abramoff, P., 127(109), *150*, 304(44), 312 (44), *332*, 363(85), *396*
Abrams, R., 77(6), 84(6), *85*
Adam, F. C., *427*
Adams, L. R., 8, *21*
Agrell, I., 138(1), *145*, 344(1), 348(1), *351*, 444(1), *448*
Ahsmann, W. B. A. M., 226(1), 236(1, 13), 238(1, 13), *258, 259*
Aiello, N., 155(6), 160(6), *168*, 209, 212, 220(2), *220*, 324(6), 325(6), *330*, 363(29), *393*
Albertson, P.-A., 226, *258*
Alfert, M., 127(2), 142, *145*, 155(1), 164, *168*, 199(1), *207*, 336(2), 351, *351, 354*, 358(1), 359(1), *391*, 447, *449*
Allen, R. D., 454(1–6), *472, 474*
Allison, J. B., *427*
Alvarez, M. R., 20(2), *21*
Amesz, J., 228(3), 229(3), *258*
Anand, A. S., 143, *145*
Anderson, E. S., 283(1), *295*
Anderson, P. J., 219(1), *220*, 243(5), 245, *258*
Anfinsen, C. B., 509(1), 512, *523*
Argauer, R. J., *426, 427*
Armstrong, J. A., 283(1), *295*
Artusi, T., 155(43), *170*
Ashburner, M., 360(2), *391*
Ashley, C. A., 488(33), *504*
Ashraf El-Bagoumi, M., *426*
Atkin, N. B., 13(16), 14(13, 14), 15(6, 18, 20), 20(3–13, 15–21, 74, 78, 80–82, 85, 90, 91), *22, 25*, 127(6), *145*, 172(1), *194*, 312(1), *330*
Atkins, N. G., 162(2), *168*
Attiya, I. R., 155(43), *170*
Atwood, K. C., 337(67), 351(67), *354*, 357

(90), 391(90), *396*
Augenstein, L., *429*
Avanzi, S., 20(22, 23, 36), *22, 23*
Axén, R., 226(61), *261*
Azumi, T., *427*

B

Bachmann, K., 312(2), *330*
Bachmann, L., 498(1), *502*
Bähr, H., 142, *145*
Bahr, G. F., 20(24), *23*, 126, 128, 130(7), 136(8), 137(8), *145*, 155(3), *168*, 245(4), *258*, 442(3), *449, 503*
Bailly, S., 87(1, 2), 88(1, 2), 89(1), 94(2), 95(2), *103*
Bairati, A., 365(3), *392*
Bajer, A., 454(7), *474*
Baker, F. N., 128(10, 31, 113, 114, 116), 130 (116), 139(11), *145, 146, 150*
Baker, M. C., 14(14), 20(19), *22*
Bakken, A. H., 20(32), *23*
Balfour, B. M., 20(25), *23*
Bane, A., 140(12), *145*
Bangham, A. D., 139(19), *146*
Barer, R., 117(1), *122*, 131(13), *145*
Barka, T., 219(1), *220*, 243(5), 245, *258*
Barnard, E. A., 20(26), *23*
Barnett, R. J., *504*
Barr, H. J., 349(60a), *354*, 358(84a), 360 (84a), *396*
Bartels, P. H., 155(3), 173(2), *168*, 192(2), *194*, 454(8), *474*
Baserga, R., 486, *504*
Bateson, R. C., 191(27), *195*
Bauer, H., 337(3), 351(3), *351*
Bayreuther, K., 337(4), 351(4), *351*, 389(4), 391(4), *392*
Bayzer, H., 110(7), *112*

527

SUBJECT INDEX

A

Abnormal chromosomes, 87, 95–102
Absorbance
  extinction of by scattering, 106
  maximum setting for, 12
  in microphotometry, 109
  of solution, 2
  of stained films, 230
Absorbancy index, 3
Absorption, phosphorescence and, 1–2
Absorption cytophotometry
  absorption errors in, 175–188
  absorption law failure in, 185–187
  distributional error, 180–185
  experimental evaluation of systematic
    errors, in 190–192
  general approaches to, 172–174
  instrument errors in, 187·
  model calculations in, 187–190
  replication error in, 193
  specimen errors in, 174–175
  stochastic errors in, 173, 192–194
  systematic errors in, 174–184
Absorption law failure, 185–187
Absorption photometry, 432–433
  see also Absorption cytophotometry
Absorption spectra, 110, 404
Absorptivity, 3
Acetic acid, chromophore fading and, 323
Acetylation, blocking by, 199
Acid phosphatase, metal salt methods for,
  243–245
Acid-Schiff procedure, 234–235
  see also Schiff reagent
Alkaline phosphatase, inhibiting of, 241
Alpha-Molykote Corp., 513
Aminocellulose films, 235–238
Aminoethylsulfuric acid, 236
Amino groups, per weight of cellulose, 236
Ascaris megalocephala
  chromosomal composition in, 57–74
  cytophotometric data on, 69

DNA content in, 64–73
  meiotic divisions in, 711
  nuclear behavior during cleavage in,
    58–62
  spermatocytes of, 70
Aufbau principle, 400
Autofluorescence, 519
Autoradiography
  data presentation in, 481–485
  defined, 478
  freeze-dried section preparation for,
    509–516
  hormone localization in, 507–523
  image formation in, 494
  quantitative, 477–502 (see also Quan-
    titative autoradiography)
  tissue preparation for, 507–523
Avian blood cells, Feulgen-staining
  procedures in, 326–327
Azo dye
  in carcenogenesis, 264
  coupling method calibration of, 249–252
Azo grouping, 1

B

Barr and Stroud, Ltd., 8
Barr and Stroud integrating microdensito-
  meter, 9–10, 155–156, 367
Barr and Stroud scanning microdensito-
  meter, 275
Basophilia, acetylation of, 200
Beer-Lambert absorption law, 172, 175, 431
Beer's law, 2–3
Bildungszentrum, in dragon fly egg, 57
Biological molecule, optical phenomena of,
  454–455
Boltzmann distribution, 404
Bombyx mori, 359
Brilliant sulfaflavine, 446–447
Bull, spermiogenesis in, 162
Bull spermatozoa, 127–130
  ejaculation rate and, 143